Meta-omics in Crop Improvement

Renu • Sanjeev Gupta • Tilak Raj Sharma
Editors

Meta-omics in Crop Improvement

Volume I: Methods and Approaches

Springer

Editors
Renu
Crop Science Division
Indian Council of Agricultural Research
New Delhi, Delhi, India

Sanjeev Gupta
Crop Science Division
Indian Council of Agricultural Research
New Delhi, Delhi, India

Tilak Raj Sharma
ICAR-National Institute for Plant Biotechnology
New Delhi, Delhi, India

ISBN 978-981-95-2293-4 ISBN 978-981-95-2294-1 (eBook)
https://doi.org/10.1007/978-981-95-2294-1

This Springer imprint is published by the registered company Springer Nature Singapore Pte Ltd.
The registered company address is: 152 Beach Road, #21-01/04 Gateway East, Singapore 189721, Singapore

Preface

Microbial communities associated with plants and soil have profound effect on soil fertility and crop productivity. To understand microbial community, culture-based traditional methods are not sufficient. The currently cultural microorganisms only account for $\leq$1.0% of the total microbial population. Also, traditional methods fail to decipher the functions of microbes in communities. The revolutionary interdisciplinary systems biology approach of meta-omics integrates various high-throughput omics disciplines for a comprehensive understanding of microbiomes. Omics technologies encompassing metagenomics, metatranscriptomics, metaproteomics, and metabolomics, which are independent of culturing the microbes, are reshaping our understanding and knowledge of soil microbial community structures, their metabolic potential, and their functional activities. There is need to provide up-to-date information on advancements in meta-omics technologies to fill the knowledge gap and facilitate the molecular microbiologists and other researchers of related fields. The current book entitled, *Meta-Omics in Crop Improvement: Methods and Approaches*, explores the advanced integrative techniques of meta-omics, including metagenomics, metatranscriptomics, metaproteomics, and metabolomics, to enhance crop productivity, resilience, and quality. This book presents a comprehensive coverage of recent advancements on the theoretical, methodological, and practical aspects of meta-omics technologies to understand the complex microbial communities and their interactions with crops. The chapters in the book offer detailed insights into techniques and tools specific to each area of the different meta-omics technologies. Insights and future perspective of application of these techniques in detection of plant pathogens and exploring plant disease ecology has been discussed in two chapters. A dedicated chapter on core plant microbiome has also been included along with a chapter on legume microbiome and nodule microbiome. Bioinformatic tools, computational analysis, and pipelines for analyzing massive data generated through these techniques have also been presented. A scientometric analysis is also attempted to have an insight over the evolution and impact of omics technologies in agricultural research and crop improvement from 2000 to 2025. We believe this compilation will immensely help students, research scholars, teachers, and researchers engaged in microbiology, plant pathology, genomics, bioinformatics, biotechnology, etc.

We are grateful to all the learned contributors for their sincere efforts in updating scientific information in this subject and sparing valuable time and knowledge in bringing out this volume. We also acknowledge and appreciate Springer Nature in terms of careful and speedy publication of this book.

New Delhi, Delhi, India

Renu
Sanjeev Gupta
Tilak Raj Sharma

Contents

Editors and Contributors

Editors

Renu is Principal Scientist (Agriculture Biotechnology), Commercial Crops, Crop Science Division at ICAR Headquarters, Krishi Bhawan, New Delhi. She received her M.Sc. in Biotechnology from Indian Veterinary Research Institute, Bareilly, and PhD in Life Sciences from Chatrapati Sahuji Maharaj University, Kanpur. She was Post-Doctoral Fellow at Plant Molecular Biology Department, National Institute for Genetic Engineering and Biotechnology (ICGEB affiliated Institute), Tehran, Iran, during 2003–2005. She joined ICAR in 1998 and worked in various capacities at Indian Agricultural Research Institute, New Delhi; Department of Science and Technology, Haryana; ICAR-National Bureau of Agriculturally Important Microorganisms, Mau. She has made significant contributions and has led numerous projects in frontier areas of microbial biotechnology, including microbe-based bioremediation of heavy metal-polluted areas, meta-omics, organic farming, etc., resulting in over 151 publications, including research papers, articles, books, technical manual, etc.; 1 patent and 1 copyright; and enriched National Agriculturally Important Microbial Culture Collection (NAIMCC) with more than 110 well-characterized microbes for public use; 40 metagenomes in JGI Gold and MgRAST; 2 metaproteomic datasets in PRIDE database; and 175 sequences registered in NCBI Genebank. Her work on microbial indicators of heavy metal pollution and microbial technologies for reduction of cadmium and arsenic uptake in spinach; nutrient management in chickpea; agrowaste management; metaproteome isolation technique and database development for agriculturally important microbes is notable. She has also headed many farmer-oriented programs funded by DST and DBT and has made first organized step leading to success in demonstration and adoption of microbe based rapid composting technology in Eastern U.P. She is recipient of several awards, Young Scientists' Award 2008, Major R.S. Yadav SRDA Award-2012, Best Women Scientist Award 2015, Scientist of the Year Award 2016, Women Scientist Award 2016 and Excellence in Research Award 2017, Outstanding Achievement Award (2018), Distinguished Scientist Award (2018), Eminent Scientist Award (2019), Women Scientist Leadership Award (2024) by Springer Nature and Society for Plant Research and is Fellow of Society of Plant Research and Fellow of Society of Applied Biotechnology.

Sanjeev Gupta has illustrious career as Scientist (Plant Breeding) at Indian Council of Agricultural Research. Presently, he is the Assistant Director General (Oilseed & Pulses) at Indian Council of Agricultural Research, New Delhi. He has more than 32 years of research experience in grain legume genetics and breeding and developed 18 high-yielding cultivars in grain legumes (Mungbean, blackgram, and common bean). He has developed world's first earliest maturing mungbean variety 'Virat' which matures in 52–55 days. He has experience in conducting several ambitious national and international projects as principal investigator on Molecular marker-assisted breeding for Mungbean Yellow Mosaic India Virus and powdery mildew resistance, Shuttle Breeding Programme among SAARC Nations, and Climate Resilient Agriculture Research. His special interest in molecular breeding led to the mapping and tagging of Yellow mosaic-resistant gene in *Vigna mungo*. He has authored nearly 165 research papers published in peer-reviewed internationally recognized journals, edited six books on pulses including *Legumes in the "Omic Era"* and *Breeding for Enhanced Nutrition and Bioactive Compounds in Food Legumes*, published by Springer Pub. (USA). He has traveled across the continents to present his research several times. He is the Fellow of National Academy of Agricultural Sciences and recipient of several awards for his research and literary contribution.

Tilak Raj Sharma, a well-known Plant Molecular Biologist, is presently working as ICAR-National Professor-BP Pal Chair at ICAR-NIPB, IARI, New Delhi. He has been former Deputy Director General (Crop Science), ICAR, Pusa Campus, New Delhi, and also acting DDG (Horticulture Science); Director and Vice Chancellor, IARI, New Delhi, IARI Assam, IARI Jharkhand; and Director, NIBSM, Raipur. Earlier he contributed in the establishment and development of the prominent DBT, GoI institutes like National Agri-Food Biotechnology Institute, Mohali (Executive Director), Center of Innovative and Applied Bioprocessing (Chief Executive Officer), and ICAR institutes like National Institute for Plant Biotechnology, New Delhi (Director), and Indian Institute of Agricultural Biotechnology Ranchi, (OSD). He obtained three Post Doctorate trainings on Plant Molecular Biology and Genome analysis and visited 15 countries for various academic visits as member of Indian delegations or for trainings. His major research interests are in the areas of genomics and plant disease resistance, and he has contributed extensively to the improvement of rice for over 35 years and identified and cloned a new rice blast resistance gene *Pi54* which has been deployed in more than 40 rice varieties in India and abroad. He was associated with the decoding of complete genomes of rice, tomato, Pigeonpea, Jute, mango, tea, and many plant pathogens. He has guided 32 PhD students and published more than 225 research papers (two in the prestigious journal *Nature*), obtained six Indian patents, and associated in the development of three blast resistant rice varieties Pusa 1637, HPR2880, and Pusa Samba1850. He is the fellow of all four national academies, viz. Indian National Science Academy, Indian Academy of Sciences, National Academy of Sciences, and National Academy of Agricultural Sciences. He has been awarded (twice) the prestigious JC Bose Fellowship of the Department of Science and Technology, Government of India. For his outstanding research contributions, he has been awarded various awards and honors, including International World Technology Award for Biotechnology (jointly to Int. Rice Genome Sequencing Team) in 2003; DBT National Bioscience Award for Career Development, 2007; NAAS Recognition Award 2008; Rafi Ahmed Kidwai Award-ICAR, 2011; Prof. K.S. Bilgrami Memorial Medal (INSA)-2018; and OP Bhasin Award in Agricultural Sciences-2022.

Contributors

Mainak Barman Department of Genetics and Plant Breeding, Bidhan Chandra Krishi Viswavidyalaya, Nadia, West Bengal, India

Pranjali V. Bhadane ICAR-Directorate of Onion and Garlic Research, Pune, Maharashtra, India

Nikita Bisht Microbial Technologies Division, CSIR-National Botanical Research Institute (CSIR-NBRI), Lucknow, Uttar Pradesh, India

Sindhunath Chakraborty ICAR-National Bureau of Agriculturally Important Microorganisms, Mau Nath Bhanjan, Uttar Pradesh, India

Tilak Chandra Division of Agricultural Bioinformatics, ICAR-Indian Agricultural Statistics Research Institute, New Delhi, Delhi, India

Shayree Chatterjee ICAR-National Institute for Plant Biotechnology, New Delhi, Delhi, India

Puneet Singh Chauhan Microbial Technologies Division, CSIR-National Botanical Research Institute (CSIR-NBRI), Lucknow, Uttar Pradesh, India

Preeti Chaurasia Department of Microbiology, SS Jain Suboth PG College, Jaipur, Rajasthan, India

Arpita Das Department of Genetics and Plant Breeding, Bidhan Chandra Krishi Viswavidyalaya, Nadia, West Bengal, India

Ritu Dixit Microbial Technologies Division, CSIR-National Botanical Research Institute (CSIR-NBRI), Lucknow, Uttar Pradesh, India

Ram Dutta ICAR-Directorate of Onion and Garlic Research, Pune, Maharashtra, India

S. P. Gawande Division of Crop Protection, ICAR- Central Institute for Cotton Research, Nagpur, India

Bishnupriya Gouda Division of Crop Improvement, ICAR-National Institute for Research on Commercial Agriculture (ICAR-NIRCA) formerly ICAR- CTRI, Rajahmundry, Andhra Pradesh, India

Sanjeev Gupta Crop Science Division, Indian Council of Agricultural Research, New Delhi, Delhi, India

Koj Haniya Division of Microbiology, ICAR-IARI, New Delhi, Delhi, India

N. S. Hiremani Division of Crop Protection, ICAR- Central Institute for Cotton Research, Nagpur, Maharashtra, India

Mir Asif Iquebal Division of Agricultural Bioinformatics, ICAR-Indian Agricultural Statistics Research Institute, New Delhi, Delhi, India

Sarika Jaiswal Division of Agricultural Bioinformatics, ICAR-Indian Agricultural Statistics Research Institute, New Delhi, Delhi, India

K. Jayalakshmi ICAR-Directorate of Onion and Garlic Research, Pune, Maharashtra, India

Girish Kumar Jha ICAR-Indian Agricultural Statistics Research Institute, New Delhi, Delhi, India

Sarala Kesamsetty Division of Crop Improvement, ICAR-National Institute for Research on Commercial Agriculture (ICAR-NIRCA) formerly ICAR- CTRI, Rajahmundry, Andhra Pradesh, India

Amrender Kumar Agricultural Knowledge Management Unit (AKMU), ICAR-Indian Agricultural Research Institute, New Delhi, Delhi, India

Ashutosh Kumar G Surgiwear Limited, Shahjahanpur, Uttar Pradesh, India

Dinesh Kumar Division of Agricultural Bioinformatics, ICAR-Indian Agricultural Statistics Research Institute, New Delhi, Delhi, India

Naina Kumari Division of Agricultural Bioinformatics, ICAR-Indian Agricultural Statistics Research Institute, New Delhi, Delhi, India

Kuldeep Kumar ICAR-Indian Institute of Pulses Research, Kanpur, Uttar Pradesh, India

Murugan Kumar ICAR-National Bureau of Agriculturally Important Microorganisms, Mau Nath Bhanjan, Uttar Pradesh, India

Shiv Kumar International Centre for Agricultural Research in the Dry Areas (ICARDA), South Asia and China Regional Programme, New Delhi, Delhi, India

Sunil Kumar Center for Agricultural Bioinformatics (CABIN), ICAR- Indian Agricultural Statistics Research Institute, New Delhi, Delhi, India

Madhubala Division of Basic Sciences, ICAR-Indian Institute of Pulses Research, Kanpur, Uttar Pradesh, India

Sharanbasappa D. Madival ICAR-Indian Agricultural Statistics Research Institute, New Delhi, Delhi, India

Pradeep Manyam Department of Plant Pathology, Sri Venkateswara Agricultural College, ANGRAU, Tirupati, Andhra Pradesh, India

Sudarshan Maurya ICAR-Indian Institute of Vegetable Research, Varanasi, Uttar Pradesh, India

M. D. Meshram Division of Crop Protection, ICAR- Central Institute for Cotton Research, Nagpur, Maharashtra, India

Dwijesh Chandra Mishra ICAR-Indian Agricultural Statistics Research Institute, New Delhi, Delhi, India

Priyanka Mishra Division of Basic Sciences, ICAR-Indian Institute of Pulses Research, Kanpur, Uttar Pradesh, India

Mousumi Murmu Department of Genetics and Plant Breeding, Bidhan Chandra Krishi Viswavidyalaya, Nadia, West Bengal, India

D. T. Nagrale Division of Crop Protection, ICAR-Central Institute for Cotton Research, Nagpur, Maharashtra, India

N. S. Nysanth Division of Microbiology, ICAR-IARI, New Delhi, Delhi, India

Sonika Pandey Division of Basic Sciences, ICAR-Indian Institute of Pulses Research, Kanpur, Uttar Pradesh, India

Kudupudi Prabhakara Rao Division of Crop Improvement, ICAR-National Institute for Research on Commercial Agriculture (ICAR-NIRCA) formerly ICAR-CTRI, Rajahmundry, Andhra Pradesh, India

Ratna Prabha Agricultural Knowledge Management Unit (AKMU), ICAR-Indian Agricultural Research Institute, New Delhi, Delhi, India

EPC, ICAR-NRC Equines, Bikaner, Rajasthan, India

Nagendra Rai ICAR-Indian Institute of Vegetable Research, Varanasi, India

B. Ramakrishnan Division of Microbiology, ICAR-IARI, New Delhi, Delhi, India

Renu Crop Science Division, Indian Council of Agricultural Research, New Delhi, Delhi, India

Soumyayan Roy Department of Genetics and Plant Breeding, Bidhan Chandra Krishi Viswavidyalaya, Nadia, India

Pramod Kumar Sahu ICAR- National Bureau of Agriculturally Important Microorganisms, Mau, Uttar Pradesh, India

Khan Mohd. Sarim Institut Ruđer Bošković, Zagreb, Croatia

Lovkush Satnami ICAR-Indian Institute of Vegetable Research, Varanasi, Uttar Pradesh, India

Murugesan Senthilkumar Division of Basic Sciences, ICAR-Indian Institute of Pulses Research, Kanpur, Uttar Pradesh, India

Nitesh Kumar Sharma ICAR-Indian Agricultural Statistics Research Institute, New Delhi, Delhi, India

Dr Rajendra Prasad Central Agricultural University Pusa, Samastipur, Bihar, India

Tilak Raj Sharma ICAR-National Institute for Plant Biotechnology, New Delhi, Delhi, India

Maganti Sheshu Madhav ICAR-National Institute for Research on Commercial Agriculture (ICAR-NIRCA) formerly ICAR-CTRI, Rajahmundry, Andhra Pradesh, India

Dhananjaya Pratap Singh ICAR-Indian Institute of Vegetable Research, Varanasi, Uttar Pradesh, India

R. Sreenath Reddy ICAR-IARI Academic Hub, The Graduate School, ICAR-CICR, Nagpur, Maharashtra, India

K. Swarnalakshmi Division of Microbiology, ICAR-Indian Agricultural Research Institute (IARI), New Delhi, Delhi, India

Manishi Tripathi Department of Microbiology, Chhatrapati Shahu Ji Maharaj University, Kanpur, Uttar Pradesh, India

Poleboina Venkata Diwakar Division of Crop Improvement, ICAR-National Institute for Research on Commercial Agriculture (ICAR-NIRCA) formerly ICAR-CTRI, Rajahmundry, Andhra Pradesh, India

Amrita Vishwakarma Division of Basic Sciences, ICAR-Indian Institute of Pulses Research, Kanpur, Uttar Pradesh, India

Recent Advances in Methods and Techniques of Meta-omics for Crop Improvement

Renu, Sanjeev Gupta, and Tilak Raj Sharma

1 Introduction

Meta-omics provide comprehensive understanding of the complex interactions between plants, microbes, and their environment thus playing crucial role in crop improvement. Microbial communities especially plant associated ones have been found to be instrumental in maintaining soil fertility, improving productivity of crop and mitigating abiotic and biotic stress. The traditional culture-based methods are not suitable for generating information on structure and functions of microbial communities. The integration of high-throughput meta-omics technologies to study the microbial communities associated with crops is made possible with recent advancements in molecular biology and opened a new frontier for sustainable agriculture. The microbiome of a crop comes with plethora of benefits, including enhancing nutrient cycling, suppressing pathogens, and improving abiotic stress tolerance in plants (Tilgam et al. 2024; Sharuddin et al. 2022; Brader et al. 2017; Khan et al. 2019).

Our ability to profile and comprehend these communities in a culture-independent manner has been enhanced by next-generation sequencing (NGS) and other omics-based technologies, including metagenomics, metatranscriptomics, metaproteomics, and metabolomics (Nilsson et al. 2018; Rai et al. 2020). The development of biofertilizers has benefitted from these techniques. With a better understanding of microbial interactions and signal transduction mechanisms, plants and their defense mechanisms become more easier (Basu et al. 2021; Lemanceau et al. 2017). By deciphering these intricate relationships between microbes, plants, and

Renu (✉) · S. Gupta
Crop Science Division, Indian Council of Agricultural Research, New Delhi, Delhi, India
e-mail: renu1.icar@icar.org.in

T. R. Sharma
ICAR-National Institute for Plant Biotechnology, New Delhi, Delhi, India

Renu et al. (eds.), *Meta-omics in Crop Improvement*,
https://doi.org/10.1007/978-981-95-2294-1_1

environment, identifying beneficial microorganisms, understanding their functions, and leverage them for crop growth, nutrient uptake and disease resistance, meta-omics technologies have strength for enabling the development of sustainable agriculture practices and improved crop varieties.

2 Survey and Sampling Strategies

Microbial community is highly diverse, and its composition depends on several biotic and environmental factors such as type of soil, root zone, plant type, season, sampling depth, etc., (Dickie et al. 2018). Hence for any meto-omics study, appropriate survey and sampling methods are the foundation step and sampling represent must be standardized against variability and reproducible across time. Soil can be gathered as bulk and rhizosphere soil, while tissue sample collection requires meticulous procedures for endophyte studies. Pre-processing steps such as sieving soil, surface-sterilizing plant tissues like leaves by swabbing or washing in phyllosphere studies (Rai et al. 2023), and avoiding contamination are essential for accurate downstream molecular analyses (Michealsamy et al. 2021).

Transport and storage systems are also critical for nucleic acids and proteins because of their high sensitivity. While DNA is stable during these processes, RNA need to be protected immediately through snap freezing with liquid nitrogen or stabilized using RNAlater and proteins/metabolites using protease inhibitors (Ranjard et al. 2009; Terrat et al. 2015). Suitable controls and replicates should be incorporated into the sampling design for either of the meta-omics studies like taxonomic or gene expression analysis or protein and metabolite identification (Huang et al. 2024).

3 Principles of Meta-omics Approaches and Extraction of Molecular Resources

Since meta-omics approaches explore microbial communities by directly analyzing their molecular resources, accurate profiling of these communities is dependent on efficient extraction of biomolecules such as DNA, RNA, proteins, and metabolites. Specific lysis and purification procedures are needed for each class of biomolecules in order to handle intricate matrices in soils and plant tissues.

3.1 Metagenomics

Metagenomics involves direct extraction and analysis of DNA from an environment without cultivating microbes providing powerful means to study the composition, diversity relative abundance, and functional capabilities of microbial communities. The effectiveness of downstream process (sequencing and functional annotation) is dependent on high quality DNA from the sample.

3.1.1 Environmental DNA Extraction

The process of DNA extraction from environmental samples such as bulk soil or rhizosphere faces challenges due to the presence of co-extracted inhibitors. Several protocols like CTAB-based methods of extraction as well as commercial kits like MoBio PowerSoil are available but each method introduces inherent biases. Adherence to ISO 11063:2012 is recommended for cross-lab standardization (Martin-Laurent et al. 2001; Vestergaard et al. 2017).

3.1.2 Methodological Strategies

3.1.2.1 Sequence-Based

Sequence based metagenomics relies on high-throughput sequencing enabling in-depth analysis of microbial communities and functional genes. 16S rRNA-based amplicon sequencing commonly introduces PCR amplification biases which is now overcame by advancement of whole-genome shotgun sequencing (WGS) (Zinicola et al. 2015; Ross et al. 2012). Sequencing technologies like Illumina and Nanopore have improved resolution in identifying taxonomic composition and functions of microbial communities in diverse agroecosystems. Tools like Illumina and PacBio help build metagenome-assembled genomes (MAGs), which represent a further advance, offering the ability to reconstruct individual microbial genomes from complex communities and reveal uncultivable taxa (Bulgarelli et al. 2015; White III et al. 2016).

3.1.2.2 Function-Based

Function-based metagenomics makes functional validation by cloning environmental DNA in suitable vectors and its screening in hosts like *E. coli* for traits like enzyme activity, stress resistance, phenotypic traits, etc. (Tchigvintsev et al. 2021; Fan et al. 2021). This method has been instrumental in discovery of novel genes, including those for xenobiotic degradation or pollutant-degrading enzymes, such as esterases and dioxygenases, and stress tolerance genes, like those for acid and heavy metal resistance (Tchigvintsev et al. 2021; Guazzaroni et al. 2015; Mirete et al. 2016).

3.2 Metatranscriptomics

Metatranscriptomics involves sequencing and analysis of environmental RNA to capture the actively expressed genes at particular point of time and offers a "real-time" assessment of microbial activities and functions. In contrast to metagenomics, metatranscriptomics is capable of distinguishing active members from inactive ones in community by providing insight into transcriptionally active population. This approach is advantageous in investigating microbiomes associated with crops and their roles under various environmental conditions like abiotic and biotic stresses, nutritional changes, etc. Disease suppressive soils studies have shown a signature pattern of highly expressed genes (polyketide synthases) in suppressive vs. oxidation stress genes in non-suppressive soils (Hayden et al. 2018). Nitrogen fixation

Genes (*nifB*, *nifZ*) were expressed abundantly in the PGPR-treated crop (Gupta et al. 2015). RNA-seq data are analyzed using packages such as SortMeRNA, MetaTrans, DESeq2 (Kopylova et al. 2012; Love et al. 2014).

3.2.1 Environmental RNA Extraction

Extraction and preservation of RNA is highly changing owing to its high sensitivity, low stability, and a short half- life. The amount of living microorganisms, humic substances contaminants and RNA loss during purification process impacts the yield of soil RNA significantly varying from tens of nanograms to micrograms per gram of soil. Significant advancements have been made in this field, especially the challenges of posed by low biomass samples, high ribosomal RNA (rRNA) abundance, and potential RNA degradation during isolation. Stringent conditions must be maintained to avoid RNA degradation for extraction. Use of guanidinium lysis buffers, followed by DNase treatment, is standard procedure. Tailored RNA extraction protocols are critical for different sample types, such as soil, aquatic samples, and host-associated environments. Pre-processing steps like microbial suspension from soil particles or rapid thawing and lysis in buffer can significantly improve RNA yield and quality. Most commonly used commercial kit is RNeasy as well as protocols employing RNAlater (Hambraeus et al. 2003; Wang et al. 2012).

3.2.2 Methodology and Challenges

Metatranscriptomics methodology involves several key steps from sample collection, its preservation, RNA extraction, mRNA enrichment, cDNA synthesis, library preparation, sequencing, and data analysis. Challenges in metatranscriptomics include RNA instability, the low proportion of mRNA (2–5% of total RNA), the absence of poly-A tails in prokaryotic mRNA, the need for efficient rRNA removal, and potential biases introduced during cDNA library construction. Advancements in RNA-seq, rRNA depletion (e.g., SortMeRNA) bioinformatics pipelines like MetaTrans, HUMAnN2 have led to improved transcript quantification and annotation of functions (Wang et al. 2009; Kopylova et al. 2012). There is need of comprehensive databases, better computational pipelines, and curated environmental transcriptome libraries for annotation of differentially expressed (DE) genes (Ijaq et al. 2019). Despite these, its ability to show active gene expression makes it crucial for understanding real-time microbial functions.

3.3 Metaproteomics

Metaproteomics is a rapidly emerging field that sheds light on functional activity of microorganisms in diverse ecosystems. It is the study of identification and quantification of whole protein complement of a microbiome within a given environment at a particular time or under specific stimuli and provides information on functional gene expression and metabolic pathways. This approach has provided insights into functionality and activity of microbial communities by identification and characterization of functional proteins like extracellular enzymes involved in nutrient cycling, organic matter decomposition, microbe driven ecosystem functions, indicators of specific biodegradation processes, host- microbe interactions, bioremediation, stress adaptation, etc. (Schneider et al. 2012; Overy et al. 2021; Waller et al. 2005; Mattarozzi et al. 2020; Wilmes and Bond 2006; Wilmes et al. 2015).

However, extracting and analyzing proteins from diverse and complex microbial communities presents significant challenges. Despite these difficulties, ongoing advancements in isolation techniques and complementary technologies are paving the way for more comprehensive and accurate metaproteome analyses.

3.3.1 Environmental Protein Extraction

The intricate structure of soil presents challenges to protein extraction from soil and rhizosphere due to co-extracted humic substances, clay particles, and organic debris that interfere with downstream analysis. Proteins get absorbed onto these particles resulting in poor yield. There are several challenges to isolate metaprotemone from plant tissue including (i) the high proportion of plant to microbial proteins in the sample, which may make it challenging to detect microbial proteins; (ii) the enormous amount of plant tissue required if techniques call for removing microbes from the plant tissue before extracting proteins (Bao et al. 2014; Knief et al. 2012); (iii) the presence of interfering compounds (phenols, carbohydrates, and pigments) in the sample (Keiblinger et al. 2012); (iv) the wide variety of microbial species with characteristics (such as cell wall thickness) that affect protein extraction differently; and, lastly, (v) the wide range of relative abundances of microbial species. Common practices include the use of SDS or urea-based buffers alongside sonication or bead-beating, followed by TCA/acetone precipitation. It is common for protein degradation to occur due to adsorption to soil (Keiblinger et al. 2016). Several protocols have been developed for soil and plant tissue samples ((Keiblinger et al. 2012; Keiblinger et al. 2016; Renu et al. 2019; Salvato et al. 2022). So far there is no commercial kit available for metaproteome isolation.

3.3.2 Methodology and Challenges

A metaproteomics workflow has multiple steps that must be tailored for every system under study, just like other meta-omics techniques. Sample preservation, protein extraction, peptide preparation, selecting liquid chromatography-tandem mass spectrometry (LC-MS/MS) techniques, and creating an appropriate protein sequence database for peptide and protein identification are some of these procedures (Kunath et al. 2019; Salvato et al. 2021). Each step has its own challenges including low protein yield in extraction step. To minimize the activity of naturally occurring proteases, processing fresh samples or storing samples at ultra-low temperatures (−80 °C for high organic matter content) is recommended. Proteins are digested (usually with trypsin) after extraction, fractionated using 2D-GE or LC-MS, and identified by peptide matching against databases such as UniProt, CAZy, and EggNOG. However, the lack of well-annotated protein reference databases continues to restrict functional annotations (Schiebenhoefer et al. 2019). Moreover, major challenges include database analysis, estimation of error rates, and the identification and annotation of peptides. The methods necessitate a high level of scientific knowledge for data computation, accurate analysis, and appropriate treatment, and the low-cost equipment.

3.4 Metabolomics

Metabolomics is comprehensive analysis including structural identification and quantification of small molecules (metabolites) like amino acids, fatty acids, sugars, vitamins, and inorganic chemical compounds in biological or environmental samples. The approach links genotypes to phenotypes by providing insight into metabolic responses to external stimuli. In studies involving soil and rhizosphere microbiomes, metabolomics offers a unique advantage by linking microbial function to plant growth, stress management, and biogeochemical cycling and provide metabolic signatures (Abiraami et al. 2020; Bonini et al. 2020; Pascale et al. 2020; Trivedi et al. 2020). Metabolomics approach can be targeted analysing specific and well characterized metabolites leading to tracking of key signalling compounds like defense molecules, phytohormones, etc. Untargeted metabolomic approach is useful for discovering novel molecules serving as bioindicators of stimuli, interaction, and state of sample like soil quality.

3.4.1 Metabolite Extraction

Metabolome isolation techniques are advancing and driven by better sensitivity, coverage, and reproducibility. Like other molecular resources, metabolome isolation and sample handling is challenging due to short half-life of metabolites, adsorption to soil particles, and variability due to environment. The extraction of metabolites is done via isolating them through varying degrees of solubility and depends on the compound's polarity. A standard solvent system used is either single-phase (methanol, methanol/acetonitrile) or two phase (methanol:chloroform:water). Microwave-assisted extraction (MAE), supercritical fluid extraction (SFE), and accelerated solvent extraction (ASE) are rapid and efficient extraction methods. In addition to it, automated and miniaturized methods like solid phase micro extraction (SFME) and online SPE-CE-MS minimize manual handling and are more reproducible (Zenati et al. 2023).

3.4.2 Methodology and Challenges

Metabolomics methodology includes steps like sample preparation, data acquisition, and analysis and interpretation. Sample preparation involves sample collection and storage, preprocessing steps to remove protein and enhance metabolite detection, metabolite extraction, and derivatization (Keiblinger et al. 2016). Data acquisition involves separation techniques (liquid chromatography or gas chromatography); ionization for converting metabolites into charged particles (electrospray ionization (ESI) or atmospheric pressure chemical ionization (APCI); and detection (MS or NMR spectroscopy). Data analysis and interpretation involves data pre-processing; feature extraction to identify and quantify metabolites; statistical analysis (ANOVA,

PCA); metabolite identification; pathway analysis; and interpretation of results (Aguiar-Pulido et al. 2016; Idle and Gonzalez 2007; Djemiel et al. 2022). Spectral matching against databases such as BioMagResBank, MassBank and Golm Metabolome Database is commonly used for metabolite annotation.

4 Bioinformatics Pipelines and Approaches

To interpret massive amount of multidimentional data generated by meta-omics platform, advancements in bioinformatics tools and pipelines are essential. Sequencing technologies have evaluated from amplicon sequencing to whole genome shotgun sequencing. Third generation sequencing platforms like Pacific Bioscience and Oxford Nanopore enable long reads up to 1 Mb. Each omics layer has dedicated tools, webservers, and softwares:

- *Metagenomics*: MG-RAST for automated annotation against SEED, KEGG, and RefSeq databases, offering insights into taxonomic and functional diversity (Meyer et al. 2008), QIIME and Mothur for sequence quality control, clustering, and OTU classification (Schloss et al. 2009; Caporaso et al. 2010), TANGO. Tools like PICRUSt predict metagenomes from 16S rRNA gene data. IMG/M, METAREP, CoMet, METAGENassist, WebMGA, MOCAT2, MGnify, MetaVelvet, IDBA-UD, and MEGAN perform *de novo* assembly and taxonomic binning with support from curated databases like SILVA, Greengenes, and *RDP.*
- *Metatranscriptomics*: SortMeRNA, Kraken, UCLUST, SOAP2, FLASH, Trimmomatic SAMSA2, TaxMapper, MetaTrans, HUMAnN, MG-RAST, *DESeq2* are used for quality control, rRNA removal, trimming, assembly, read mapping, and gene expression analysis. Tools rely on curated repositories like *COG*, *KEGG*, and *SEED* for gene annotation, and functional comparisons.
- *Metaproteomics*: iMetaLab, Galaxy-P, MetaGOmics, MetaProteomeAnalyzer, Prophane, PeptideShaker, and Unipept are used for taxonomic classification, functional enrichment, and visualization, comparing peptide mass fingerprints against databases like eggnog, MetaCyc, Pfam, UniProtKB, CAZy, and NCBI. metaQuantome is used for quantitative comparisons across conditions (Mehta et al. 2021).
- Metabolomics: XCMS, MetaboAnalyst and *MZmine* for data cleaning and statistical analysis; BioMagResBank, Golm Metabolome Database, and MassBank for spectral comparison and compound identification (Aguiar-Pulido et al. 2016; Roessner and Bowne 2009).

Multi-omics integration tools are required for cross validation and systems level modeling of functionality of microbes. Tools such as Cytoscape, KEGG Mapper, and HUMAnN2 enable cross-omics pathway mapping (Abubucker et al. 2012; Liu et al. 2021).

4.1 Challenges in Bioinformatics

Despite advancements in computational field pertaining to meta-omics data which is massive in size and is heterogeneous, there are several limitations which need to be addressed. Challenges continue to persist in standardization, data normalization, computational efficiency and interpretability due to incomplete coverage, short read lengths especially from second generation sequencing platforms, high polymorphism or heterogeneity complicating assembly, noise in raw reads, and the high percentage of reads not form known source or trait due to limited reference genomes. The specialized assemblers like MetaVelvet-SL and Meta-IDBA are tools to address chimeric reads and varying coverage depths among different species. Besides this the dearth of integrating approaches across multi-omics datasets remain significant (Hu et al. 2022).

5 Conclusion and Future Outlook

Microorganisms play a key role in agricultural sustainability by affecting soil and plant health and plant productivity. Limitations of traditional cultivation-dependent methods for studying microbes and their communities have been circumvented by meta-omics approaches. Meta-omics has transformed and dramatically advanced agricultural microbiology by providing unparalleled insights into microbial community structure, function, and interactions within agro-ecosystems. By decoding potential of microbial communities in rhizosphere, endosphere, and phyllosphere, meta-omics techniques offer transformative solutions for sustainable agriculture. However, single-omics analysis has its own limitations and insufficient to fully characterize the complex physiological and biochemical phenomena of microbial communities. For comprehensive understanding of microbial diversity and system-level understanding of soil and plant microbiomes, integrating multi-omics data is crucial which will also help in overcoming biases of diverse omics technologies and achieve cross-validation. Future efforts should focus on sample standardization, advanced culturomics and synthetic microbial communities (SynComs) to validate meta-omics findings in vivo, SIP-omics for sorting active microbial members, AI-powered analytics for predictive ananysis and bioindicator development, portable real-time sequencing, and microbiome engineering for sustainable farming by creating need based tailored bioformulations (Nikoloudaki et al. 2024; Schwartz et al. 2016; Saad et al. 2020; Mitter et al. 2021). Despite the challenges in field of meta-oimcs technologies, there is unprecedented potential through research advances to underpin the next generation of sustainable, microbiome-informed agriculture by understanding and manipulating plant-microbe interactions in better way.

Conflict of Interest The authors declare no conflict of interest.

References

Abiraami TV, Singh S, Nain L (2020) Soil metaproteomics as a tool for monitoring functional microbial communities: promises and challenges. Rev Environ Sci Biotechnol 19:73–102

Abubucker S, Segata N, Goll J, Schubert AM, Izard J, Cantarel BL et al (2012) Metabolic reconstruction for metagenomic data and its application to the human microbiome. PLoS Comput Biol 8(6):e1002358

Aguiar-Pulido V, Huang W, Suarez-Ulloa V, Cickovski T, Mathee K, Narasimhan G (2016) Metagenomics, metatranscriptomics, and metabolomics approaches for microbiome analysis. Evol Bioinforma 12:5–16

Bao Z, Okubo T, Kubota K, Kasahara Y, Tsurumaru H, Anda M, Ikeda S, Minamisawa K (2014) Metaproteomic identification of diazotrophic methanotrophs and their localization in root tissues of field-grown rice plants. Appl Environ Microbiol 80:5043–5052. https://doi.org/10.1128/AEM.00969-14

Basu A, Prasad P, Das SN, Kalam S, Sayyed RZ, Reddy MS, El-Enshasy HA (2021) Plant growth promoting rhizobacteria (PGPR) as green bioinoculants: recent developments, constraints, and prospects. Sustainability 13(3):1140

Bonini P, Rouphael Y, Moreno MBM, Lee B, Cardarelli M, Erice G, Cirino V, Lucini L, Colla G (2020) Microbial-based biostimulant enhances sweet pepper performance by metabolic reprogramming of phytohormone profile and secondary plant metabolism. Front Plant Sci 11:567388

Brader G, Compant S, Mitter B, Trognitz F, Sessitsch A (2017) Metabolic potential of endophytic bacteria. Curr Opin Biotechnol 43:82–90

Bulgarelli D, Garrido-Oter R, Münch PC, Weiman A, Dröge J, Pan Y et al (2015) Structure and function of the bacterial root microbiota in wild and domesticated barley. Cell Host Microbe 17(3):392–403

Caporaso JG, Kuczynski J, Stombaugh J, Bittinger K, Bushman FD, Costello EK, Fierer N, Peña AG, Goodrich JK, Gordon JI, Huttley GA, Kelley ST, Knights D, Koenig JE, Ley RE, Lozupone CA, McDonald D, Muegge BD, Pirrung M, Reeder J, Sevinsky JR, Turnbaugh PJ, Walters WA, Widmann J, Yatsunenko T, Zaneveld J, Knight R (2010) QIIME allows analysis of high-throughput community sequencing data. Nat Methods 7(5):335–336. https://doi.org/10.1038/nmeth.f.303. Epub 2010 Apr 11. PMID: 20383131; PMCID: PMC3156573.

Dickie IA et al (2018) Towards robust and repeatable sampling methods in eDNA-based studies. Mol Ecol Resour 18(5):940–952

Djemiel C et al (2022) Meta-omics to provide modern microbial indicators for monitoring soil quality and securing food production. Front Microbiol 13:889788

Fan Y et al (2021) Functional screening of thermophilic esterases from metagenomic library of urban basin soil. Biodegradation 32(2):147–160

Guazzaroni ME, Silva-Rocha R, Ward RJ (2015) Synthetic biology approaches for bioremediation. Curr Opin Biotechnol 33:142–148

Gupta G, Parihar SS, Ahirwar NK, Snehi SK, Singh V (2015) Plant growth promoting rhizobacteria (PGPR): current and future prospects. J Microb Biochem Technol 7:96–102

Hambraeus G, von Wachenfeldt C, Hederstedt L (2003) Genome-wide survey of mRNA half-lives in Bacillus subtilis identifies extremely stable mRNAs. Mol Gen Genomics 269:706–714

Hayden HL, Savin KW, Wadeson J, Gupta VVSR, Mele PM (2018) Comparative metatranscriptomics of wheat rhizosphere microbiomes in disease suppressive and non-suppressive soils for rhizoctonia solani AG8. Front Microbiol 9:859. https://doi.org/10.3389/fmicb.2018.00859

Hu B, Canon S, Eloe-Fadrosh EA, Anubhav, Babinski M, Corilo Y, Davenport K, Duncan WD, Fagnan K, Flynn M, Foster B, Hays D, Huntemann M, Jackson EKP, Kelliher J, Li P-E, Lo C-C, Mans D, McCue LA, Mouncey N, Mungall CJ, Piehowski PD, Purvine SO, Smith M, Varghese NJ, Winston D, Xu Y, Chain PSG (2022) Challenges in bioinformatics workflows for processing microbiome omics data at scale. Front Bioinform 1:826370. https://doi.org/10.3389/fbinf.2021.826370

Huang Y, Hu H, Zhang T, Wang W, Liu W, Tang H (2024) Meta-omics assisted microbial gene and strain resources mining in contaminant environment. Eng Life Sci 24(1):e2300207

Idle JR, Gonzalez FJ (2007) Metabolomics. Cell Metab 6(5):348–351

Ijaq J, Malik G, Kumar A, Kumar A, Das PS, Meena N et al (2019) A model to predict the function of hypothetical proteins through a nine-point classification scoring schema. BMC Bioinform 20:14. https://doi.org/10.1186/s12859-018-2554-y

Keiblinger KM, Wilhartitz IC, Schneider T et al (2016) Soil metaproteomics—comparative evaluation of protein extraction protocols. Soil Biol Biochem 54:14–24

Keiblinger KM, Wilhartitz IC, Schneider T, Roschitzki B, Schmid E, Eberl L, Riedel K, Zechmeister-Boltenstern S (2012) Soil metaproteomics—comparative evaluation of protein extraction protocols. Soil Biol Biochem 54:14–24. https://doi.org/10.1016/j.soilbio.2012.05.014

Khan A, Sayyed RZ, Seifi S (2019) Rhizobacteria: legendary soil guards in abiotic stress management. In: Sayyed RZ, Arora NK, Reddy MS (eds) Plant growth promoting rhizobacteria for sustainable stress management: abiotic stress management, vol 1. Springer, Singapore, pp 27–342

Knief C, Delmotte N, Chaffron S, Stark M, Innerebner G, Wassmann R, von Mering C, Vorholt JA (2012) Metaproteogenomic analysis of microbial communities in the phyllosphere and rhizosphere of rice. ISME J 6:1378–1390. https://doi.org/10.1038/ismej.2011.192

Kopylova E, Noé L, Touzet H (2012) SortMeRNA: filtering rRNAs in metatranscriptomic data. Bioinformatics 28(24):3211–3217

Kunath B, Minniti G, Skaugen M, Hagen L, Vaaje-Kolstad G, Eijsink V, Pope P, Arntzen M (2019) Metaproteomics: sample preparation and methodological considerations. In: Capelo-Martínez JL (ed) Emerging sample treatments in proteomics, Advances in experimental medicine and biology, vol 1073. Springer, Cham. https://doi.org/10.1007/978-3-030-12298-0_8

Lemanceau P, Blouin M, Muller D, Moënne-Loccoz Y (2017) Let the core microbiota be functional. Trends Plant Sci 22(7):583–595

Liu Z, Ma A, Mathé E, Merling M, Ma Q, Liu B (2021) Network analyses in microbiome based on high-throughput multi-omics data. Brief Bioinform 22(2):1639–1655

Love MI, Huber W, Anders S (2014) Moderated estimation of fold change and dispersion for RNA-seq data with DESeq2. Genome Biol 15:550

Martin-Laurent F, Philippot L, Hallet S, Chaussod R, Germon JC, Soulas G et al (2001) DNA extraction from soils: old bias for new microbial diversity analysis methods. Appl Environ Microbiol 67:2354–2359. https://doi.org/10.1128/AEM.67.5.2354-2359.2001

Mattarozzi M, Di Zinno J, Montanini B, Manfredi M, Marengo E, Fornasier F et al (2020) Biostimulants applied to maize seeds modulate the enzymatic activity and metaproteome of the rhizosphere. Appl Soil Ecol 148:103480. https://doi.org/10.1016/j.apsoil.2019.103480

Mehta S, Kumar P, Crane M, Johnson JE, Sajulga R, Nguyen DDA, McGowan T, Arntzen MØ, Griffin TJ, Jagtap PD (2021) Updates on metaQuantome software for quantitative metaproteomics. J Proteome Res 20(4):2130–2137. https://doi.org/10.1021/acs.jproteome.0c00960

Meyer F, Paarmann D, D'Souza M et al (2008) The metagenomics RAST server – a public resource for the automatic phylogenetic and functional analysis of metagenomes. BMC Bioinform 9:386. https://doi.org/10.1186/1471-2105-9-386

Michealsamy A, Thangamani L, Manivel G, Kumar P, Sundar S, Piramanayagam S, Natarajan J (2021) Current research and applications of meta-omics stratagems in bioremediation: a bird's-eye view. J Appl Biotechnol Rep 8(2):109–115

Mirete S, López-Pérez M, González-Pastor JE (2016) Functional metagenomics of extreme environments. Curr Opin Biotechnol 38:143–149

Mitter B, Pfaffenbichler N, Sessitsch A (2021) Microbiome management to improve crop productivity: a future perspective. Agronomy 11(1):251

Nikoloudaki O, Aheto F, Di Cagno R, Gobbetti M (2024) Synthetic microbial communities: a gateway to understanding resistance, resilience, and functionality in spontaneously fermented food microbiomes. Food Res Int 192:114780. https://doi.org/10.1016/j.foodres.2024.114780

Nilsson RH, Anslan S, Bahram M, Wurzbacher C, Baldrian P, Tedersoo L (2018) Mycobiome diversity: high-throughput sequencing and identification of fungi. Nat Rev Microbiol 17(2):95–109

Overy DP, Bell MA, Habtewold J, Helgason BL, Gregorich EG (2021) Omics' technologies for the study of soil carbon stabilization: a review. Front Environ Sci 9:116

Pascale A, Proietti S, Pantelides IS, Stringlis IA (2020) Modulation of the root microbiome by plant molecules: the basis for targeted disease suppression and plant growth promotion. Front Plant Sci 10:1741

Rai S, Solanki AK, Anal AKD (2020) Modern biotechnological tools: an opportunity to discover complex phytobiomes of horticulture crops. In: Solanki MK, Kashyap PL, Ansari RA, Kumari B (eds) Microbiomes and plant health: panoply and their applications. Elsevier, United States, pp 85–124

Rai S, Ayman O, Rehan M, Al-Turki A, Sagar A, Ilyas N, Sayyed R, Hasanuzzaman M (2023) Crop microbiome: their role and advances in molecular and omic techniques for the sustenance of agriculture. Planta 257:27. https://doi.org/10.1007/s00425-022-04052-5

Ranjard L et al (2009) Sampling strategy for assessing microbial diversity in soil. Appl Soil Ecol 42:150–159

Renu, Gupta SK, Rai AK, Sarim KM, Sharma A, Budhlakoti N et al (2019) Metaproteomic data of maize rhizosphere for deciphering functional diversity. Data Brief 27:104574. https://doi.org/10.1016/j.dib.2019.104574

Roessner U, Bowne J (2009) What is metabolomics all about? BioTechniques 46(5):363–365

Ross EM, Petrovski S, Moate PJ, Hayes BJ (2012) Metagenomics of rumen bacteriophage from thirteen lactating dairy cattle. BMC Microbiol 13:242

Saad MM, Eida AA, Hirt H (2020) Tailoring plant-associated microbial inoculants in agriculture: a roadmap for successful application. J Exp Bot 71(15):3870–3885

Salvato F, Hettich RL, Kleiner M (2021) Five key aspects of metaproteomics as a tool to understand functional interactions in host-associated microbiomes. PLoS Pathog (2):e1009245. https://doi.org/10.1371/journal.ppat.1009245. PMID: 33630960; PMCID: PMC7906368.

Salvato F, Vintila S, Finkel OM, Dangl JL, Kleiner M (2022) Evaluation of protein extraction methods for metaproteomic analyses of root-associated microbes. Mol Plant-Microbe Interact 35(11):977–988. https://doi.org/10.1094/MPMI-05-22-0116-TA

Schiebenhoefer H et al (2019) Comparative evaluation of functional annotation in metaproteomics. Expert Rev Proteomics 16(5):375–390

Schloss PD, Westcott SL, Ryabin T, Hall JR, Hartmann M, Hollister EB, Lesniewski RA, Oakley BB, Parks DH, Robinson CJ, Sahl JW, Stres B, Thallinger GG, Van Horn DJ, Weber CF (2009) Introducing mothur: open-source, platform-independent, community-supported software for describing and comparing microbial communities. Appl Environ Microbiol 75(23):7537–7541. https://doi.org/10.1128/AEM.01541-09. Epub 2009 Oct 2. PMID: 19801464; PMCID: PMC2786419.

Schneider T et al (2012) Who is who in litter decomposition? ISME J 6:1749–1762

Schwartz E, Van Horn DJ, Hermann M et al (2016) Stable isotope probing to investigate microbial growth and death in complex communities. Curr Opin Biotechnol 41:14–18

Sharuddin SS, Ramli N, Yusoff MZM, Muhammad NAN, Ho LS, Maeda T (2022) Advancement of metatranscriptomics towards productive agriculture and sustainable environment: a review. Int J Mol Sci 23(7):3737

Tchigvintsev A, Tran H, Popovic A et al (2021) Functional metagenomics reveals novel biocatalysts in cold marine environments. Front Microbiol 12:632

Terrat S et al (2015) Improving soil microbial community description: a meta-analysis. Soil Biol Biochem 84:31–40

Tilgam J, Jayaswal D, Ashajyothi M, Singh JP, Kumar A, Chakdar H (2024) Meta-omics approaches for understanding and exploring soil microbial communities for sustainable agriculture. In: Applications of metagenomics. Academic Press, pp 3–22. https://doi.org/10.1016/B978-0-323-98394-5.00013-4

Trivedi P, Leach JE, Tringe SG, Sa T, Singh BK (2020) Plant–microbiome interactions: from community assembly to plant health. Nat Rev Microbiol 18(11):607–621

Vestergaard G et al (2017) ISO standard DNA extraction methods. Soil Biol Biochem 115:122–130

Waller F, Achatz B, Baltruschat H, Fodor J, Becker K, Fischer M et al (2005) The endophytic fungus *Piriformospora indica* reprograms barley to salt-stress tolerance, disease resistance, and higher yield. Proc Natl Acad Sci 102(38):13386–13391

Wang Z, Gerstein M, Snyder M (2009) RNA-Seq: a revolutionary tool for transcriptomics. Nat Rev Genet 10(1):57–63. https://doi.org/10.1038/nrg2484

Wang Y, Chen Y, Zhou Q et al (2012) A culture-independent approach to unravel uncultured bacteria and functional genes in a complex microbial community. PLoS One 7(10):e47530

White RA III, Bottos EM, Roy Chowdhury T, Zucker JD, Brislawn CJ, Nicora CD et al (2016) Moleculo long-read sequencing facilitates assembly and genomic binning from complex soil metagenomes. Msystems 1(3):e00045–e00016

Wilmes P, Bond PL (2006) Metaproteomics: studying functional gene expression in microbial ecosystems. Trends Microbiol 14(2):92–97

Wilmes P, Heintz-Buschart A, Bond PL (2015) A decade of metaproteomics: where we stand and what the future holds. Proteomics 15(20):3409–3417

Zenati N, Hachani M, Aissani L, Bacha MK, Djemaa I (2023) Comparison between 5 extractions methods in either plasma or serum for metabolomics analysis using LC-HRMS. Cell Mol Biol Lett 28(1):1–14

Zinicola M et al (2015) Shotgun metagenomic sequencing reveals microbiome and functional genes associated with bovine digital dermatitis. PLoS One 10(7):e0133674

Metagenomics as a Tool to Decipher Rhizosphere Microbial Community Shift Due to Climate Change

K. Jayalakshmi, Pramod Kumar Sahu, Pranjali V. Bhadane, Pradeep Manyam, and Ram Dutta

1 Introduction

Climate change can affect many biological communities. These communities are combination of many complex form of organisms coexist alongside thousands of different species, some of which are beneficial, some are pathogenic, and rest with insignificant impact on environment (Vandenkoornhuyse et al. 2015). All these climate changes can affect soil microflora or microbiota either directly, or indirectly through affecting plant physiological processes and plant community composition (Bardgett et al. 2013). It is not clear whether natural communities will react to climate change in the same manner because they are made up of creatures with wildly varied physiological features and dispersal abilities.

Relationships between plants and soil microbes not yet received as much attention as those between plants and animals under global climate change. Changes in the interactions between soil microorganisms and plants may have impact on microbial communities and functioning of ecosystems, because these microbes govern plant populations, feed nutrients to plants, permit coexistence among neighbors, and regulate nutrient transformations. According to Schimel et al. (2007) and de Vries et al. (2012), soil communities demonstrate changes in species interactions by climate change cascade to changes in biodiversity and the function of terrestrial

K. Jayalakshmi (✉) · P. V. Bhadane · R. Dutta
ICAR-Directorate of Onion and Garlic Research, Pune, Maharashtra, India

P. K. Sahu
ICAR-National Bureau of Agriculturally Important Microorganisms, Mau, Uttar Pradesh, India

P. Manyam
Department of Plant Pathology, Sri Venkateswara Agricultural College, ANGRAU, Tirupati, Andhra Pradesh, India

Renu et al. (eds.), *Meta-omics in Crop Improvement*,
https://doi.org/10.1007/978-981-95-2294-1_2

ecosystems. Indeed, interactions between soil microbes and plants can influence landscape patterns of plant and animal abundance, diversity, and composition (van der Putten et al. 2013). Particularly in the light of climate change, microbial communities are climate-sensitive; climate change can alter their composition and function. Metagenomics enables monitoring of microbial shifts occurred due to climate change for example it detects the changes in community structure, gene abundance as well as metabolic potential under different climate scenarios.

Metagenomics, the study of genetic material from environmental samples, has emerged as a powerful tool to investigate these shifts. It allows researchers to identify and analyze bacterial, archaeal, fungal, and viral species present in the rhizosphere without culturing. Identification of the genes and metabolic processes can be performed in the microbial communities due to global climate change. It also helps to calculate the relative abundances of different microbial species in various climate conditions and to understand how microbial communities have changed genetically over time in response to the pressures brought on by the climate. Metagenomics also reveals microbial role in key biogeochemical cycles like nitrogen cycle, carbon cycle and helps to predict how microbial communities can work for climate change, e.g., methane production and carbon sequestration. Metagenomic data can serve as indicators of ecosystem resilience or vulnerability, bioindicator to assess the resilience of ecosystems to climate stressors like drought, heatwaves, or flooding. By using metagenomics predictive models can be developed which anticipate responsible responses to climate shifts. Thompson et al. (2017) discusses the Earth Microbiome Project (EMP) and its role in generating a global baseline of microbial diversity which may help long term and global monitoring for detecting environmental and climate induced shifts in microbial communities over time. This chapter focuses on metagenomics and its pivotal role in understanding the effect of climate change on microbial communities throughout different ecosystems.

2 Rhizosphere Microbial Community Structure and Function

2.1 Diversity and Composition of Rhizosphere Microbiome

Rhizosphere microbiome diversity and microbiome composition are directly influenced by climate change, and metagenomic approaches offer powerful tool to unravel these shifts both taxonomic and functional level. The rhizosphere is an evolving interface between plant roots and soil and works as a hotspot for complex interactions between both (Korenblum et al. 2020; Liu et al. 2020). These interactions play a pivotal role in a complex dynamic of microbial communities in the rhizosphere, have already be studied in different ecosystems like forests (Chen et al. 2018b), grasslands (Birgander et al. 2017), and croplands (Simonin et al. 2020). Different levels of complexity and functionality found at different ecosystems, which plays an important role in the growth and development of plants (Huang et al. 2014).

Rhizosphere microbes contribute as a major factor in maintaining plant health by enhancing nutrient acquisition and providing resilience against many biotic and abiotic stresses. For example, they help plants combat diseases (Song et al. 2023), tolerate high salinity (Schmitz et al. 2022), and tolerant to drought (De Vries et al. 2020), and adapt to the changing environmental conditions (Berendsen et al. 2012; Trivedi et al. 2020). In exchange, plants also supply essential nutrients to the soil microbiota such as carbon by creating a symbiotic relationship with microbes (Bais et al. 2006). The study between the living communities, plants and rhizosphere microbes has advanced significantly, yet limitations persist in the traditional cultivation techniques, which have been criticized for their bias and inability to represent the full diversity of microbial communities (Torsvik et al. 1996; Van Elsas and Boersma 2011).

Metagenomics works on genetic data derived from cultured strains for understanding microbial physiology and functions (Giovannoni and Stingl 2007; Nichols 2007). While high-throughput sequencing has revolutionized microbial ecology, bridging it with cultivation-based approaches remains essential for a holistic understanding of rhizosphere dynamics. This integrated strategy not only reveals the intricacies of plant-microbe interactions but also paves the way for sustainable agricultural innovations.

2.2 Microbial Interactions: Plant-Microbe and Microbe-Microbe Associations

Plants have evolved in a microbial world where their survival, growth, and establishment are profoundly influenced by interactions with soil microbial communities. These plant-microbe interactions are fundamental, specifically in the event of climate change effect on soil microbial populations could determine the plant species ability to establish and grow in desired communities, thereby shaping ecosystem (Walther et al. 2002; Parmesan and Yohe 2003). Plants migrations, driven by climate change, are well-documented, yet the cascading effects of these shifts on plant-soil-microbe interactions remain less understood (Fischer et al. 2014; Mohan et al. 2014). The rhizosphere which is a the narrow soil region directly influenced by plant roots plays a crucial role in plant growth, nutrition, and productivity (Mendes et al. 2013; Lakshmanan et al. 2014). It is a highly dynamic niche enriched with root exudates, including amino acids, organic acids, sugars, and mucilage, which collectively account for a significant proportion of a plant's fixed carbon. These exudates drive microbial activity, altering the composition, structure, and functions of the soil microbial community, which, in turn, impacts plant health and development (Fierer et al. 2007b). Changing temperature and precipitation patterns, as well as CO_2 levels shifts the structure of rhizosphere microbial communities, impacting plant health and nutrient uptake (Singh et al. 2021).

The interplay between plants and their rhizosphere microbiome exemplifies a co-evolved system involving symbiotic, parasitic, mutualistic, and antagonistic relationships (Mendes et al. 2013). Agronomic practices such as crop rotation, intercropping, irrigation, fertilization, and pesticide use can further influence plant-microbial interactions (Liu et al. 2015; Xie et al. 2022). These practices not only affect the abundance and diversity of rhizosphere microbes but also modulate their role in mitigating plant stress and enhancing growth. For instance, microbial communities can buffer plants against biotic stresses like pathogens and abiotic stresses such as drought and salinity, thus contributing to ecosystem stability (Ahmad et al. 2008; Cattelan et al. 1999).

The rhizosphere can be conceptualized as a "holobiont," where the plant and its microbiome form an evolutionary unit of selection (Rosenberg et al. 2007; Zilber-Rosenberg and Rosenberg 2008). This concept is supported by evidence that microbial diversity in the rhizosphere significantly impacts plant fitness and that changes in either the plant genome or the microbial community can influence the holobiont's functionality (Manter et al. 2010; Uroz et al. 2010). However, the transmission of genetic variation in the rhizosphere differs from traditional inheritance, as the microbial genetic pool in soil is vast, with a single gram containing millions of cells from over 10,000 taxa (Fierer et al. 2007a,b). Plants may selectively enrich functional microbial groups that support growth under specific conditions, reflecting a high degree of co-adaptation. Environmental changes and plant stress can also alter the symbiotic nature of plant-microbe interactions. For example, the relationship between mycorrhizal fungi and plant hosts is not always mutualistic and can shift depending on environmental factors or plant stress levels (Johnson et al. 2008). These variations highlight the complexity of interactions within the rhizosphere and the potential for indirect effects of climate change, mediated by shifts in plant communities, to counteract or differ from direct impacts on microbial communities (Kardol et al. 2010).

Metagenomic sequencing data helps to understand microbial networks, revealing both taxonomic diversity and functional gene content in the rhizosphere as reported by Mendes et al. (2013). Competitive and cooperative interactions among microbes influence nutrient cycling, pathogen suppression, and ecosystems responses to stressors like drought or warming (Faust and Raes 2012). Symbiotic relationships critical for nutrient acquisition and stress tolerance are disrupted by climate extremes, and metagenomics can detect shifts in symbionts abundance and function (Van der Heijden et al. 2016).

2.3 Factors Influencing Rhizosphere Microbial Assemblages

The formation of the rhizosphere microbiome assemblage is a highly intricate process shaped by various factors, such as the genetic characteristics of the host plant, the soil's physical and chemical attributes, and the diversity and composition of the native soil microbial community (Haichar et al. 2008; İnceoğlu et al. 2012). These factors interact dynamically to shape the microbial communities that drive critical ecological processes. Microbial responses to climate change often exhibit phylogenetic clustering, with traits underlying these responses being conserved within microbial lineages. For instance, elevated CO_2, warming, drought, flooding, and snow cover changes consistently affect the relative abundance of both bacteria and fungi (Schimel et al. 2007; De Vries and Shade 2013). These groups demonstrate higher stress resistance, enabling them to persist under fluctuating moisture conditions and other climate-induced stresses. Moreover, the dominance of fungal-dominated soil food webs under climate change conditions has been linked to significant carbon (C) and nitrogen (N) losses, underscoring the functional impact of these microbial shifts on ecosystem processes (Gordon et al. 2008; de Vries et al. 2012).

Climate change also affects plant-rhizosphere interactions by influencing root phenology, a process less explored in phenology studies compared to shoot phenology (Iversen et al. 2015). Root phenology, unlike that of shoots, exhibits significant variation across species and ecosystems, influenced by abiotic factors such as temperature, moisture, and the availability of stored carbon and nutrients (Finzi et al. 2015). These differences in root and shoot phenology can impact rhizosphere interactions, potentially leading to seasonal shifts in microbial community composition (Waldrop and Firestone 2006). Root-associated symbionts, including rhizobia bacteria and mycorrhizal fungi, are crucial for plant productivity as they regulate plant nutrient acquisition (De Bello et al. 2010). Mycorrhizal fungi, in particular, significantly influence soil microbial decomposition activity by modifying plant litter quality and carbon inputs, thereby playing a key role in carbon cycling (Clemmensen et al. 2015).

3 Climate Change and Its Effects on Rhizosphere Microbial Dynamics

3.1 Temperature Variability and Microbial Responses

Rising global temperatures significantly impact soil respiration rates, firstly due to the temperature sensitivity of soil microorganisms and then biochemical processes they mediate. This phenomenon contributes to a positive feedback loop in climate change, as elevated soil respiration releases additional carbon dioxide (CO_2) into the atmosphere, thereby intensifying global warming (Bradford et al. 2008). However, field studies reveal that these temperature-induced increases in microbial activity, soil organic matter decomposition, and carbon turnover are often transient. Over time, acclimation of microbial communities or resource limitations appears to moderate these effects.

Temperature and soil management practices also play an important role in the composition of soil microbial communities. Notably, drought and warming tend to favor certain microbial taxa, such as Acidobacteria, which are adapted to low-nutrient environments (oligotrophic conditions). Conversely, other taxa, including Actinobacteria and Proteobacteria, which thrive in nutrient-rich environments (copiotrophic conditions), often decline under such stressors. These shifts in microbial community structure can be attributed to intrinsic traits like oligotrophy and copiotrophy, which influence the resilience and functional adaptability of microbial populations under changing environmental conditions (DeAngelis et al. 2015). These findings underscore the intricate connections between microbial responses to warming, soil carbon dynamics, and broader climate feedback mechanisms. They highlight the necessity of understanding microbial processes to accurately predict ecosystem responses to climate change. Furthermore, this knowledge is vital for developing soil microbial communities' management strategies to make the effective soil health aimed at mitigating the adverse effects of climate change by enhancing soil carbon storage and maintaining ecosystem stability.

3.2 Drought and Salinity

Drought and fluctuations in soil moisture, often increased by climate change, it can lead to change the soil microbial communities and their functional dynamics. The most adverse effect of drought is reduction in microbial diversity, just due to the low moisture and high temperature coupled with an increase in taxa that are well-adapted to desiccation. Metagenomic analyses have revealed genes associated with osmotic stress tolerance and alternative nutrient acquisition pathways, highlighting microbial strategies to cope with water stress. For instance, gram-positive bacteria, due to their thick cell walls, spore-forming abilities, and slow growth rates, exhibit higher resistance to drought compared to gram-negative bacteria (De Vries and Shade 2013).

Fungi, with their thick cell walls and slower growth strategies, are also more drought-resistant than bacteria, reported by (Schimel et al. 2007; Briones 2014) on the basis of temperature sensitivity between these groups under drought conditions. While bacterial communities respond quickly to moisture pulses, fungi often lag due to their slower growth dynamics (Bell et al. 2008; Cregger et al. 2012). Taxonomic shifts under drought conditions often show an increase in Acidobacteria and Alphaproteobacteria, whereas Actinobacteria may decrease temporarily but recover rapidly (Thomson et al. 2010; Bouskill et al. 2013; Amend et al. 2016). Drought-induced changes in microbial communities are further influenced by soil pH and nutrient availability, often altered by factors such as acid rain or decomposition rates. Functional metagenomics has identified microbial genes involved in acid tolerance and nutrient cycling, emphasizing the role of microbes in maintaining ecosystem functions under stress. However, extreme climatic events, such as drought, have been shown to affect ecosystem functioning more profoundly than gradual changes in CO_2 concentrations or temperature (Reichstein et al. 2013). In some findings underscore the importance of understanding moisture-temperature interactions, as temperature-induced microbial activity can be constrained by diffusion and microbial access to substrates under dry conditions (Zak et al. 1999). Microbial community responses to nitrogen addition and drought reveal functional adaptations. In case of bacteria, genes coding for the degradation of various carbon compounds are linked to nitrogen addition, while drought responses are associated with specific genes, such as those involved in xylan degradation (Amend et al. 2016). These findings suggest that microbial functional traits are key to understanding ecosystem responses to combined climatic stressors. Drought, coupled with temperature fluctuations and soil management factors, exerts complex and often taxon-specific effects on microbial communities. These type of climate changes not only influence microbial diversity but also have cascading effects on malfunctioning of ecosystems processes and decomposition, carbon cycling, nutrient dynamics.

To comprehensively understand the impacts on microbial biomass and activity in saline soils, it is essential to consider the soil's water potential, which is the sum of the osmotic potential and the matrix potential. As salt concentration in the soil solution increases, the water content decreases, leading to changes in the water potential. These changes significantly affect microbial communities, as microorganisms are subjected to varying osmotic and water potentials. While the electrical conductivity (EC) of the soil may show only minor fluctuations, the alterations in water availability and the resulting osmotic stress can have a considerable influence on microbial function. This highlights the importance of considering both water potential and salinity when evaluating microbial behaviour in saline environments, as microorganisms respond not only to the concentration of salts but also to the reduced water availability that accompanies salinization. Salinization refers to over accumulation of several salts including K+, Mg^{2+}, Ca^{2+}, Cl^-, SO_4^{2-}, CO_3^{2-}, HCO_3^-, and Na^+. Sodium accumulation, known as sodification, is particularly harmful as it replaces other essential cations, degrading soil structure and reducing water infiltration. This leads to poor soil permeability, nutrient imbalances, and diminished plant growth. Salinization can occur due to natural factors, such as in arid regions, or through human activities like improper irrigation practices, posing significant risks to agriculture and water resources.

Globally, the productivity of approximately 400 million hectares of arable land is severely constrained by soil salinity (Bot et al. 2000). Salinity adversely impacts plant growth and development, ultimately reducing crop yields, and has become an increasingly widespread issue in agricultural soils worldwide. In temperate and humid climates, salinity is less extensive but is observed in specific areas such as saltwater marshes, roadside environments affected by salts, and regions receiving saline effluents (Pathak and Rao 1998; Qadir et al. 2000; Wichern et al. 2006). Microbial responses to environmental stressors, including high concentrations of salts, heavy metals, fire, and variations in soil moisture, are crucial. These stresses can harm sensitive microorganisms and reduce the activity of survivors due to the metabolic burden of stress tolerance mechanisms (Schimel et al. 2007; Yuan et al. 2007; Ibekwe et al. 2010; Chowdhury et al. 2011).

In fungi, reduced osmotic potential inhibits spore germination, slows hyphal growth, alters morphology (Juniper and Abbott 2006), and modifies gene expression (Liang et al. 2007). These changes often lead to the development of thick-walled spores (Mandeel 2006). However, synthesizing osmolytes to counteract osmotic stress is energy-intensive, requiring 30 to 110 ATP molecules, compared to the 30 ATPs needed for cell wall synthesis (Oren 1999). This high energy demand imposes a significant metabolic burden, limiting the energy available for microbial growth.

3.3 Elevated CO_2 and Its Impact on Microbial Interactions

Climate change leads to change in atmospheric CO_2 levels which effects soil microbial communities indirectly by modifying plant physiological processes, growth, and community composition. Additionally, temperature changes impact these microbial communities both indirectly, through plant-mediated effects, and directly, independent of plant involvement. For instance, numerous studies have reported increased fungal abundance under elevated temperature conditions (Castro et al. 2010; Deslippe et al. 2012; Yergeau et al. 2012). Understanding the effects of elevated CO_2 on microbial community composition is critical, as these communities play a central role in regulating the balance between the storage and release of greenhouse gases such as CO_2 and N_2O, which can either exacerbate or mitigate climate change. Klironomos and Kendrick (1996) observed that under low nitrogen (N) conditions, elevated CO_2 levels led to an increase in the abundance of both decomposer and arbuscular mycorrhizal (AM) fungi, as determined through microscopic hyphal counts. Similarly, Jones and Edwards (1998) reported that elevated CO_2 stimulated cellulose decomposers, which in turn accelerated decomposition rates. These findings highlight the significant impact of CO_2 enrichment on soil microbial dynamics, particularly under varying nitrogen availability. The rhizosphere microbial populations may also be indirectly affected by elevated CO_2 due to potential changes in root exudation patterns. Studies employing metagenomics have revealed shifts in microbial processes associated with phosphorus solubilization, carbon sequestration, and nitrogen fixation under elevated CO_2 conditions. Despite evidence that elevated CO_2 alters fungal and bacterial community composition by influencing the quantity and quality of plant carbon (C) allocation to the soil, its effects on fungal and bacterial diversity remain inconclusive (Ebersberger et al. 2004; Weber et al. 2011; Hayden et al. 2012). Microbial communities exhibit various responses to warming and other environmental perturbations. These responses include resistance, facilitated by microbial trait plasticity, and resilience, demonstrated by the community's ability to return to its original composition after the stress subsides (Allison and Martiny 2008). Understanding these dynamics is essential for predicting microbial contributions to ecosystem functioning in a changing climate.

4 Metagenomics Approaches for Assessing Rhizosphere Microbial Shifts

Metagenomics serves as a powerful tool to explore the functional potential and taxonomic composition of microbial communities (Ye et al. 2019). However, these approaches are often costly and demand significant quantities of microbial starting material. Moreover, many metagenomic techniques lack the ability to differentiate between DNA from transient sources and that originating from living microorganisms (Cheng et al. 2019).

5 Metagenomics Tools and Different Sequencing Technologies

High-Throughput Sequencing (HTS), commonly known as Next-Generation Sequencing (NGS), represents a technological breakthrough revolutionizing the field of genomics. This cutting-edge approach allows researchers to rapidly and efficiently sequence DNA and RNA molecules on an unparalleled scale.

HTS delivers high-quality sequencing data with minimal errors allowing large-scale sequencing of DNA or RNA in a single setup, making it ideal for comprehensive genome and transcriptome studies requiring deep coverage. It generates massive data volumes rapidly, expediting research timelines. Metagenome sequencing analyzes all DNA in a sample, offering a comprehensive view of species diversity, gene composition, and functional roles, including discovering uncultivated species. 16S rRNA sequencing specifically targets the conserved 16S rRNA gene in bacteria and archaea, enabling species identification and diversity quantification like alpha and beta diversity analysis. These approaches provide valuable insights into microbial composition, dynamics, and functions in various ecosystems. Some sequencing platforms are explained as follows:

5.1 Illumina Sequencing

Illumina sequencing, a transformative technology in high-throughput sequencing, has revolutionized genomic research with its exceptional scalability, accuracy, and cost-efficiency. Known as *sequencing-by-synthesis*, this method involves fragmenting DNA samples, attaching sequencing adapters, and amplifying the fragments through Polymerase Chain Reaction (PCR). The process begins with bridge amplification, where DNA fragments ligated to adaptors undergo clonal amplification on a glass slide (Bentley et al. 2008). Sequencing occurs via a cyclic reversible termination method, which sequentially determines each nucleotide in the DNA template. This involves incorporating fluorescently labeled nucleotides, washing away unincorporated bases, imaging, and chemically cleaving the fluorescent label and blocking group to prepare for the next cycle. The method uses modified nucleotides, specifically fluorescently labeled 3′-O-azidomethyl-dNTPs, to temporarily halt the incorporation process. This controlled pausing allows precise imaging of the newly added nucleotide (Guo et al. 2008). A CCD camera scans the flow cell to capture the fluorescent signals, after which the blocking group and fluorescent label are removed, enabling the next nucleotide to be added and read.

This approach allows Illumina platforms to sequence millions of DNA fragments simultaneously, making them ideal for applications such as whole-genome sequencing, exome sequencing, and RNA-seq for transcriptome analysis.

5.2 Oxford Nanopore

Oxford Nanopore sequencing represents a paradigm shift in genomics by offering real-time sequencing capabilities with long-read potential. This technology operates by passing DNA or RNA molecules through nanoporous structures embedded in a membrane. As these molecules traverse the nanopores, changes in the electrical current are recorded and converted into DNA sequence data. Oxford Nanopore's current technology utilizes sequencing flow cells containing hundreds of micro-wells, each equipped with biological nanopores. Distinct from other sequencing methods, Oxford Nanopore sequencing does not require pre-amplification or fragmentation, allowing for swift, mobile, and direct sequencing of genetic material. Library preparation for this process involves fragmenting DNA and attaching adapters, which can be done without the need for clonal amplification. One adapter is linked to a proprietary motor enzyme, while the other is a hairpin oligonucleotide that is connected by a HP motor protein. This setup allows for precise sequencing of both strands of DNA from a single molecule, enabling highly accurate and long-read sequencing. It has proven transformative, especially for applications requiring long-read sequencing, such as de novo genome assembly, structural variant detection, and real-time pathogen surveillance.

5.3 Thermofisher's Ion Torrent

Ion Torrent by ThermoFisher follows a template preparation and clonal amplification process similar to Roche/454 pyrosequencing (Margulies et al. 2005), using emulsion-PCR to amplify adapter-ligated DNA fragments on beads. These beads are then placed in microwells for sequencing-by-synthesis. Unlike pyrosequencing, which detects nucleotides based on light production via luciferase, Ion Torrent employs semiconductor sequencing, relying on ion sensing technology (Rothberg et al. 2011). This method measures pH changes that occur during DNA extension as hydrogen ions are released. The pH variations are detected by an ion sensor in the microwells, which then converts them into a voltage signal, proportional to the number of incorporated bases. This process is repeated for each nucleotide, enabling base discrimination without the need for optical scanning. This advantage accelerates sequencing runs and reduces costs.While its read lengths may not match those of other platforms, the speed and simplicity of Ion Torrent sequencing make it a valuable tool for routine sequencing tasks. Its user-friendly nature and cost-effectiveness contribute to its widespread adoption in various genomic research areas, expanding the scope of high-throughput sequencing technologies.

5.4 Pacific Biosciences Sequencing: Single-Molecule Real-Time (SMRT) Technology

Pacific Biosciences (PacBio) sequencing, based on Single-Molecule Real-Time (SMRT) technology, offers long-read sequencing capabilities with high accuracy. Unlike Illumina and Oxford Nanopore sequencing, which use different methods to read DNA, PacBio sequencing directly observes DNA synthesis in real-time through the use of fluorescently labeled nucleotides. This allows for the sequencing of long DNA fragments, making it particularly useful for identifying structural variants, genomic rearrangements, and epigenetic modifications. SMRT sequencing employs single-stranded hairpin adapters (SMRT-bell) that enable the sequencing of the original DNA molecule multiple times, improving accuracy and eliminating the need for clonal amplification. The sequencing process occurs in zero-mode waveguides (ZMWs), nanoscale wells with a single immobilized polymerase, allowing for real-time monitoring of nucleotide incorporation and continuous polymerization. Although PacBio sequencing typically has lower throughput than Illumina sequencing, its ability to generate long, high-fidelity reads makes it invaluable for applications in genome assembly, metagenomics, and transcriptomics, where detailed analysis of complex genomic regions and full-length transcripts is essential.

6 Bioinformatics and Data Analysis Tools: Details Regarding Data Processing Pipelines and Functional Annotation Tools

6.1 Bioinformatic Approaches for NGS Data Analysis

Next-Generation Sequencing (NGS) technologies generate vast quantities of data, making bioinformatics indispensable for managing, processing, and interpreting this information. The core bioinformatic approaches in NGS data analysis encompass multiple stages, from initial data preprocessing to advanced integrative analyses.

6.1.1 Tools for Data Processing

Before meaningful biological insights can be derived, raw sequencing data must undergo preprocessing to ensure quality and accuracy.

Quality Check of Sequences

I. *FastQC* evaluates the quality of sequencing data by analyzing various parameters such as per-base quality scores, GC content, and adapter contamination. It provides visual reports to help identify potential issues in the raw reads (Andrews 2010).

II. *FASTX-toolkit* focuses on pre-processing reads, allowing users to evaluate and improve the quality of sequencing data through tasks like filtering, trimming, and quality assessment (Gordon and Hannon 2010).
III. *MultiQC* aggregates output from various bioinformatics tools like FastQC into a single, unified report, simplifying the evaluation of multiple datasets (Ewels et al. 2016).

Trimming of Adapters and Low-Quality Base

I. *Trimmomatic* removes low-quality reads and adapter sequences, enabling high-quality data for downstream analysis. It offers flexibility in setting quality thresholds and handling paired-end reads (Bolger et al. 2014).
II. *Cutadapt* focuses on detecting and removing adapter sequences from sequencing data efficiently. It supports custom adapter sequences and is widely used for small RNA sequencing (Martin 2011).
III. *fastp* is an all-in-one tool for quality control and read preprocessing. It integrates adapter trimming, quality filtering, and basic QC metrics in a fast and user-friendly manner (Chen et al. 2018a,b).

Alignment of Sequence Reads to Reference Genome

I. BWA (Burrows-Wheeler Aligner) maps short reads to a reference genome using efficient indexing methods, balancing speed and memory usage. It is highly reliable for analyzing large-scale sequencing data (Li and Durbin 2009).
II. Bowtie is a lightweight and ultra-fast tool for aligning short reads to large genomes. It uses a Burrows-Wheeler transform-based approach to achieve high speed while maintaining alignment accuracy (Langmead et al. 2009).
III. dragMAP is a newer alignment tool designed for cloud-based environments. It efficiently handles large datasets, optimizing cost and time for alignment tasks (DragMAP team 2021).

Report Visualization

MultiQC consolidates reports from multiple bioinformatics tools, providing a comprehensive overview of sequencing quality, alignment metrics, and other analyses. It simplifies comparative assessments across multiple samples or datasets (Ewels et al. 2016).

6.2 16S rRNAseq

The 16S rRNA gene is the "gold standard" for prokaryotic studies due to its conserved regions and variable domains. Tools like QIIME and Mothur are widely used for processing sequencing data, including tasks like denoising, operational taxonomic unit (OTU) classification, and taxonomic lineage inference. Databases such as Greengenes, Silva, and the Ribosomal Database Project (RDP) provide references for identifying microbial taxa.

6.2.1 16S rRNAseq Analysis Pipelines

I. *QIIME2* (Quantitative Insights Into Microbial Ecology) is a comprehensive and extensible platform for microbiome analysis, providing tools for sequence quality control, diversity analyses, and taxonomic classification (Bolyen 2019).
II. *mothur* is an open-source software package designed for processing and analyzing microbial community sequencing data, including clustering and classification of 16S rRNA sequences (Schloss et al. 2009).
III. *USEARCH* is a fast and efficient pipeline for clustering, denoising, and taxonomic assignment of 16S rRNA sequences, known for its high-performance tools like OTU picking and chimera detection (Edgar 2010).

6.2.2 Ribosomal RNA Databases

I. *Greengenes* is a curated database for 16S rRNA gene sequences, offering reference data for taxonomic classification and phylogenetic studies (DeSantis et al. 2006).
II. *Silva* database is a robust resource containing high-quality ribosomal RNA sequence data, supporting microbial taxonomic assignments and phylogenetic analyses (Quast 2013).
III. *RDP (Ribosomal Database Project)* provides curated ribosomal RNA sequences and analytical tools for taxonomic classification, phylogeny construction, and diversity analyses (Cole et al. 2014).

6.3 Shotgun Metagenomics

Shotgun metagenomics enables the analysis of complete protein-coding genes and operons, providing valuable functional insights. To achieve this, assembling short reads into contigs and scaffolds is crucial. Early assembly tools, originally designed for single genomes, struggled with metagenomic data. However, modern tools have been optimized for handling multiple genomes, making them more effective for constructing a comprehensive view of the sequenced community.

6.3.1 Taxonomic Classification

I. *MetaPhlAn4* (Metagenomic Phylogenetic Analysis) is a computational tool that uses marker genes to provide high-resolution taxonomic profiling of microbial communities from metagenomic sequencing data (Blanco-Míguez 2023).
II. *Kaiju* is a sensitive and fast taxonomic classification tool for metagenomic reads, based on protein-level comparisons, enabling the identification of microbial communities, including viruses (Menzel 2016).

III. *Kraken* uses k-mer-based exact matching to classify metagenomic sequences against a database of genomes for rapid and accurate taxonomic identification (Wood et al. 2014).

6.3.2 Assembly of Metagenomic Reads

I. *metaSPAdes* is a metagenome-specific assembler that optimally reconstructs genomes from complex microbial communities with varying abundance levels (Nurk et al. 2017).
II. *metaIDBA* is a tool designed for de novo assembly of metagenomic sequences, accounting for uneven sequencing depths in microbial communities (Peng et al. 2011).

6.3.3 Protein Databases for Taxonomic Classification

I. *NCBI non-redundant protein database* This is a comprehensive database of curated protein sequences, widely used in metagenomic studies for taxonomic and functional classification (Pruitt 2012).

6.3.4 Gene Annotation

I. *Prokka* is a rapid prokaryotic genome annotation tool that predicts and annotates genes, focusing on bacterial and archaeal genomes (Seemann 2014).
II. *MetaGeneMark* is an efficient tool for identifying protein-coding genes in metagenomic sequences, even in unassembled reads (Zhu et al. 2010).

6.3.5 Databases for Functional Annotation of Genes

I. *COG (Clusters of Orthologous Groups)* database classifies genes into functional categories based on orthologous relationships, aiding in the understanding of gene functions in microbial communities (Tatusov et al. 2000).
II. *KEGG (Kyoto Encyclopedia of Genes and Genomes)* is a database that maps genes to metabolic pathways and functional networks, facilitating the study of cellular processes and environmental systems (Kanehisa et al. 2017).
III. *GO (Gene Ontology)* provides a structured vocabulary for gene functions, biological processes, and cellular components, helping annotate metagenomic datasets (Ashburner et al. 2000).

6.3.6 Pipelines for Metagenomics Analysis

Metagenomics analysis has become a cornerstone for exploring the microbial diversity and functional potential within environmental or host-associated communities. To effectively handle and interpret the massive datasets generated from metagenomics studies, bioinformatics pipelines provide streamlined frameworks for processing, analyzing, and visualizing data. Several robust tools and platforms have been developed to cater to the diverse needs of metagenomics research:

I. *CloVR-Metagenomics*: This pipeline is a cloud-based platform designed to facilitate metagenomic analysis. It integrates multiple tools and automates workflows for taxonomic classification and functional annotation of metagenomic data. CloVR's compatibility with cloud computing resources makes it accessible to researchers with varying computational resources (Angiuoli et al. 2011).

II. *Galaxy Platform (Metagenomics Pipeline)*: The Galaxy platform offers a user-friendly, web-based interface for metagenomics data analysis. It enables researchers to construct custom pipelines for tasks such as sequence quality control, taxonomic profiling, and functional annotation. The platform supports reproducible research by maintaining detailed records of analysis steps (Blankenberg et al. 2010; Afgan et al. 2018).

III. *IMG/M*: The Integrated Microbial Genomes with Microbiomes system (IMG/M) is a comprehensive platform for analyzing and visualizing metagenomic datasets. It facilitates the annotation and comparison of microbial genomes, with an emphasis on functional and phylogenetic insights (Markowitz et al. 2014; Chen et al. 2021).

IV. *MetAMOS*: MetAMOS is an open-source software suite that integrates various tools for metagenomic assembly and annotation. Its modular design allows for flexibility in tailoring workflows to specific research needs, making it a versatile option for researchers working with diverse datasets (Treangen et al. 2013).

V. *MG-RAST*: The Metagenomics Rapid Annotation using Subsystem Technology (MG-RAST) pipeline is widely used for processing and annotating metagenomic data. It provides extensive taxonomic and functional analysis features and includes a public repository for data sharing and comparative studies. MG-RAST's user-friendly interface and robust analytical capabilities make it a popular choice (Meyer et al. 2008).

VI. *RAMMCAP*: This pipeline offers a comprehensive framework for analyzing shotgun metagenomics data. It integrates tools for assembly, annotation, and statistical analysis, streamlining the entire metagenomics workflow. Its efficient processing capabilities are especially useful for large-scale studies (Arndt et al. 2012).

VII. *SmashCommunity*: A modular metagenomics analysis pipeline, SmashCommunity emphasizes scalability and flexibility. It is particularly effective for large datasets, offering tools for assembly, annotation, and comparative analysis (Arndt et al. 2012).

These pipelines collectively address the complexities of metagenomic data, offering tailored solutions for tasks like taxonomic profiling, functional annotation, assembly, and comparative analysis. Their adaptability and efficiency have revolutionized how researchers explore microbial communities, enabling more comprehensive and accurate insights into microbial ecology and function.

7 Linking Taxonomic Shifts to Functional Changes

Identifying key indicator taxa and Connecting metagenomics data to ecological functions.

The study of microbial communities has gained significant importance due to their critical roles in human health, industrial processes, and environmental ecosystems. Understanding these communities involves examining both their taxonomic composition and functional capacities, which provide complementary insights into community structure and function. Taxonomic composition is often derived from targeted 16S rRNA sequencing, a method that utilizes hypervariable regions of the 16S rRNA gene to identify microbes and estimate their relative abundances. This taxonomic data reveals interactions within microbial communities or between microbes and their host, aiding in the detection of ecological shifts or disease associations (Caporaso et al. 2010; Schloss et al. 2009). In contrast, functional capacities are assessed through whole metagenome shotgun sequencing, which provides detailed gene-level insights, including the potential metabolic capabilities of microbial communities and how they respond to environmental changes (Langille et al. 2013; Kanehisa et al. 2017).

The connection between taxonomic composition and functional capacities is called as the taxa-function relationship, which is fundamental in understanding microbial ecology. Taxonomic shifts directly influence functional profiles because the community's genes are distributed across the genomes of its members, with their relative abundances reflecting taxonomic changes. This relationship has been effectively utilized in tools like PICRUSt and Tax4Fun, which predict community-level functional potential based on taxonomic profiles and reference genome databases (Langille et al. 2013; Abhauer et al. 2015). Such predictions help identify taxonomic factors that drivers the functional shifts linked to diseases or environmental changes, enabling a deeper understanding of how community metabolism and functional pathways are altered.

The structure-to-function landscape can be used to visualize the relationship between microbial taxa and their functions, similar to that of fitness landscape of evolutionary biology. This framework illustrates how variations in microbial community composition correspond to changes in functional capabilities. A key feature of this landscape is the concept of taxa-function robustness, which reflects the resilience of a community's functional traits despite taxonomic fluctuations. For instance, when a specific gene or metabolic pathway is unique to a single species within the community, any alteration in the abundance of that species can lead to substantial shifts in the community's overall functional profile. Conversely, if multiple species encode the same genes or pathways, their functional impact is more stable due to compensatory effects among community members (Louca et al. 2018).

Moreover, the co-occurrence of genes across microbial genomes further influences taxa-function robustness. Genes that are distributed among multiple taxa within a community are less susceptible to functional perturbation, even when taxonomic shifts occur. These dynamics highlight the importance of identifying key indicator taxa-species that disproportionately contribute to functional changes. Understanding these relationships enables researchers to connect metagenomic data to ecological functions, offering mechanistic insights into how microbial communities adapt to environmental changes or perturbations (Louca et al. 2016).

8 Rhizosphere Microbial Shifts Under Different Climate Change Scenarios

Knowledge regarding the interactions between plant growth-promoting bacteria (PGPB), plants, and the soil microbiome is essential for enhancing strain selection, developing effective inoculant formulations, and optimizing bioproduct application technologies (Volpiano et al. 2022). These efforts can also help refine soil microbiome management through improved agricultural practices. However, current knowledge about the rhizobiome largely focuses on microbial community composition, with limited understanding of how PGPB and native soil microbial communities' function in association with host plants (Castellano-Hinojosa and Strauss 2021; Wang et al. 2021).

The effectiveness and persistence of PGPB applied in natural seed- and soil-associated microbial communities depend significantly on their interactions with the native microbiota. This dynamic raises concerns about the reliability of PGPB in promoting plant growth under field conditions, which are heavily affected by edaphoclimatic factors as reported by Orozco-Mosqueda et al. (2021).

Recent advancements in omics technologies, coupled with bioinformatics and statistical tools, have enabled deeper exploration of microbial communities' ecological and metabolic contributions to host plants. These tools facilitate the identification of efficient bacterial strains and their functional roles in enhancing plant growth. For instance, studies on microbial community responses to elevated CO_2 have revealed specific changes, such as increased abundance of alpha-proteobacteria, beta-proteobacteria, and Acidobacteria, as well as their association with shifts in microbial functionality (Lipson et al. 2006). These findings underscore the importance of integrating ecological, molecular, and functional analyses to better harness the potential of PGPB for sustainable agriculture.

9 Challenges in Metagenomics-Based Studies for Studying Microbial Dynamics

Despite its transformative potential, metagenomics faces several challenges and limitations when applied to studying rhizosphere microbial shifts driven by climate change. A primary issue is the complexity of rhizosphere ecosystems, which harbor highly diverse microbial communities. Capturing the full extent of this diversity requires deep sequencing, which can be cost-prohibitive and computationally demanding. Additionally, current databases are often incomplete, limiting the ability to annotate and interpret the vast number of microbial genes identified.

Another challenge lies in the lack of standardization in sampling and data analysis protocols. Variations in methodologies can lead to inconsistent results, making it difficult to compare findings across studies. Furthermore, while metagenomics provides valuable insights into microbial composition and functional potential, it cannot directly link these to active processes without complementary techniques like metatranscriptomics or metabolomics.

Environmental factors, such as soil heterogeneity and seasonal variability, further complicate the interpretation of metagenomic data. These factors can obscure the detection of climate-induced microbial shifts. Additionally, while metagenomics identifies microbial taxa and genes, it often fails to establish causal relationships between microbial changes and ecosystem functions.

To address these limitations, integrating metagenomics with other omics approaches, advancing bioinformatics tools, and improving reference databases will be essential. Overcoming these challenges will enhance the utility of metagenomics in understanding and mitigating the impacts of climate change on rhizosphere microbial communities.

10 Conclusion and Future Prospects

Metagenomics is poised to revolutionize our understanding of rhizosphere microbial responses to climate change. Future studies should aim to integrate metagenomics with other advanced omics techniques, such as transcriptomics and metabolomics, to provide a comprehensive view of microbial functional changes in response to environmental shifts. This integration can uncover how microbial communities adapt to alterations in temperature, moisture, and soil conditions, shedding light on their roles in maintaining ecosystem stability.

Standardizing methodologies for sample collection, sequencing, and data interpretation will be crucial to ensure consistent and comparable results across studies. The application of advanced bioinformatics and machine learning tools can further enhance predictions of microbial behavior, identifying critical microbial taxa and functional genes that support plant and soil resilience under stress.

Metagenomics also offers practical applications, such as identifying and harnessing beneficial rhizosphere microbes to enhance crop resilience to climate-induced stress. These findings can drive the development of tailored microbial bioinoculants and inform sustainable agricultural practices. Long-term monitoring of rhizosphere microbial communities using metagenomics can help evaluate the impact of these interventions and guide adaptive management strategies. By unraveling microbial dynamics in the rhizosphere, metagenomics provides valuable insights to sustain soil health, promote crop productivity, and mitigate the impacts of climate change on agricultural ecosystems.

Conflict of Interest The authors declare no conflict of interest.

References

Abhauer KP, Wemheuer B, Daniel R, Meinicke P (2015) Tax4Fun: predicting functional profiles from metagenomic 16S rRNA data. Bioinformatics 31(17):2882–2884. https://doi.org/10.1093/bioinformatics/btv287

Afgan E, Baker D, Batut B (2018) The Galaxy platform for accessible, reproducible, and collaborative biomedical analyses: 2018 update. Nucleic Acids Res 46(W1):W537–W544. https://doi.org/10.1093/nar/gky379

Ahmad F, Ahmad I, Khan MS (2008) Screening of free-living rhizospheric bacteria for their multiple plant growth promoting activities. Microbiol Res 163(2):173–181. https://doi.org/10.1016/j.micres.2006.04.001

Allison SD, Martiny JB (2008) Resistance, resilience, and redundancy in microbial communities. Proc Natl Acad Sci 105(suppl_1):11512–11519. https://doi.org/10.1073/pnas.0801925105

Amend AS, Martiny AC, Allison SD, Berlemont R, Goulden ML, Lu Y, Martiny JB (2016) Microbial response to simulated global change is phylogenetically conserved and linked with functional potential. ISME J 10(1):109–118. https://doi.org/10.1038/ismej.2015.96

Andrews S (2010) FastQC: a quality control tool for high throughput sequence data. Available online at: http://www.bioinformatics.babraham.ac.uk/projects/fastqc

Angiuoli SV, Matalka M, Gussman A (2011) CloVR: A virtual machine for automated and portable sequence analysis from the desktop using cloud computing. BMC Bioinform 12(1):356. https://doi.org/10.1186/1471-2105-12-356

Arndt D, Xia J, Liu Y (2012) METAGENassist: A comprehensive web server for comparative metagenomics. Nucleic Acids Res 40(W1):W88–W95. https://doi.org/10.1093/nar/gks497

Ashburner M, Ball C, Blake (2000) Gene ontology: tool for the unification of biology. Nat Genet 25:25–29. https://doi.org/10.1038/75556

Bais HP, Weir TL, Perry LG, Gilroy S, Vivanco JM (2006) The role of root exudates in rhizosphere interactions with plants and other organisms. Annu Rev Plant Biol 57(1):233–266. https://doi.org/10.1146/annurev.arplant.57.032905.105159

Bardgett RD, Manning P, Morriën E, De Vries FT (2013) Hierarchical responses of plant–soil interactions to climate change: consequences for the global carbon cycle. J Ecol 101(2):334–343. https://doi.org/10.1111/1365-2745.12043

Bell TH, El-Din Hassan S, Lauron-Moreau A, Al-Otaibi F, Hijri M, Yergeau E, St-Arnaud M (2008) Linkage between bacterial and fungal rhizosphere communities in hydrocarbon-contaminated soils is related to plant phylogeny. ISME J 8(2):331–343. https://doi.org/10.1038/ismej.2013.149

Bentley DR, Balasubramanian S, Swerdlow HP, Smith GP, Milton J, Brown CG et al (2008) Accurate whole human genome sequencing using reversible terminator chemistry. Nature 456(7218):53–59. https://doi.org/10.1038/nature07517

Berendsen RL, Pieterse CM, Bakker PA (2012) The rhizosphere microbiome and plant health. Trends Plant Sci 17(8):478–486

Birgander J, Rousk J, Olsson PA (2017) Warmer winters increase the rhizosphere carbon flow to mycorrhizal fungi more than to other microorganisms in a temperate grassland. Glob Chang Biol 23(12):5372–5382. https://doi.org/10.1111/gcb.13803

Blanco-Míguez A (2023) MetaPhlAn 4: a comprehensive taxonomic profiler for metagenomics. Nat Biotechnol 41(1):50–58. https://doi.org/10.1038/s41587-023-01688-w

Blankenberg D, Von Kuster G, Bouvier E et al (2010) Dissemination of scientific software with Galaxy ToolShed. Genome Biol 11(1):R19. https://doi.org/10.1186/gb4161

Bolger AM, Lohse M, Usadel B (2014) Trimmomatic: a flexible read trimming tool for Illumina NGS data. Bioinformatics 30(15):2114–2120. https://doi.org/10.1093/bioinformatics/btu170

Bolyen E (2019) Reproducible, interactive, scalable and extensible microbiome data science using QIIME 2. Nat Biotechnol 37:852–857. https://doi.org/10.1038/s41587-019-0209-9

Bot AJ, Nachtergaele FO, Young A (2000) Land resource potential and constraints at regional and country levels (World Soil Resources Report No. 90). Food and Agriculture Organization of the United Nations

Bouskill NJ, Lim HC, Borglin S, Salve R, Wood TE, Silver WL, Brodie EL (2013) Pre-exposure to drought increases the resistance of tropical forest soil bacterial communities to extended drought. ISME J 7(2):384–394. https://doi.org/10.1038/ismej.2012.113

Bradford MA, Davies CA, Frey SD, Maddox TR, Melillo JM, Mohan JE et al (2008) Thermal adaptation of soil microbial respiration to elevated temperature. Ecol Lett 11(12):1316–1327. https://doi.org/10.1111/j.1461-0248.2008.01251.x

Briones MJI (2014) Soil fauna and soil functions: a jigsaw puzzle. Front Environ Sci 2:7. https://doi.org/10.3389/fenvs.2014.00007

Caporaso JG, Kuczynski J, Stombaug J (2010) QIIME allows analysis of high-throughput community sequencing data. Nature Meth 7(5):335–336. https://doi.org/10.1038/nmeth.f.303

Castellano-Hinojosa A, Strauss SL (2021) Insights into the taxonomic and functional characterization of agricultural crop core rhizobiomes and their potential microbial drivers. Sci Rep 11:10068. https://doi.org/10.1038/s41598-021-89569-7

Castro HF, Classen AT, Austin EE, Norby RJ, Schadt CW (2010) Soil microbial community responses to multiple experimental climate change drivers. Appl Environ Microbiol 76(4):999–1007. https://doi.org/10.1128/AEM.02874-09

Cattelan AJ, Hartel PG, Fuhrmann JJ (1999) Screening for plant growth–promoting rhizobacteria to promote early soybean growth. Soil Sci Soc Am J 63(6):1670–1680. https://doi.org/10.2136/sssaj1999.6361670x

Chen IA, Chu K, Palaniappan K et al (2021) The IMG/M data management and analysis system v.6.0: new tools and advanced capabilities. Nucleic Acids Res 49(D1):D751–D763. https://doi.org/10.1093/nar/gkaa939

Chen S, Zhou Y, Chen Y, Gu J (2018a) fastp: an ultra-fast all-in-one FASTQ preprocessor. Bioinformatics 34(17):1884–1890. https://doi.org/10.1093/bioinformatics/bty560

Chen L, Ren C, Zhang B, Wang Z, Xi Y (2018b) Estimation of forest above-ground biomass by geographically weighted regression and machine learning with sentinel imagery. Forests 9(10):582. https://doi.org/10.3390/f9100582

Cheng L, Zhu J, Abraham J, Trenberth KE, Fasullo JT, Zhang B, Song X (2019) 2018 continues record global ocean warming. https://doi.org/10.1007/s00376-019-8276-x

Chowdhury N, Marschner P, Burns R (2011) Response of microbial activity and community structure to decreasing soil osmotic and matric potential. Plant Soil 344:241–254. https://doi.org/10.1007/s11104-011-0743-9

Clemmensen KE, Finlay RD, Dahlberg A, Stenlid J, Wardle DA, Lindahl BD (2015) Carbon sequestration is related to mycorrhizal fungal community shifts during long-term succession in boreal forests. New Phytol 205(4):1525–1536. https://doi.org/10.1111/nph.13208

Cole JR et al (2014) Ribosomal database project: data and tools for high throughput rRNA analysis. Nucleic Acids Res 42(D1):D633–D642. https://academic.oup.com/nar/article/42/D1/D633/1063201

Cregger MA, Schadt CW, McDowell NG, Pockman WT, Classen AT (2012) Response of the soil microbial community to changes in precipitation in a semiarid ecosystem. Appl Environ Microbiol 78(24):8587–8594. https://doi.org/10.1128/aem.02050-12

De Bello F, Lavorel S, Díaz S, Harrington R, Cornelissen JH, Bardgett RD et al (2010) Towards an assessment of multiple ecosystem processes and services via functional traits. Biodivers Conserv 19:2873–2893. https://doi.org/10.1007/s10531-010-9850-9

De Vries FT, Shade A (2013) Controls on soil microbial community stability under climate change. Front Microbiol 4:265. https://doi.org/10.3389/fmicb.2013.00265

de Vries FT, Bloem J, Quirk H, Stevens CJ, Bol R, Bardgett RD (2012) Extensive management promotes plant and microbial nitrogen retention in temperate grassland. PLoS One 7(12):e51201. https://doi.org/10.1371/journal.pone.0051201

De Vries FT, Griffiths RI, Knight CG, Nicolitch O, Williams A (2020) Harnessing rhizosphere microbiomes for drought-resilient crop production. Science 368(6488):270–274. https://doi.org/10.1126/science.aaz5192

DeAngelis KM, Pold G, Topçuoğlu BD, van Diepen LT, Varney RM, Blanchard JL, Frey SD (2015) Long-term forest soil warming alters microbial communities in temperate forest soils. Front Microbiol 6:104. https://doi.org/10.3389/fmicb.2015.00104

DeSantis TZ et al (2006) Greengenes, a chimera-checked 16S rRNA gene database and workbench compatible with ARB. Appl Environ Microbiol 72(7):5069–5072. https://doi.org/10.1128/AEM.03006-05

Deslippe JR, Hartmann M, Simard SW, Mohn WW (2012) Long-term warming alters the composition of Arctic soil microbial communities. FEMS Microbiol Ecol 82(2):303–315. https://doi.org/10.1111/j.1574-6941.2012.01350.x

DragMAP Team (2021) dragMAP documentation

Ebersberger D, Wermbter N, Niklaus PA, Kandeler E (2004) Effects of long term CO 2 enrichment on microbial community structure in calcareous grassland. Plant Soil 264:313–323. https://doi.org/10.1023/B:PLSO.0000047768.89268.8c

Edgar RC (2010) Search and clustering orders of magnitude faster than BLAST. Bioinformatics 26(19):2460–2461. https://doi.org/10.1093/bioinformatics/btq461

Ewels P, Magnusson M, Lundin S, Kaller M (2016) MultiQC: summarize analysis results for multiple tools and samples in a single report. Bioinformatics 32(19):3047–3048. https://doi.org/10.1093/bioinformatics/btw354

Faust K, Raes J (2012) Microbial interactions: from networks to models. Nat Rev Microbiol 10(1): 538–550. https://doi.org/10.1038/nrmicro2832

Fierer N, Bradford MA, Jackson RB (2007a) Toward an ecological classification of soil bacteria. Ecology 88(6):1354–1364. https://doi.org/10.1890/05-1839

Fierer N, Morse JL, Berthrong ST, Bernhardt ES, Jackson RB (2007b) Environmental controls on the landscape-scale biogeography of stream bacterial communities. Ecology 88(9):2162–2173. https://doi.org/10.1890/06-1746.1

Finzi AC, Abramoff RZ, Spiller KS, Brzostek ER, Darby BA, Kramer MA, Phillips RP (2015) Rhizosphere processes are quantitatively important components of terrestrial carbon and nutrient cycles. Glob Chang Biol 21(5):2082–2094. https://doi.org/10.1111/gcb.12816

Fischer RA, Byerlee D, Edmeades G (2014) Crop yields and global food security. ACIAR: Canberra, ACT, pp 8–11

Giovannoni S, Stingl U (2007) The importance of culturing bacterioplankton in the'omics' age. Nat Rev Microbiol 5(10):820–826. https://doi.org/10.1038/nrmicro1752

Gordon A, Hannon G (2010) FASTX-Toolkit

Gordon H, Haygarth PM, Bardgett RD (2008) Drying and rewetting effects on soil microbial community composition and nutrient leaching. Soil Biol Biochem 40(2):302–311. https://doi.org/10.3389/fmicb.2023.1129471

Guo ZT, Sun B, Zhang ZS, Peng SZ, Xiao GQ, Ge JY et al (2008) A major reorganization of Asian climate by the early Miocene. Clim Past 4(3):153–174. https://doi.org/10.5194/cp-4-153-2008

Haichar FEZ, Marol C, Berge O, Rangel-Castro JI, Prosser JI, Balesdent J et al (2008) Plant host habitat and root exudates shape soil bacterial community structure. ISME J 2(12):1221–1230. https://doi.org/10.1038/ismej.2008.80

Hayden HL, Mele PM, Bougoure DS, Allan CY, Norng S, Piceno YM, Brodie EL, DeSantis TZ, Andersen GL, Williams AL, Hovenden MJ (2012) Changes in the microbial community structure of bacteria archaea and fungi in response to elevated CO 2 and warming in an Australian native grassland soil. Summary Environl Microbiol 14(12):3081–3096. https://doi.org/10.1111/emi.2012.14

Huang X-F, Chaparro JM, Reardon KF, Zhang R, Shen Q, Vivanco JM (2014) Rhizosphere interactions: root exudates, microbes, and microbial communities. Botany 92(4):267–275. https://doi.org/10.1139/cjb-2013-0225

Ibekwe AM, Kennedy AC, Stubbs TL (2010) An assessment of environmental conditions for control of downy brome by Pseudomonas fluorescens D7. Int J Environ Technol Manag 12(1):27–46. https://doi.org/10.1504/IJETM.2010.029959

İnceoğlu Ö, Falcão Salles J, van Elsas JD (2012) Soil and cultivar type shape the bacterial community in the potato rhizosphere. Microb Ecol 63(2):460–470. https://doi.org/10.1007/s00248-011-9930-8

Iversen CM, Victoria LS, Patrick FS, Eugenie SE, David McGuire, Richard, JN, Anthony PW, Jeffrey MW, Stan DW (2015) The unseen iceberg: plant roots in arctic tundra. New Phytol 205(1):34–58. https://doi.org/10.1111/nph.13003

Johnson NC, Graham JH, Smith FA (2008) Functioning of mycorrhizal associations along the mutualism parasitism continuum. New Phytol 135(4):575–585. https://doi.org/10.1046/j.1469-8137.1997.00729.x

Jones DL, Edwards AC (1998) Influence of sorption on the biological utilization of two simple carbon substrates. Soil Biol Biochem 30(14):1895–1902. https://doi.org/10.1016/S0038-0717(98)00060-1

Juniper S, Abbott LK (2006) Soil salinity delays germination and limits growth of hyphae from propagules of arbuscular mycorrhizal fungi. Mycorrhiza 16:371–379. https://doi.org/10.1007/s00572-006-0046-9

Kanehisa M, Sato Y, Morishima K (2017) KEGG: new perspectives on genomes, pathways, diseases and drugs. Nucleic Acids Res 45(D1):D353–D361. https://doi.org/10.1093/nar/gkw1092

Kardol P, Cregger MA, Campany CE, Classen AT (2010) Soil ecosystem functioning under climate change: plant species and community effects. Ecology 91(3):767–781. https://doi.org/10.1890/09-0135.1

Klironomos JN, Kendrick WB (1996) Palatability of microfungi to soil arthropods in relation to the functioning of arbuscular mycorrhizae. Biol Fertil Soils 21:43–52. https://doi.org/10.1007/BF00335992

Korenblum E, Dong Y, Szymanski J, Panda S, Jozwiak A, Massalha H, Aharoni A (2020) Rhizosphere microbiome mediates systemic root metabolite exudation by root-to-root signaling. Proc Natl Acad Sci 117(7):3874–3883. https://doi.org/10.1073/pnas.1912130117

Lakshmanan V, Selvaraj G, Bais HP (2014) Functional soil microbiome: belowground solutions to an aboveground problem. Plant Physiol 166(2):689–700. https://doi.org/10.1104/pp.114.245811

Langille MGI, Zaneveld J, Caporaso JG (2013) Predictive functional profiling of microbial communities using 16S rRNA marker gene sequences. Nat Biotechnol 31(9):814–821. https://doi.org/10.1038/nbt.2676

Langmead B, Trapnell C, Pop M, Salzberg SL (2009) Ultrafast and memory-efficient alignment of short DNA sequences to the human genome. Genome Biol 10(3):R25. https://doi.org/10.1186/gb-2009-10-3-r25

Li H, Durbin R (2009) Fast and accurate short read alignment with Burrows–Wheeler transform. Bioinformatics 25(14):1754–1760. https://doi.org/10.1093/bioinformatics/btp324

Li M, Ma Z, Wu P, Liu J, Lv M, Yang Q, Han Y (2022) Ecological response to climate change across China from combined soil temperature and moisture changes. Earth Space Sci 9(12):e2022EA002640. https://doi.org/10.1029/2022EA002640

Liang YU, Chen HU, Tang M, Shen S (2007) Proteome analysis of an ectomycorrhizal fungus Boletus edulis under salt shock. Mycological Res 111(8):939–946. https://doi.org/10.1016/j.mycres.2007.06.005

Lipson DA, Wilson RF, Oechel WC (2006) Effects of elevated atmospheric C02 on soil microbial biomass, activity, and diversity in a chaparral ecosystem. Appl Environ Microbiol 71:8573–8580. https://doi.org/10.1128/AEM.71.12.8573-8580

Liu JK, Hyde KD, Jones EG, Ariyawansa HA, Bhat DJ, Boonmee S et al (2015) Fungal diversity notes 1–110: taxonomic and phylogenetic contributions to fungal species. Fungal Divers 72:1–197. https://doi.org/10.1007/s13225-015-0324-y

Liu W, Huang L, Komorek R, Handakumbura PP, Zhou Y, Hu D, Zhu Z (2020) Correlative surface imaging reveals chemical signatures for bacterial hotspots on plant roots. Analyst 145(2):393–401. https://doi.org/10.1039/C9AN01954E

Louca S, Parfrey LW, Doebeli M (2016) Decoupling function and taxonomy in the global ocean microbiome. Science 353(6305):1272–1277. https://doi.org/10.1126/science.aaf4507

Louca S, Polz MF, Mazel F (2018) Function and functional redundancy in microbial systems. Nat Ecol Evol 2(6):936–943. https://doi.org/10.1038/s41559-018-0519-1

Mandeel QA (2006) Biodiversity of the genus Fusarium in saline soil habitats. J Basic Microbiol 46(6):480–494. https://doi.org/10.1002/jobm.200510128

Manter DK, Delgado JA, Holm DG, Stong RA (2010) Pyrosequencing reveals a highly diverse and cultivar-specific bacterial endophyte community in potato roots. Microb Ecol 60:157–166. https://doi.org/10.1007/s00248-010-9658-x

Margulies M, Egholm M, Altman WE, Attiya S, Bader JS, Bemben LA, Rothberg JM (2005) Genome sequencing in microfabricated high-density picolitre reactors. Nature 437(7057):376–380. https://doi.org/10.1038/nature03959

Markowitz VM, Chen IM, Palaniappan K (2014) IMG 4 version of the integrated microbial genomes comparative analysis system. Nucleic Acids Res 42(D1):D560–D567. https://doi.org/10.1093/nar/gkt963

Martin M (2011) Cutadapt removes adapter sequences from high-throughput sequencing reads. EMBnet. Journal 17(1):10. https://doi.org/10.14806/ej.17.1.200

Mendes R, Garbeva P, Raaijmakers JM (2013) The rhizosphere microbiome: significance of plant beneficial, plant pathogenic, and human pathogenic microorganisms. FEMS Microbiol Rev 37(5):634–663. https://doi.org/10.1111/1574-6976.12028

Menzel P (2016) Kaiju: fast and sensitive taxonomic classification for metagenomics. Nat Commun 7:11257. https://doi.org/10.1038/ncomms11257

Meyer F, Paarmann D, D'Souza M (2008) The metagenomics RAST server–a public resource for the automatic phylogenetic and functional analysis of metagenomes. BMC Bioinform 9(1):386. https://doi.org/10.1186/1471-2105-9-386

Mohan JE, Cowden CC, Baas P, Dawadi A, Frankson PT, Helmick K et al (2014) Mycorrhizal fungi mediation of terrestrial ecosystem responses to global change: mini-review. Fungal Ecol 10:3–19. https://doi.org/10.1016/j.funeco.2014.01.005

Nichols D (2007) Cultivation gives context to the microbial ecologist. FEMS Microbiol Ecol 60(3):351–357. https://doi.org/10.1111/j.1574-6941.2007.00332.x

Nurk S et al (2017) metaSPAdes: a new versatile metagenomic assembler. Genome Res 27:824–834. https://doi.org/10.1101/gr.213959.116

Oren A (1999) Bioenergetic aspects of halophilism. Microbiol Mol Biol Rev 63(2):334–348. https://doi.org/10.1128/mmbr.63.2.334-348.1999

Orozco-Mosqueda M, Flores A, Rojas-Sánchez B, Urtis-Flores CA, Morales-Cedeño LR, Valencia-Marin MF (2021) Plant growth-promoting bacteria as bioinoculants: attributes and challenges for sustainable crop improvement. Agronomy 11(6):1167. https://doi.org/10.3390/agronomy11061167

Parmesan C, Yohe G (2003) A globally coherent fingerprint of climate change impacts across natural systems. Nature 421(6918):37–42. https://doi.org/10.1038/nature01286

Pathak H, Rao DLN (1998) Carbon and nitrogen mineralization from added organic matter in saline and alkali soils. Soil Biol Biochem 30(6):695–702. https://doi.org/10.1016/S0038-0717(97)00208-3

Peng Y et al (2011) IDBA-UD: a de novo assembler for single-cell and metagenomic sequencing data with highly uneven depth. Bioinformatics 27(8):823–830. https://doi.org/10.1093/bioinformatics/bts174

Pruitt KD (2012) NCBI Reference Sequences (RefSeq): current status, new features, and genome annotation policy. Nucleic Acids Res 40(D1):D130–D135. https://doi.org/10.1093/nar/gkr1079

Qadir M, Ghafoor A, Murtaza G (2000) Amelioration strategies for saline soils: a review. Land Degrad Dev 11(6):501–521. https://doi.org/10.1002/1099-145X(200011/12)11:6<501::AID-LDR405>3.0.CO;2-S

Quast C (2013) The SILVA ribosomal RNA gene database project: improved data processing and web-based tools. Nucleic Acids Res 41(D1):D590–D596. https://doi.org/10.1093/nar/gkr1079

Reichstein M, Bahn M, Ciais P, Frank D, Mahecha MD, Seneviratne SI, Wattenbach M (2013) Climate extremes and the carbon cycle. Nature 500(7462):287–295. https://doi.org/10.1038/nature12350

Rosenberg E, Koren O, Reshef L, Efrony R, Zilber-Rosenberg I (2007) The role of microorganisms in coral health, disease and evolution. Nat Rev Microbiol 5(5):355–362. https://doi.org/10.1038/nrmicro1635

Rothberg JM, Hinz W, Rearick TM, Schultz J, Mileski W, Davey M et al (2011) An integrated semiconductor device enabling non-optical genome sequencing. Nature 475(7356):348–352. https://doi.org/10.1038/nature10242

Schimel J, Balser TC, Wallenstein M (2007) Microbial stress-response physiology and its implications for ecosystem function. Ecology 88(6):1386–1394. https://doi.org/10.1890/06-0219

Schloss PD, Westcott SL, Ryabin T (2009) Introducing mothur: open-source, platform-independent, community-supported software for describing and comparing microbial communities. Appl Environ Microbiol 75(23):7537–7541. https://doi.org/10.1128/AEM.01541-09

Schmitz L, Yan Z, Schneijderberg M, De Roij M, Pijnenburg R, Zheng Q, Franken C, Dechesne A, Trindade LM, Van Velzen R, Bisseling T, Geurts R, Cheng X (2022) Synthetic bacterial community derived from a desert rhizosphere confers salt stress resilience to tomato in the presence of a soil microbiome. ISME J 16(8):1907–1920. https://doi.org/10.1038/s41396-022-01238-3

Seemann T (2014) Prokka: rapid prokaryotic genome annotation. Bioinformatics 30(14):2068–2069. https://doi.org/10.1093/bioinformatics/btu153

Simonin M, Dasilva C, Terzi V, Ngonkeu ELM, Diouf D, Kane A, Béna G, Moulin L (2020) Influence of plant genotype and soil on the wheat rhizosphere microbiome: evidences for a core microbiome across eight African and European soils. FEMS Microbiol Ecol 96(6):fiaa067. https://doi.org/10.1093/femsec/fiaa067

Singh A, Shourie A (2021) Climate change impacts on plant–microbe interactions. In Climate change and the microbiome: sustenance of the ecosphere Cham: Springer International Publishing, pp 155–186. https://doi.org/10.1007/978-3-030-76863-8_8

Song W, Shao H, Zheng A, Zhao L, Xu Y (2023) Advances in roles of salicylic acid in plant tolerance responses to biotic and abiotic stresses. Plants 12(19):3475. https://doi.org/10.3390/plants12193475

Weedon JT, Kowalchuk G, Aerts R, van Hal J, van Logtestijn R, Taş N et al (2012) Summer warming accelerates sub-arctic peatland nitrogen cycling without changing enzyme pools or microbial community structure. Glob Chang Biol 18(1):138–150. https://doi.org/10.1111/j.1365-2486.2011.02548.x

Tatusov RL et al (2000) The COG database: a tool for genome-scale analysis of protein functions and evolution. Nucleic Acids Res 28(1):33–36. https://doi.org/10.1093/nar/28.1.33

Thomson BC, Ostle NJ, McNamara NP, Whiteley AS, Griffiths RI (2010) Effects of sieving, drying and rewetting upon soil bacterial community structure and respiration rates. J Microbiol Methods 83(1):69–73. https://doi.org/10.1016/j.mimet.2010.07.021

Thompson LR, Sanders JG, McDonald D, Amir A, Ladau J, Locey KJ, Knight R (2017) A communal catalogue reveals earth's multiscale microbial diversity. Nature 551(7681):457–463. https://doi.org/10.1038/nature24621

Torsvik TH, Smethurst MA, Meert JG, Van der Voo R, McKerrow WS, Brasier MD, Walderhaug HJ (1996) Continental break-up and collision in the Neoproterozoic and Palaeozoic—a tale of Baltica and Laurentia. Earth Sci Rev 40(3–4):229–258. https://doi.org/10.1016/0012-8252(96)00008-6

Treangen TJ, Koren S, Sommer DD et al (2013) MetAMOS: a modular and open-source metagenomic assembly and analysis pipeline. Genome Biol 14(1):R2. https://doi.org/10.1186/gb-2013-14-1-r2

Trivedi P, Leach JE, Tringe SG, Sa T, Singh BK (2020) Plant–microbiome interactions: from community assembly to plant health. Nat Rev Microbiol 18(11):607–621. https://doi.org/10.1038/s41579-020-0412-1

Uroz S, Buée M, Murat C, Frey-Klett P, Martin F (2010) Pyrosequencing reveals a contrasted bacterial diversity between oak rhizosphere and surrounding soil. Environ Microbiol Rep 2(2):281–288. https://doi.org/10.1111/j.1758-2229.2009.00117.x

Van der Putten WH, Bardgett RD, Bever JD, Bezemer TM, Casper BB, Fukami T, Wardle DA (2013) Plant-soil feedbacks: the past, the present and future challenges. J Ecol 101 (2):265–276. https://doi.org/10.1111/1365-2745.12054

Van Der Heijden MG, Hartmann M (2016) Networking in the plant microbiome. PloS Biol 14: e1002378. https://doi.org/10.1371/journal.pbio.1002378

Van Elsas JD, Boersma FGH (2011) A review of molecular methods to study the microbiota of soil and the mycosphere. Eur J Soil Biol 47(2):77–87. https://doi.org/10.1016/j.ejsobi.2010.11.010

Vandenkoornhuyse P, Quaiser A, Duhamel M, Le Van A, Dufresne A (2015) The importance of the microbiome of the plant holobiont. New Phytol 206(4):1196–1206. https://doi.org/10.1111/nph.13312

Volpiano CG, Lisboa BB, Jose JFBS, Beneduzi A, Granada CE, Vargas LK (2022) Soil-plant-microbiota interactions to enhance plant growth. Rev Bras Cienc Solo 46:e0210098. https://doi.org/10.36783/18069657rbcs20210098

Waldrop MP, Firestone MK (2006) Response of microbial community composition and function to soil climate change. Microb Ecol 52:716–724. https://doi.org/10.1007/s00248-006-9103-3

Walther GR, Post E, Convey P, Menzel A, Parmesan C, Beebee TJ, Bairlein F (2002) Ecological responses to recent climate change. Nature 416(6879):389–395. https://doi.org/10.1038/416389a

Wang H, Liu R, You MP, Barbetti MJ, Chen Y (2021) Pathogen biocontrol using plant growth-promoting bacteria (PGPR): role of bacterial diversity. Microorganisms 9(9):1988. https://doi.org/10.3390/microorganisms9091988

Weber KA, Thrash JC, Van Trump JI, Achenbach LA, Coates JD (2011) Environmental and taxonomic bacterial diversity of anaerobic uranium (IV) bio-oxidation. Appl Environ Microbiol 77(13):4693–4696. https://doi.org/10.1128/AEM.02539-10

Wichern J, Wichern F, Joergensen RG (2006) Impact of salinity on soil microbial communities and the decomposition of maize in acidic soils. Geoderma 137(1–2):100–108. https://doi.org/10.1016/j.geoderma.2006.08.001

Wood DE et al (2014) Kraken: ultrafast metagenomic sequence classification using exact alignments. Genome Biol 15:R46. https://doi.org/10.1186/gb-2014-15-3-r46

Xie H, Wang X, van der Hooft JJ, Medema MH, Chen ZY, Yue X et al (2022) Fungi population metabolomics and molecular network study reveal novel biomarkers for early detection of aflatoxigenic Aspergillus species. J Hazard Mater 424:127173. https://doi.org/10.1016/j.jhazmat.2021.127173

Ye SH, Siddle KJ, Park DJ, Sabeti PC (2019) Benchmarking metagenomics tools for taxonomic classification. Cell 178:779–794. https://doi.org/10.1016/j.cell.2019.07.010

Yergeau E, Bokhorst S, Kang S, Zhou J, Greer CW, Aerts R, Kowalchuk GA (2012) Shifts in soil microorganisms in response to warming are consistent across a range of Antarctic environments. ISME J 6(3):692–702. https://doi.org/10.1038/ismej.2011.124

Yuan BC, Li ZZ, Liu H, Gao M, Zhang YY (2007) Microbial biomass and activity in salt affected soils under arid conditions. Appl Soil Ecol 35(2):319–328. https://doi.org/10.1016/j.apsoil.2006.07.004

Zak DR, Holmes WE, MacDonald NW, Pregitzer KS. (1999) Soil temperature, matric potential, and the kinetics of microbial respiration and net N mineralization in northern hardwood forests. Soil Sci Soc Am J 63:575–584. https://doi.org/10.2136/sssaj1999.03615995006300030021x

Zhu W et al (2010) Ab initio gene identification in metagenomic sequences. Nucleic Acids Res 38(12):e132. https://doi.org/10.1093/nar/gkq275

Zilber-Rosenberg I, Rosenberg E (2008) Role of microorganisms in the evolution of animals and plants: the hologenome theory of evolution. FEMS Microbiol Rev 32(5):723–735. https://doi.org/10.1111/j.1574-6976.2008.00123.x

Metagenomics in the Era of Genome Editing

Ratna Prabha, Dhananjaya Pratap Singh, Renu, Kuldeep Kumar, and Amrender Kumar

1 Introduction

Genome editing is a powerful technology that allows for deliberate but precise modifications in the DNA sequence of an organism (Yan et al. 2019; Gopika et al. 2024). This can include the insertion, deletion, or replacement of specific DNA fragments (Shi et al. 2021). The technology depends on engineered nucleases to create double-stranded breaks at targeted locations in the genome, which are then repaired by the cell's natural DNA repair mechanisms, such as non-homologous end joining or homologous recombination (Shi et al. 2021; Yan et al. 2019). Genome editing can impact genetic diversity by enabling precise, specific, and accurate changes in the genetic material of an organism, altering both phenotypic characteristics and genotypic composition (Pande et al. 2019).

R. Prabha (✉)
Agricultural Knowledge Management Unit (AKMU), ICAR-Indian Agricultural Research Institute, New Delhi, Delhi, India

EPC, ICAR-NRC Equines, Bikaner, Rajasthan, India

D. P. Singh
ICAR-Indian Institute of Vegetable Research, Varanasi, Uttar Pradesh, India

Renu
Crop Science Division, Indian Council of Agricultural Research, New Delhi, Delhi, India

K. Kumar
ICAR-Indian Institute of Pulses Research, Kanpur, Uttar Pradesh, India

A. Kumar
Agricultural Knowledge Management Unit (AKMU), ICAR-Indian Agricultural Research Institute, New Delhi, Delhi, India

Renu et al. (eds.), *Meta-omics in Crop Improvement*,
https://doi.org/10.1007/978-981-95-2294-1_3

Table 1 Comparison of key features between different genome editing techniques

Feature	CRISPR-Cas9	TALENs	ZFNs	Base editors	Prime editing
Target recognition	RNA-guided	Protein-based	Protein-based	RNA-guided	RNA-guided
Efficiency	High	Moderate	Moderate	High	Moderate
Specificity	Moderate-high	High	High	High	Very high
Ease of design	Easy	Complex	Complex	Moderate	Complex
Cost	Low	High	High	Moderate	Moderate
Size	~4.2 kb	~3 kb each	~1 kb each	~5 kb	~6.4 kb
Multiplexing	Yes	Limited	Limited	Limited	Limited
DNA cleavage	Yes	Yes	Yes	No	No

2 Different Approaches of Genome Editing

There are three main genome editing techniques: Zinc Finger Nucleases (ZFNs), Transcription Activator-Like Effector Nucleases (TALENs), Clustered Regularly Interspaced Short Palindromic Repeats (CRISPR)-associated Cas9 (CRISPR/Cas9) system, though CRISPR/Cas9 are most widely used and versatile tool (Kanwal et al. 2024). Recently, several modifications in the CRISPR/Cas9 system such as Base Editors, and Prime Editors are developed which are more precise in creating point as well as targeted mutation. A comparative overview of these techniques is given in Table 1. Further, a brief description of these genome editing techniques is provided.

2.1 Zinc Finger Nucleases (ZFNs)

ZFNs are composed of a DNA-binding domain (zinc fingers) and a DNA-cleaving domain (FokI nuclease) that create double-strand breaks (DSBs) at specific genomic locations (Garcia and Lamba 2017). It is used in gene therapy and basic research to introduce targeted genetic modifications (Garcia and Lamba 2017). Zinc Finger Nucleases (ZFNs) are one of the earliest genome editing tools, which use engineered zinc finger proteins to target specific DNA sequences.

2.2 Transcription Activator-Like Effector Nucleases (TALENs)

Transcription Activator-Like Effector Nucleases (TALENs) are designed to bind to specific DNA sequences and introduce breaks (Yamamoto et al. 2014). TALENs consist of a DNA-binding domain derived from transcription activator-like effectors and a FokI nuclease that induces DSBs at target sites (Garcia and Lamba 2017). Similar to ZFNs, TALENs are used for precise genome modifications in various organisms (Garcia and Lamba 2017).

2.3 CRISPR/Cas9

CRISPR/Cas9 uses a guide RNA to direct the Cas9 enzyme to a specific DNA sequence, where it introduces a DSBs break (Shi et al. 2021; Yamamoto et al. 2014). CRISPR/Cas9 are widely used due to its simplicity and versatility in gene editing, including gene knockouts, insertions, and corrections (Garcia and Lamba 2017).

2.4 Base Editing

It allows for the substitution of one DNA base pair into another without introducing DSBs (Sakuma and Yamamoto 2018; Gopika et al. 2024). It is used for precise point mutations, which can be crucial for correcting genetic disorders (Sakuma and Yamamoto 2018; Gopika et al. 2024).

2.5 Prime Editing

Comparatively a newer technique that can introduce various types of mutations, including insertions, deletions, and all 12 possible base-to-base conversions, without DSBs (Marzec and Hensel 2020). It is adapted for use in both human cells and plants, offering a high degree of precision (Marzec and Hensel 2020).

Apart from the above mentioned approaches, Programmable Addition via Site-specific Targeting Elements (PASTE), Epigenome Editing, and RNA Editing Systems are also emerging approaches. Programmable Addition via Site-specific Targeting Elements (PASTE) involves site-specific integration of new genetic material without DSBs (Gopika et al. 2024). It is useful for adding new genes or regulatory elements to the targeted genome (Gopika et al. 2024). In case of Epigenome Editing, epigenetic modifiers are added to specific genomic loci to alter gene expression without changing the DNA sequence (Sakuma and Yamamoto 2018). RNA Editing Systems targets RNA molecules for editing, allowing for modifications at the transcript level (Sakuma and Yamamoto 2018) and is useful for transient and reversible modifications, particularly in gene expression studies (Sakuma and Yamamoto 2018).

Among these techniques, CRISPR-Cas9 is known for its extraordinary efficiency compared to traditional gene-editing tools, ease of use, and low cost (Kanwal et al. 2024; Jeon et al. 2017). TALENs consist of endonucleases fused to a DNA-binding domain, while the CRISPR/Cas9 system uses guide RNAs to target the bacterial Cas9 endonuclease to the desired genomic location (Pellagatti et al. 2016). CRISPR/Cas9 has high specificity and efficiency, making it a popular and successful genome editing system (Rao et al. 2022; Bhat et al. 2020). A comparative study found that paired Cas9 nucleases induced targeted genomic deletion more efficiently and precisely than two TALEN pairs (He et al. 2016). When concurrently supplied with a plasmid template spanning the DNA double-strand breaks, TALENs stimulated homology directed repair more efficiently than CRISPR/Cas9 and caused fewer targeted genomic deletions (He et al. 2016).

Prime editing allows the introduction of all mutation types, including insertions, deletions, and various base-to-base conversions (Marzec and Hensel 2020). Base editing and prime editing have the potential to further advance the precision and versatility of genome editing technologies, enabling targeted modifications with high accuracy (Marzec and Hensel 2020). Different databases and tools are available in the area of genome editing (Table 2).

Table 2 Details of different genome editing tools and databases

Tool name	Description	Weblink
CRISPR design and analysis tools		
CHOPCHOP	Web tool for CRISPR design	https://chopchop.cbu.uib.no/
CRISPOR	Guide RNA design and analysis	https://crispor.gi.ucsc.edu/
E-CRISP	CRISPR target design tool	http://www.e-crisp.org/
Cas-Designer	CRISPR guide RNA design	http://www.rgenome.net/cas-designer/
CRISPR-ERA	gRNA design platform	http://crispr-era.stanford.edu/
Off-target analysis tools		
Cas-OFFinder	Off-target site prediction	http://www.rgenome.net/cas-offinder/
COSMID	CRISPR search with mismatches, insertions and/or deletions	https://crispr.bme.gatech.edu/
CROP-IT	Ranked off-target prediction	http://www.adlilab.org/CROP-IT/homepage.html
CCTop	CRISPR target predictor	http://crispr.cos.uni-heidelberg.de
CRISPR databases		
CRISPRCasdb	CRISPR array collection	https://crisprcas.i2bc.paris-saclay.fr/
CRISPR-RT	Reference targets database	http://bioinfolab.miamioh.edu/CRISPR-RT
GenomeCRISPR	CRISPR screening data	http://genomecrispr.org/
Base editing tools		
BE-Designer	Base editor design	http://www.rgenome.net/be-designer/
BE-Analyzer	Base editing analysis	http://www.rgenome.net/be-analyzer/#!
EditR	Editing efficiency calculation	http://baseeditr.com/
Prime editing tools		
PrimeDesign	pegRNA design	https://github.com/pinellolab/PrimeDesign
PE-Designer	Prime editing design	http://www.rgenome.net/pe-designer/
General genome editing resources		
Addgene	Plasmid repository	https://www.addgene.org/
CRISPResso2	Editing outcome analysis	http://crispresso2.pinellolab.org/submission
TIDE	Editing efficiency analysis	https://tide.nki.nl/
ICE	Editing analysis tool	https://www.synthego.com/products/bioinformatics/analysis
CRISPR-GA	Genome analysis tool	https://www.cropbioengineering.iastate.edu/cgat/

3 Role of CRISPR-Cas9 in Genome Editing

CRISPR-Cas9 is a powerful and efficient genome editing tool that has revolutionized the field of genetic engineering. The CRISPR/Cas9 system is derived from a bacterial immune system and functions as an active immune system to prevent the bacteria from viruses and plasmids (Nitish et al. 2023). It is a highly specific and efficient technique for genome editing or DNA alteration, using an RNA molecule as the main functional part to recognize and cleave foreign viral elements (Nitish et al. 2023). CRISPR/Cas9 has been successfully used to edit genomes of living organisms in many fields, working in vitro and in vivo in germline and somatic cells to knock-out harmful mutated genes or knock-in beneficial genes (Khouzam and Tivakaran 2021). CRISPR-Cas9 is known for its ease of use and high efficiency compared to other genome editing techniques like zinc finger nucleases (ZFNs) and transcription activator-like effector nucleases (TALENs) (Li et al. 2017; Rezaei et al. 2020; Qi et al. 2017). CRISPR-Cas9 is much easier to use and is highly efficient, with the capacity for multiplex editing of multiple genes simultaneously (Li et al. 2017; Rezaei et al. 2020). It has been shown to be easy to establish in new species and boasts high efficiency and specificity due to the use of an RNA molecule to target the gene or sequence of interest (Wilson and Wingfield 2020).

CRISPR-Cas9 has revolutionized genome editing due to its high precision and efficiency compared to other techniques like zinc-finger nucleases (ZFNs) and transcription activator-like effector nucleases (TALENs) (Ma and Liu 2016; Li et al. 2017; Rezaei et al. 2020; Okoli et al. 2018; Demirci et al. 2018). CRISPR-Cas9 is simpler to design and implement. It uses RNA molecules to guide the Cas9 protein to the target DNA, making it more straightforward than the protein-based targeting of ZFNs and TALENs (Ma and Liu 2016; Li et al. 2017; Rezaei et al. 2020; Demirci et al. 2018). CRISPR-Cas9 demonstrates higher editing efficiency. It allows for multiplexing, meaning multiple genes can be edited simultaneously, which is more challenging with ZFNs and TALENs (Ma and Liu 2016; Li et al. 2017; Rezaei et al. 2020; Okoli et al. 2018; Demirci et al. 2018). CRISPR-Cas9 offers high specificity, although off-target effects can occur. Advances in Cas9 variants have reduced these off-target effects, albeit sometimes at the cost of lower on-target efficiency (Khoshandam et al. 2024; Lee et al. 2016; Zheng et al. 2018). The *Neisseria meningitidis* (Nme) CRISPR-Cas9 system, for instance, shows reduced off-target effects compared to the *Streptococcus pyogenes* (Spy) system (Lee et al. 2016). CRISPR-Cas9 is versatile, being used in various fields from plant genome editing to treating human diseases, which highlights its broad applicability and robustness.

Techniques like multiplex genome editing allow for the simultaneous targeting of multiple genes, which can enhance the plant's ability to withstand various stresses by knocking out negative regulatory genes or introducing beneficial traits (Behera et al. 2023). By enabling precise and targeted modifications of stress-responsive genes, accelerating the development of stress-tolerant varieties, and integrating with advanced delivery systems, genome editing can significantly enhance agricultural productivity and sustainability in the face of climate change.

4 Metagenomics

Metagenomics is the study of genetic information at the community level, providing a more complete view of a microbial community than is obtained by attempting to culture the target organisms (Haynes 2018). Metagenomics includes the genomic analysis of microorganisms by extracting and cloning DNA from a group of microorganisms, or the direct use of purified DNA or RNA for sequencing, bypassing the usual protocol of isolating and culturing individual microbial species (de Bruijn 2011a; de Bruijn 2011b). It allows for the study of microorganism diversity and the isolation of novel medical and industrial compounds (de Bruijn 2011a). Commonly used approach for metagenomic data analysis includes assembly of the data into longer contiguous sequence and classification of sequenced metagenomic data into taxonomic groups with the aim of correctly identifying species presented in the metagenomic sample (Marić and Šikić 2019). Metagenomic analysis also includes the identification, and functional and evolutionary analysis of the genomic sequences of a community of organisms, not limited to taxonomy but also by the encoded metabolic potential (Haynes 2018). The computational analysis of a meta-genome or meta-transcriptome addresses questions such as "Who is out there?" and "What are they doing?" (Mitra et al. 2011). A typical metagenomics data analysis flowchart is given in Fig. 1. Computational methods commonly used in metagenomic data analysis include high-throughput systems like MG-RAST, customizable web servers like WebMGA, and standalone tools like MEGAN. Different tools and databases are available for metagenomics data analysis and studies, a few of them are given in Table 3. Metagenomic analysis differs from genomic data analysis in that it provides a more complete view of a microbial community and includes the study of genetic information at the community level. Metagenomic samples often contain hundreds of genomes from different microbial species, leading to high genomic diversity and complexity (Glass and Meyer 2012; Su et al. 2014; Su et al. 2011).

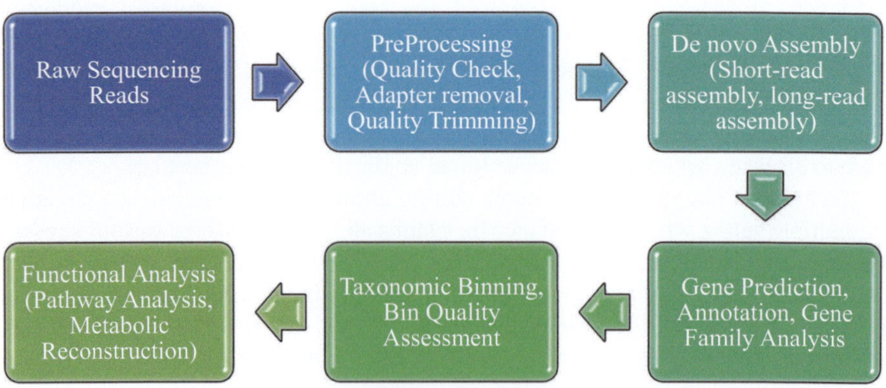

Fig. 1 A basic metagenomics data analysis pipeline

Table 3 Details of different tools and databases available for metagenomics data analysis

Name	Description	Availability
Quality control and pre-processing tools		
FastQC	Quality control	https://github.com/s-andrews/FastQC
Trimmomatic	Read trimming, adapter removal, quality filtering	http://www.usadellab.org/cms/?page=trimmomatic
BBTools	Read processing, filtering, trimming, error correction	https://github.com/kbaseapps/BBTools
Cutadapt	Flexible adapter trimming, quality filtering	https://cutadapt.readthedocs.io/en/stable/
KneadData	Contamination removal host sequence removal, quality control	https://github.com/biobakery/kneaddata
Assembly tools		
MEGAHIT	De novo assembly, memory-efficient, fast assembly	https://github.com/voutcn/megahit
MetaSPAdes	De novo assembly, specialized for metagenomes	http://cab.spbu.ru/software/spades
IDBA-UD	De novo assembly, multiple k-mer strategy	https://github.com/loneknightpy/idba
Flye	Long-read assembly, PacBio/Nanopore data assembly	https://github.com/mikolmogorov/Flye
SoapDeNovo	De novo assembly, short read assembly	https://github.com/aquaskyline/SOAPdenovo2
Binning tools		
MetaBAT2	Contig binning	https://bitbucket.org/berkeleylab/metabat/src/master/
CONCOCT	Binning	https://github.com/BinPro/CONCOCT
MaxBin2	Automated binning	https://sourceforge.net/projects/maxbin2/
DAS Tool	Bin aggregation	https://github.com/cmks/DAS_Tool
CheckM	Bin quality assessment	https://github.com/Ecogenomics/CheckM
Taxonomic classification tools		
Kraken2	Read classification	https://ccb.jhu.edu/software/kraken2/
MetaPhlAn	Taxonomic profiling	https://huttenhower.sph.harvard.edu/metaphlan/
DIAMOND	Sequence alignment	https://github.com/bbuchfink/diamond
Kaiju	Protein-level classification	https://bioinformatics-centre.github.io/kaiju/
MEGAN	Taxonomic analysis	https://software-ab.cs.uni-tuebingen.de/download/megan6/welcome.html
Functional analysis tools		
Prodigal	Gene prediction	https://github.com/hyattpd/Prodigal
eggNOG-mapper	Functional annotation	http://eggnog-mapper.embl.de/
InterProScan	Protein domain analysis	https://www.ebi.ac.uk/interpro/search/sequence/

(continued)

Table 3 (continued)

Name	Description	Availability
MinPath	Pathway prediction, parsimonious pathway reconstruction	https://omics.informatics.indiana.edu/MinPath/
HUMAnN	Functional profiling	https://huttenhower.sph.harvard.edu/humann/
Metagenomics associated databases and servers		
SILVA	Ribosomal RNA databases	https://www.arb-silva.de/documentation/silva-taxonomy/
RDP	Ribosomal database project	https://www.glbrc.org/data-and-tools/glbrc-data-sets/ribosomal-database-project
IMG/M	Integrated microbial genomes	https://img.jgi.doe.gov/
KEGG	Pathway database metabolic pathways and functions	https://www.genome.jp/kegg/
MetaCyc	Metabolic pathways	https://metacyc.org/
COG	Clusters of orthologous groups	https://www.ncbi.nlm.nih.gov/research/cog/
SEED	Subsystem-based annotation	https://rast.nmpdr.org/
CAZy	Carbohydrate-active enzymes	https://www.cazy.org/
CARD	Antibiotic resistance database	https://card.mcmaster.ca/
MGnify	Metagenome analysis	https://www.ebi.ac.uk/metagenomics
MG-RAST	Metagenome analysis	https://www.mg-rast.org/

5 Potential of Metagenomics in Genome Editing

Metagenomics, which involves sequencing all the genomic nucleic acids from a biome or consortium of microorganisms, provides a more complete view of a microbial community than traditional culture-based methods (Haynes 2018). It allows for the analysis of a metagenome not only by taxonomy but also by the encoded metabolic potential, making it valuable for retrieving and modifying valuable microorganisms for production, particularly in marine metagenomics (Kodzius and Gojobori 2015).

The integration of metagenomics with genome editing holds significant potential. Metagenomics can uncover novel genes, pathways, and genomes from uncultivated organisms, which can then be targeted for genome editing to enhance biological and chemical production (Kodzius and Gojobori 2015; Aliaga Goltsman et al. 2020). For example, novel CRISPR-associated proteins discovered through metagenomic studies can be developed into new gene-editing tools (Aliaga Goltsman et al. 2020; Zhang et al. 2014; Weissman and Johnson 2020). Metagenomic data has significantly expanded the diversity of known CRISPR-Cas systems, identifying new families of CRISPR-associated genes that can be harnessed for genome editing (Zhang et al. 2014; Weissman and Johnson 2020; Podlevsky et al. 2020). These discoveries provide new tools for precise genetic modifications and have applications in gene therapy, agriculture, and biotechnology (Pande et al. 2019; Tay Fernandez et al. 2022). Genome editing of genes discovered through metagenomics can lead to the development of crops with improved traits, such as disease resistance and stress tolerance (Tay Fernandez et al. 2022). Metagenomics provides a powerful tool for exploring the genetic diversity of microbial communities, while genome editing offers precise methods for manipulating this genetic information. Together, they hold great promise for advancing research and applications across various fields.

6 Development in the Field of Metagenomics and Its Relevance to Genome Editing

Metagenomics and genome editing are increasingly intersecting fields, particularly in the context of enhancing biological and chemical production, developing novel bioresources, and improving industrial strain efficiency. Metagenomics significantly contributes to the discovery of novel genetic elements for genome editing tools by enabling the exploration of genetic material from non-cultivable microorganisms. This approach is particularly valuable as it allows researchers to identify new genes, pathways, and genomes that are otherwise inaccessible through traditional culturing methods (Kodzius and Gojobori 2015). Metagenomics aids in the development of CRISPR-based tools for various biotechnological applications, such as improving agricultural productivity, developing new therapeutics, and creating bioresources for industrial use (Srivastava and Verma 2024; Kodzius and Gojobori 2015).

Metagenomics has been instrumental in identifying new CRISPR-Cas systems, which are essential for genome editing. Large-scale metagenomic analyses have led to the discovery of novel CRISPR-Cas systems, such as Cas12a and CoCas9, which are derived from uncultivated organisms. These systems exhibit unique properties and high efficiency in gene editing, expanding the CRISPR toolbox for therapeutic and biotechnological applications (Aliaga Goltsman et al. 2020; Pedrazzoli et al. 2024; Aliaga Goltsman et al. 2022; Al-Shayeb et al. 2022; Zhang et al. 2014). The exploration of metagenomes has led to the identification of diverse CRISPR-associated proteins and novel PAM sequences, expanding the toolkit available for genome editing. This allows for more tailored and efficient editing strategies (Zhang et al. 2014; Ciciani et al. 2022). Metagenomics has led to the identification of new CRISPR-Cas systems from uncultivated organisms. These systems, such as CRISPR-CasX and CRISPR-CasY, expand the toolkit available for genome editing, offering more compact and efficient options (Burstein et al. 2017). Tools like digIS, which utilize profile hidden Markov models, are specifically designed to detect distant and novel insertion sequence (IS) elements in metagenomic data. These elements are crucial for understanding genome organization and evolution, which can be harnessed for genome editing purposes (Puterová and Martínek 2021). CasCollect and metaCRISPR tools are developed to efficiently identify CRISPR arrays and Cas genes from metagenomic data. These tools improve the detection and annotation process, making it feasible to discover novel CRISPR systems from complex metagenomic datasets (Podlevsky et al. 2020; Lei and Sun 2016).

7 Potential Applications of Novel Genetic Elements Discovered Through Metagenomics in the Genome Editing Tools

Metagenomics has significantly contributed to the discovery of novel genetic elements, which can be harnessed to enhance genome editing tools. These advancements have several potential applications. Novel genetic elements identified through metagenomics can be engineered to improve the precision and efficiency of genome editing tools like CRISPR/Cas9, base editing, and prime editing. These improvements can lead to more accurate modifications with fewer off-target effects (Fu et al. 2023; Arif et al. 2023). Metagenomics-derived genetic elements can be used to develop crops with enhanced traits such as disease resistance and improved yield. Additionally, these elements can be applied in industrial biotechnology to create more efficient bio-production processes (Kaur et al. 2023; Kumari 2020; Kodzius and Gojobori 2015). Novel genes and pathways discovered through metagenomics can be utilized to engineer microorganisms for environmental protection and the production of bioresources, such as biofuels and pharmaceuticals, thereby addressing resource depletion issues (Kodzius and Gojobori 2015).

8 Conclusion

The intersection of metagenomics and genome editing represents a powerful frontier in genetic research and biotechnology. Metagenomics has proven invaluable in expanding our understanding of genetic diversity by enabling the exploration of genetic material from non-cultivable microorganisms, leading to the discoveries of novel genes, pathways, and genomes. This expanded knowledge base has directly contributed to advancing genome editing technologies, particularly in the development of new CRISPR-Cas systems and other editing tools. The integration of these fields has significant implications across multiple sectors, including agriculture, medicine, and industrial biotechnology. Through metagenomics, researchers have discovered novel CRISPR-associated proteins and systems like CasX and CasY, which have enhanced the precision and versatility of genome editing techniques. These discoveries are particularly valuable for developing improved crops with enhanced traits, creating more efficient bio-production processes, and advancing therapeutic applications.

Looking forward, the synergy between metagenomics and genome editing holds promising potential for addressing current limitations in genome editing, particularly in predicting and defining unintended mutations. This advancement is crucial for developing more reliable and precise genome-editing therapies and applications. As these fields continue to evolve together, they will likely play an increasingly important role in addressing global challenges in agriculture, medicine, and environmental conservation. The future prospects of metagenomics in genome editing research lie in the development of new approaches for predicting and defining the probable function of unintended genome-editing mutations, which will inspire confidence in the next generation of promising genome-editing therapies.

Acknowledgement The authors are thankful to Director, ICAR-IARI, New Delhi, for their guidance and support. RP is thankful for the support received under the Institute Project "Soil and Crop health monitoring through sensing, modelling, and big data analytics for resource efficient smart agriculture at field to regional scale (CRSCIARISIL20210034319)."

Conflict of Interest The authors declare no conflict of interest.

References

Aliaga Goltsman DS, Alexander LM, Devoto AE, Albers JB, Liu J, Butterfield CN, Brown CT, Thomas BC (2020) Novel type V-A CRISPR effectors are active nucleases with expanded targeting capabilities. CRISPR J 3(6):454–461. https://doi.org/10.1089/crispr.2020.0043

Aliaga Goltsman DS, Alexander LM, Lin JL, Fregoso Ocampo R, Freeman B, Lamothe RC, Perez Rivas A, Temoche-Diaz MM, Chadha S, Nordenfelt N, Janson OP, Barr I, Devoto AE, Cost GJ, Butterfield CN, Thomas BC, Brown CT (2022) Compact Cas9d and HEARO enzymes for genome editing discovered from uncultivated microbes. Nat Commun 13(1):7602. https://doi.org/10.1038/s41467-022-35257-7

Al-Shayeb B, Skopintsev P, Soczek KM, Stahl EC, Li Z, Groover E, Smock D, Eggers AR, Pausch P, Cress BF, Huang CJ, Staskawicz B, Savage DF, Jacobsen SE, Banfield JF, Doudna JA (2022) Diverse virus-encoded CRISPR-Cas systems include streamlined genome editors. Cell 185(24):4574–4586.e16

Arif T, Farooq A, Ahmad FJ, Akhtar M, Choudhery MS (2023) Prime editing: a potential treatment option for β-thalassemia. Cell Biol Int 47(4):699–713. https://doi.org/10.1002/cbin.11972

Behera L, Pawan KA, Sawant SB, Sanghamitra S, Parmeswaran C, Kailash CS, Sahoo B, Nayak I (2023) Genome editing for multiple stress tolerance. In: Multiple abiotic stress tolerances in higher plants: addressing the growing challenges. CRC Press, p p19

Bhat MA, Bhat MA, Kumar V, Wani IA, Bashir H, Shah AA, Rahman S, Jan AT (2020) The era of editing plant genomes using CRISPR/Cas: a critical appraisal. J Biotechnol 324:34–60. https://doi.org/10.1016/j.jbiotec.2020.09.013

Burstein D, Harrington LB, Strutt SC, Probst AJ, Anantharaman K, Thomas BC, Doudna JA, Banfield JF (2017) New CRISPR-Cas systems from uncultivated microbes. Nature 542(7640):237–241. https://doi.org/10.1038/nature21059

Ciciani M, Demozzi M, Pedrazzoli E, Visentin E, Pezzè L, Signorini LF, Blanco-Miguez A, Zolfo M, Asnicar F, Casini A, Cereseto A, Segata N (2022) Automated identification of sequence-tailored Cas9 proteins using massive metagenomic data. Nat Commun 13(1):6474. https://doi.org/10.1038/s41467-022-34213-9

de Bruijn FJ (2011a) Handbook of molecular microbial ecology I: metagenomics and complementary approaches. https://doi.org/10.1002/9781118010518

de Bruijn FJ (2011b) Handbook of molecular microbial ecology II: metagenomics in different habitats. https://doi.org/10.1002/9781118010549

Demirci Y, Zhang B, Unver T (2018) CRISPR/Cas9: an RNA-guided highly precise synthetic tool for plant genome editing. J Cell Physiol 233(3):1844–1859. https://doi.org/10.1002/jcp.25970

Fu Y, He X, Gao XD, Li F, Ge S, Yang Z, Fan X (2023) Prime editing: current advances and therapeutic opportunities in human diseases. Sci Bull (Beijing) 68(24):3278–3291. https://doi.org/10.1016/j.scib.2023.11.015

Garcia T, Lamba DA (2017) CRISPR-based genome engineering in human stem cells, Patient-specific stem cells. CRC Press

Glass EM, Meyer F (2012) Analysis of metagenomics data. In: Rodríguez-Ezpeleta N, Hackenberg M, Aransay A (eds) Bioinformatics for high throughput sequencing. Springer, New York, NY. https://doi.org/10.1007/978-1-4614-0782-9_13

Gopika AB, Anand AV et al (2024) Gene editing using CRISPR/Cas9 system: methods and applications. CRISPR and Plant Functional Genomics. ISBN 9781032469492

Haynes M (2018) Metagenomics. Encyclopedia of ecology

He Z, Proudfoot C, Whitelaw CB, Lillico SG (2016) Comparison of CRISPR/Cas9 and TALENs on editing an integrated EGFP gene in the genome of HEK293FT cells. Springerplus 5(1):814. https://doi.org/10.1186/s40064-016-2536-3

Jeon S, Lim JM, Lee HG et al (2017) Current status and perspectives of genome editing technology for microalgae. Biotechnol Biofuels 10:267. https://doi.org/10.1186/s13068-017-0957-z

Kanwal F, Deabel RA, Alqahtani A et al (2024) Evolution of genome editing technologies. In: CRISPRized horticulture crops: genome modified plants and microbes in food and agriculture

Kaur R, Singh AK, Singh DK, Singh S (2023) Modern tools of genome engineering and their applications. In: Sobti R, Kuhad RC, Lal R, Rishi P (eds) Role of microbes in sustainable development. Springer, Singapore. https://doi.org/10.1007/978-981-99-3126-2_9

Khoshandam M, Soltaninejad H, Mousazadeh M, Hamidieh AA, Hosseinkhani S (2024) Clinical applications of the CRISPR/Cas9 genome-editing system: delivery options and challenges in precision medicine. Genes Dis 11(1):268–282

Khouzam JPS, Tivakaran VS (2021) CRISPR-Cas9 applications in cardiovascular disease. Curr Probl Cardiol 46(3):100652

Kodzius R, Gojobori T (2015) Marine metagenomics as a source for bioprospecting. Mar Genomics 24(1):21–30

Kumari PS (2020) Genome editing tools in Livestock. Res J Biotechnol 15(8):145–151

Lee CM, Cradick TJ, Bao G (2016) The Neisseria meningitidis CRISPR-Cas9 system enables specific genome editing in mammalian cells. Mol Ther 24(3):645–654. https://doi.org/10.1038/mt.2016.8

Lei J, Sun Y (2016) Assemble CRISPRs from metagenomic sequencing data. Bioinformatics 32(17):i520–i528. https://doi.org/10.1093/bioinformatics/btw456

Li Y, Song YH, Liu B, Yu XY (2017) The potential application and challenge of powerful CRISPR/Cas9 system in cardiovascular research. Int J Cardiol 227:191–193. https://doi.org/10.1016/j.ijcard.2016.11.177

Ma X, Liu YG (2016) CRISPR/Cas9-based multiplex genome editing in monocot and dicot plants. Curr Protoc Mol Biol 115(1):31.6.1–31.6.21

Marić J, Šikić M (2019) Approaches to metagenomic classification and assembly. In: International convention on information and communication technology, electronics and microelectronics (MIPRO), pp 348–356. https://doi.org/10.23919/MIPRO.2019.8756644

Marzec M, Hensel G (2020) Prime editing: game changer for modifying plant genomes. Trends Plant Sci 25(8):722–724. https://doi.org/10.1016/j.tplants.2020.05.008

Mitra S, Rupek P, Richter DC et al (2011) Functional analysis of metagenomes and metatranscriptomes using SEED and KEGG. BMC Bioinform 12(Suppl 1):S21. https://doi.org/10.1186/1471-2105-12-S1-S21

Nitish B, Jebi S, Mangla H, Sharma A, Dave S, Pathak H (2023) RNA-mediated nuclease genome editing system type II: a mini review. Appl Environ Biotechnol 8(1):25–31. https://doi.org/10.26789/AEB.2023.01.004

Okoli A, Okeke MI, Tryland M, Moens U (2018) CRISPR/Cas9-advancing orthopoxvirus genome editing for vaccine and vector development. Viruses 10(1):50. https://doi.org/10.3390/v10010050

Pande A, Bhat RAH, Saxena A, Tyagi M (2019) Concepts and potential applications of gene editing in aquaculture. In: Genomics and biotechnological advances in veterinary, poultry, and fisheries

Pedrazzoli E, Demozzi M, Visentin E, Ciciani M, Bonuzzi I, Pezzè L, Lucchetta L, Maule G, Amistadi S, Esposito F, Lupo M, Miccio A, Auricchio A, Casini A, Segata N, Cereseto A

(2024) CoCas9 is a compact nuclease from the human microbiome for efficient and precise genome editing. Nat Commun 15(1):3478. https://doi.org/10.1038/s41467-024-47800-9
Pellagatti A, Dolatshad H, Yip BH, Valletta S, Boultwood J (2016) Application of genome editing technologies to the study and treatment of hematological disease. Adv Biol Regul 60:122–134
Podlevsky JD, Hudson CM, Timlin JA, Williams KP (2020) CasCollect: targeted assembly of CRISPR-associated operons from high-throughput sequencing data. NAR Genom Bioinform 2(3):lqaa063. https://doi.org/10.1093/nargab/lqaa063
Puterová J, Martínek T (2021) digIS: towards detecting distant and putative novel insertion sequence elements in prokaryotic genomes. BMC Bioinform 22(1):258. https://doi.org/10.1186/s12859-021-04177-6
Qi X, Zhang J, Zhao Y, Chen T, Xiang Y, Hui J, Cai D, Liu Y, Xia L, Yu T, Li G (2017) The applications of CRISPR screen in functional genomics. Brief Funct Genomics 16(1):34–37
Rao Y, Yang X, Pan C, Wang C, Wang K (2022) Advance of clustered regularly interspaced short palindromic repeats-Cas9 system and its application in crop improvement. Front Plant Sci 13:839001. https://doi.org/10.3389/fpls.2022.839001
Rezaei H, Khadempar S, Farahani N, Hosseingholi EZ, Hayat SMG, Sathyapalan T, Sahebkar AH (2020) Harnessing CRISPR/Cas9 technology in cardiovascular disease. Trends Cardiovasc Med 30(2):93–101. https://doi.org/10.1016/j.tcm.2019.03.005
Sakuma T, Yamamoto T (2018) Acceleration of cancer science with genome editing and related technologies. Cancer Sci 109(12):3679–3685. https://doi.org/10.1111/cas.13832
Shi M, Shen Z, Zhang N, Wang L, Yu C, Yang Z (2021) CRISPR/Cas9 technology in disease research and therapy: a review. Shengwu Gongcheng Xuebao/Chin J Biotechnol 37(4):1205–1228
Srivastava A, Verma D (2024) Environmental metagenomics and CRISPR-Cas. In: CRISPR-Cas system in translational biotechnology. Academic Press, pp 175–190. https://doi.org/10.1016/B978-0-323-91808-4.00014-6
Su X, Xu J, Ning K (2011) Parallel-META: a high-performance computational pipeline for metagenomic data analysis. In: IEEE international conference on systems biology (ISB), pp 173–178. https://doi.org/10.1109/ISB.2011.6033151
Su X, Pan W, Song B, Xu J, Ning K (2014) Parallel-META 2.0: enhanced metagenomic data analysis with functional annotation, high performance computing and advanced visualization. PLoS ONE 9(3):e89323
Tay Fernandez CG, Nestor BJ, Danilevicz MF, Marsh JI, Petereit J, Bayer PE, Batley J, Edwards D (2022) Expanding gene-editing potential in crop improvement with pangenomes. Int J Mol Sci 23(4):2276. https://doi.org/10.3390/ijms23042276
Weissman JL, Johnson PLF (2020) Network-based prediction of novel CRISPR-associated genes in metagenomes. mSystems 5:e00752–e00719
Wilson AM, Wingfield BD (2020) CRISPR-Cas9-mediated genome editing in the filamentous ascomycete Huntiella omanensis. J Vis Exp 160. https://doi.org/10.3791/61367
Yamamoto T, Sakamoto N, Sakuma T (2014) Genome editing with programmable site-specific nucleases. Uirusu 64(1):75–82
Yan Y, Zhu J, Xie C, Liu C (2019) Regulatory framework of genome-edited products—a review. Shengwu Gongcheng Xuebao/Chin J Biotechnol 35(6):921–930
Zhang Q, Doak TG, Ye Y (2014) Expanding the catalog of cas genes with metagenomes. Nucleic Acids Res 42(4):2448–2459. https://doi.org/10.1093/nar/gkt1262
Zheng L, Shi J, Mu Y (2018) Dynamics changes of CRISPR-Cas9 systems induced by high fidelity mutations. Phys Chem Chem Phys 20:27439–27448

Tools and Techniques of Meta-Transcriptomic Studies and Its Applications in Crop Improvement

Naina Kumari, Tilak Chandra, Sarika Jaiswal, Mir Asif Iquebal, and Dinesh Kumar

1 Introduction

The growing global population and the impacts of climate change pose significant threats to agriculture by adversely affecting crop health and yield. However, plant microbiomes offer promising solutions for enhancing resilience and mitigating these challenges (Thingujam et al. 2024). Advances in next-generation sequencing (NGS) technologies have greatly expanded our ability to uncover hidden information within molecular sequences, including DNA, coding and noncoding RNA, and various proteins (Kumar and Yadav 2024). These molecular footprints have been integral to numerous studies aimed at resolving complex biological questions, including those related to the microbiome. The microbiome, encompassing all the genetic material within microbiota, refers to the entire community of microorganisms inhabiting a specific environment (Kaul et al. 2021). Present in all living organisms and ecosystems, it interacts with and regulates the environment in which it exists (Braga et al. 2016). The microbiome plays a crucial role in the health of plants and crop ecosystems, influencing interactions within complex environments such as soil and water bodies, where microorganisms engage in a wide range of community-based operations (Mukherjee and Reddy 2020). Traditional culture-based methods only capture about 1% of an ecosystem's microbiome, as they are limited to culturable microorganisms, leaving over 99% of the microbial community uncharacterized (Garza and Dutilh 2015; Nath et al. 2021). The development of advanced technologies and tools has significantly advanced microbiological research, driving paradigm shifts in our understanding of agricultural productivity and environmental sustainability (Sharuddin et al. 2022). High-throughput sequencing platforms have

Naina Kumari and Tilak Chandra contributed equally with all other contributors.

N. Kumari · T. Chandra · S. Jaiswal · M. A. Iquebal (✉) · D. Kumar
Division of Agricultural Bioinformatics, ICAR-Indian Agricultural Statistics Research Institute, New Delhi, India

Renu et al. (eds.), *Meta-omics in Crop Improvement*,
https://doi.org/10.1007/978-981-95-2294-1_4

revolutionized the quantification of transcript expression in both known organisms and microbial communities. These methods enable detailed annotation and quantification of known and novel transcripts and variants, thus facilitating more comprehensive and efficient research (Shakya et al. 2019; Kumar and Yadav 2024).

Most microbiome studies employ either a targeted approach or shotgun sequencing. The targeted approach involves high-throughput DNA sequencing of specific regions, such as 16S rRNA and 18S rRNA, to characterize microbial communities, while shotgun sequencing involves metagenomic analysis, which sequences the entire DNA content from a sample (Shakya et al. 2019). Amplicon sequencing is a cost-effective method commonly used in microbiome research; however, it primarily identifies microorganisms without providing insight into their functional roles. In contrast, metagenomics can identify both microbes and their functional profiles, but it does not distinguish between active and inactive members of the microbial community (Shakya et al. 2019). Metatranscriptomics, however, focuses on the genes expressed by the entire microbial community, offering a snapshot of gene expression at a specific time and under particular environmental conditions. For example, agricultural soils provide valuable insights into the microbial community structure, soil health, and crop performance (Sharma and Sharma 2019), while climate change has been shown to impact the rhizosphere microbiomes in European grassland ecosystems (Bei et al. 2019), among other studies. Despite its broad applications in agriculture and its role in identifying the active microbial community under varying conditions, the tools and technologies used for metatranscriptomic analysis have not been comprehensively discussed. This chapter aims to explore the tools, techniques, and algorithms to analysis and visualization of metatranscriptomic data within the context of their application in agriculture.

2 Established Pipelines for Metatranscriptomic Data Analysis

With the availability of numerous bioinformatics tools and methodologies for data analysis, several pipelines have been developed to analyze metatranscriptomic data from various environmental samples. However, a comprehensive guide for selecting appropriate metatranscriptomic tools can be found in a paper published by Terron-Camero et al. (2022). The first pipeline dedicated to comprehensive metatranscriptome analysis was developed and validated using an RNA-seq dataset from the human small intestine microbiota (Leimena et al. 2013). Since then, a variety of other metatranscriptomic pipelines have been introduced, including MetaTrans (Martinez et al. 2016), FMAP (Kim et al. 2016), IMP (Narayanasamy et al. 2016), SAMSA2 (Westreich et al. 2018), COMAN (Ni et al. 2016), SQUEEZEMETA (Tamames and Puente-Sanchez 2019), CoMW (Anwar et al. 2019), and MOSCA (Sequeira et al. 2019). Two of the most recent pipelines, MetaPro (Taj et al. 2023) and AsaiM-MT (Mehta et al. 2021), have also been developed. While the ASaiM-MT (v1) workflow was originally designed for both metagenomics and metatranscriptomics data analysis, it was initially validated only for metagenomics datasets. In contrast, the tools in ASaiM-MT (v2) have been specifically optimized and validated for analyzing metatranscriptomic data. Additionally, several metagenomic analysis pipelines, such as MG-RAST and HumAnN2, have also been applied to metatranscriptomic data. A detailed overview and applied methodologies to the algorithm and tools utilized in various metatranscriptomic pipelines is given in Table 1, Fig. 1.

3 Processing and Workflow

3.1 Data Quality Check and Trimming

The first and most crucial step in analyzing any next-generation sequencing (NGS) data is the removal of erroneous and spurious reads. Most analysis pipelines use FASTQC to assess data quality and remove unwanted reads, either through quality control tools or custom scripts (Andrews 2010). The Trimmomatic tool is employed by pipelines such as SAMSA2, IMP, and MOSCA, while BM-Tagger, Kraken, and CutAdapt are used in the FMAP, MetaTrans, and ASaiM-MT pipelines, respectively. Trimmomatic utilizes two approaches for quality filtering of Illumina reads: sliding window quality filtering and maximum information quality filtering. Both methods evaluate the quality score of each base position to identify the optimal location for trimming the read. The 5′ portion of the sequence is retained, while the read is cut at the identified position on the 3′ end, as Illumina data typically exhibit lower quality towards the 3′ end (Bolger et al. 2014). To detect technical sequences, Trimmomatic offers two modes: "simple mode" and "palindrome mode." The simple mode locates a close match between the read and the user-defined technical sequence, making it effective for detecting various types of technical sequences, including adapter and primer sequences.

The palindrome mode is specifically designed to detect "adapter read-through," where the sequenced DNA fragment is shorter than the read length, leading to adapter contamination at the ends of the reads. This mode employs a "seed and extend" alignment approach, similar to the simple mode, and applies global alignment scoring to ensure an end-to-end match across the entire overlap (Bolger et al. 2014). CutAdapt utilizes an algorithm comparable to BWA (Burrows-Wheeler Alignment) for quality trimming of reads, employing a semi-global (free shift) alignment method to identify and remove adapter sequences (Martin 2011). The Kraken suite uses three standalone tools for pre-processing NGS reads: (i) Reaper, which employs Smith-Waterman alignments for adapter sequence identification and trimming, demultiplexes reads by recognizing different barcode sequences, and generates comprehensive summary statistics; (ii) Tally, a C program that tallies redundant sequences in memory and optimizes memory usage through efficient allocation, hashing techniques, and sequence compression; and (iii) Minion, a small utility that uses a De Bruijn graph to detect sequence fragments with characteristics indicative of adapter sequences, such as high abundance towards the 3′ end of the reads (Devis et al. 2013).

Table 1 Tabulation of metatranscriptomics data analysis pipelines and various features such as pre-processing, filtering of rRNA, assembly, taxonomy, putative genes and their functional annotation, differential expression analysis and advantage and limitation

Pipelines	Pre-processing	Filtering of rRNA	Assembly	Taxonomy	Putative genes	Functional annotation	Differential expression	Advantage	Limitation
Leimena—2013	Avg. Phred score ≥ 10 as a cutoff	SortMeRNA and BLASTN (NCBI and SILVA ribosomal dastabase)	Not performed	MegaBLAST and BLASTN (NCBI)	–	BLAST and KAAS (COG and KEGG)	–	Clustering of identical sequences to avoid inefficient and repetitive mapping	Assembly of reads not performed. DE analysis not available in the pipeline
SAMSA2—2018	Trimmomatic	SortMeRNA	Not performed		–	DIAMOND (NCBI RefSeq and SEED subsystems)	R scripts (Deseq2)	95% accuracy at annotation step and 87.5% accuracy for species level taxonomy Ability to customize index database Less time and computational resources are required	Require large computing resource (10 CPUs/20 GB RAM even for small sequence files)
MetaTrans—2016	KRAKEN	SortMeRNA	No	16S rRNA and16S DNA by SOAP2	FragGeneScan	SOAP2	DESeq2		Uses the METAHIT database optimized only for gut microbiome

Pipelines	Pre-processing	Filtering of rRNA	Assembly	Taxonomy	Putative genes	Functional annotation	Differential expression	Advantage	Limitation
COMAN—2016	In-house perl script	BLASTN	No	DIAMOND	–	RPS-BLAST and DIAMOND, KOBAS	Wilcoxon rank sum test, GAGE	Includes biosynthetic gene clusters (BGCs) for identifying microbial secondary metabolites and co-expression network analysis as well Also facilitate the researchers, who may lack the expertise on computational resources	
MOSCA—2019	Trimmomatic	SortMeRNA	MetaSPAdes and Megahit	Uniprot's ID mapping service and DIAMOND	FragGeneScan	Uniprot's ID mapping service and DIAMOND	DeSeq2	Integration of metagenomics and metatranscriptomics analysis	Relies on the reference database for annotation Can be computationally intensive User expertise in bioinformatics is needed

(continued)

Table 1 (continued)

Pipelines	Pre-processing	Filtering of rRNA	Assembly	Taxonomy	Putative genes	Functional annotation	Differential expression	Advantage	Limitation
SqueezeMeta—2019	Trimmomatic, Prinseq	Barnap	CD-HIT	DIAMOND, LCA algorithm	Prodigal	DIAMOND HMMER3	Does not perform	Provides analysis of multiple metagenomes allowing for co-assembly of related metagenomes Resource efficiency Designed to work on a moderately-sized computational infrastructure Flexible web interface: Allows users to interact with the stored data easily	May not be suitable for analysing extremely large and complex metagenomes Dependency on sequencing technology: It is adapted to MinION sequences but may not support other sequencing technologies or formats.
IMP—2016	Trimmomatic	SortMeRNA	MegaHit	MetaQUAST	–	Prokka	Not performed	Reference-independent analysis: Reducing the reliance on external reference databases It offers reproducibility and modularity allowing the users to customize analysis components Employs an iterative co-assembly approach enhancing data usage and generating large number of contigs	Computational demand may be high for large and complex dataset Customization complexity: Users may need to have some expertise in bioinformatics as incorrect customization may result in suboptimal output

Pipelines	Pre-processing	Filtering of rRNA	Assembly	Taxonomy	Putative genes	Functional annotation	Differential expression	Advantage	Limitation
FMAP—2016	BM-Tagger	–	–	DIAMOND or USEARCH	–	DIAMOND or USEARCH	(1) MetagenomeSeq (2) Kruskal-Wallis rank-sum tests (3) quasi-Poisson	Provides operon level analysis Capable of processing large datasets Optimized reference database, which is a functionally annotated filtered UniRef90	Reference database may introduce bias towards well studied organisms and functions Operon analysis challenges: The accuracy of operon predictions may vary depending on the dataset
CoMW—2019			Trinity				Deseq2		
AsaiM-MT—2021	MultiQC, CutAdapt	SortMeRNA	No	MetaPhlan2	HUMAnN2	HUMAnN2	Not performed		

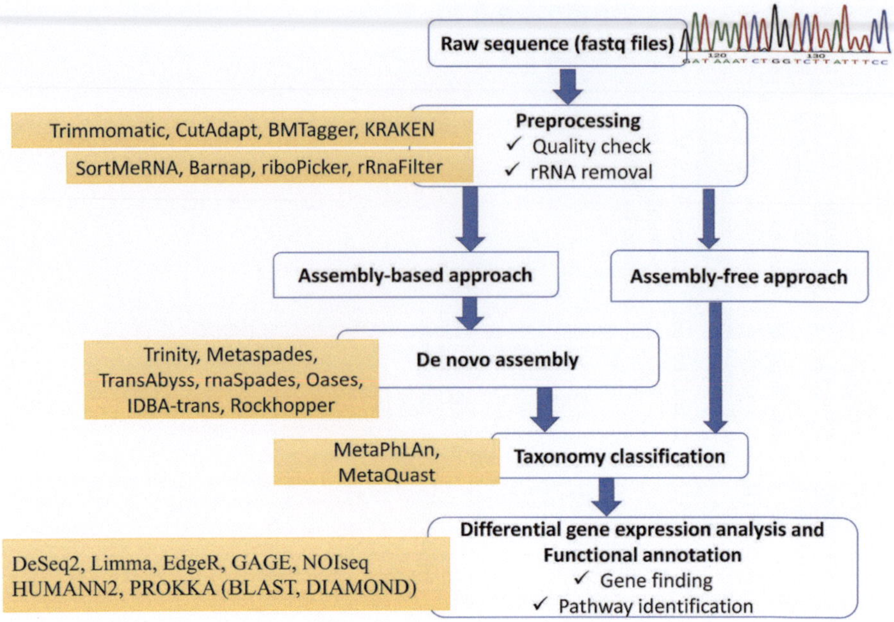

Fig. 1 The workflow model for metatranscriptomic studies involves the use of multiple pipelines and algorithms to analyze microbiome communities and their functional roles within their respective habitats

3.2 Identification of rRNA and Its Removal

The removal of rRNA sequences is a critical step in the pre-processing of metatranscriptomic data (Koseoglu 2024). Although most rRNA reads are eliminated during the mRNA enrichment step in sequencing, a significant number of rRNA reads often remain in the data and must be identified and removed before expression analysis (Wolf 2013). However, many pipelines, such as MetaTrans, retain removed rRNA sequences separately for taxonomic classification of bacterial communities. Typically, pipelines align query sequences against an rRNA database using BLASTn to identify rRNA sequences. However, most pipelines, including Leimena-2013, SAMSA2, MOSCA, MetaTrans, IMP, and ASaiM-MT, use SortMeRNA, which remains the most popular and preferred rRNA removal tool to date. SortMeRNA's algorithm scans each read with a sliding window to search for short, similar regions between the read and the rRNA database (Kopylova et al. 2012). Reads that exceed a specified threshold number of matching windows are accepted, and errors (substitutions, insertions, or deletions) are allowed between the read and the rRNA database for a given window. The rRNA database is stored in a burst trie data structure combined with a lookup table, enabling efficient access. This approach accelerates access to the burst trie and leverages conserved regions in rRNA sequences to optimize performance (Kopylova et al. 2012).

3.3 Assembly/Classification

After the pre-processing step, two general approaches are typically followed. In the assembly-free approach, quality-filtered reads are directly used for taxonomic profiling, whereas in the assembly-based approach, reads are assembled into contigs and supercontigs before proceeding to downstream analysis. The assembly step in metatranscriptomic data analysis is considered optional, and many pipelines, such as SAMSA2, MetaTrans, FMAP, and ASaiM-MT, do not include this step. However, pipelines like SqueezeMeta, MOSCA, IMP, and CoMW follow the assembly-based approach for metatranscriptomic analysis. Previous studies have shown that assembling metatranscriptome datasets can significantly improve read annotation and gene expression quantification (Celaj et al. 2014; Terron-Camero et al. 2022). Additionally, RNA-Seq analysis of complex microbial communities is challenging because different species may be represented at vastly different abundance levels (Celaj et al. 2014; Terron-Camero et al. 2022; Kumar and Yadav 2024). Given the lack of reference genomes for many microorganisms, de novo assembly (i.e., assembly without reference) is commonly used for datasets derived from non-model biological systems. De novo assemblers such as MegaHIT, CD-HIT, and Trinity are used by the IMP, SqueezeMeta, and CoMW pipelines, respectively, while MOSCA includes both MegaHIT and MetaSpades for read assembly. MegaHIT and MetaSpades were primarily designed for metagenomic datasets, whereas Trinity was specifically developed for assembling transcriptomic reads. Most de novo assemblers utilize the De Bruijn graph approach for constructing transcriptome assemblies.

Comparative studies have shown that Trinity outperforms other transcriptome assemblers, such as Oases, Rockhopper, IDBA-trans, and Trans-Abyss, in read assembly (Grabherr et al. 2011). Trinity consists of three modules: Inchworm, Chrysalis, and Butterfly, which together facilitate the efficient analysis of large volumes of RNA-Seq reads. The Inchworm module uses a k-mer-based approach to quickly and efficiently assemble reads into unique transcript sequences. Chrysalis clusters related contigs by constructing de Bruijn graphs, corresponding to segments of alternatively spliced transcripts or unique portions of paralogous genes. The Butterfly module then analyzes the paths taken by reads and read pairings, reporting all probable transcript sequences and resolving alternatively spliced isoforms and transcripts from paralogous genes (Grabherr et al. 2011; Celaj et al. 2014). Pincho is a de novo transcriptomic analysis workflow that uses a multi-assembly approach, comparing nine de novo assemblers in combination. It produces a consensus assembly using the TransRate tool and employs CD-HIT to isolate transcripts under a specified length threshold. Redundancy reduction and alignment are performed through HISAT2, BUSCO, and TransRate. The authors recommend combining Trans-ABySS, rnaSPAdes, and Translig assemblers for de novo transcriptome assembly, as this combination provides the highest metric scores. After filtering, clean mRNA reads undergo taxonomic identification and functional assignment, either on a read-based or contig-based approach (Ortiz et al. 2021).

Some pipelines use only mRNA reads, while others retain rRNA reads for taxonomic classification, as 16S rRNA sequences have proven to be excellent markers for identifying bacterial species. For instance, the MetaTrans pipeline maps rRNA/tRNA reads against the GreenGenes 16S rRNA database for taxonomic classification, randomly selecting a high number of reads (~1 million) to expedite the classification process. Most pipelines align filtered reads to prokaryotic databases using tools such as BLASTN, MEGABLAST, or DIAMOND. In the Leimena-2013 pipeline, both MegaBLAST and BLASTN are used in iterative steps to align reads to a prokaryotic database, with alignment bit-scores employed for taxonomic classification at various levels. Specifically, mRNA reads with bit-scores above a threshold of 148 and 110 are used for phylogenetic assignments at the genus and family levels, respectively, while reads with bit-scores below 110 are categorized as unclassified (Leimena et al. 2013).

COMAN, SqueezeMeta, and FMAP utilize the DIAMOND tool for taxonomic classification, which is highly sensitive and significantly faster than BLAST for aligning the millions of reads present in NGS datasets. MOSCA uses the UniProt ID mapping service to automatically retrieve taxonomic and functional information for annotated genes and provide cross-references to other databases. In contrast, ASaiM-MT employs MetaPhlan2 for taxonomic classification of metatranscriptomic (MT) reads. MetaPhlan2 uses a clade-specific marker gene database comprising approximately 17,000 reference genomes of microorganisms. It scans each read against this database to determine the presence and coverage of clade-specific markers, enabling taxonomic clade identification in microbiome samples. The inclusion of subspecies markers allows for strain-level analysis, while quasi-markers help detect viruses and eukaryotic microbes, thus improving classification accuracy (Truong et al. 2015).

3.4 Gene Identification with Its Expression

The identification and annotation of genes to characterize the function of transcripts is a crucial step in metatranscriptomic data analysis, as the expressed transcripts serve as proxies for the actual phenotype (Shakya et al. 2019). Detection of putative genes is typically performed using tools such as FragGeneScan and Prodigal. FragGeneScan employs a hidden Markov model that integrates sequencing error models and codon usage patterns to enhance the prediction of protein-coding regions in short reads (Rho et al. 2010). Functional annotation of genes generally involves aligning them against functional databases using tools like BLAST, RPS-BLAST, and DIAMOND, which are also used for taxonomic profiling. Additional tools for functional annotation include KOBAS, HUMANN2, KAAS, PROKKA, and SOAP2. While HUMANN2 is itself a comprehensive pipeline for metagenomics and metatranscriptomics data analysis, some pipelines, such as ASaiM-MT, integrate HUMANN2 specifically for functional gene annotation. The choice of functional databases is a key factor in gene characterization. Many pipelines, such as SqueezeMeta, COMAN, and FMAP, primarily rely on KEGG (Kyoto

Encyclopedia of Genes and Genomes) and COG (Clusters of Orthologous Groups) databases for functional annotation.

SqueezeMeta employs a classical best-hit approach for gene annotation against COG and KEGG IDs. For each COG and KEGG ID, the pipeline selects the first hit that exceeds an identity threshold, and the bitscores are averaged. The ORF (open reading frame) is assigned to the highest-scoring COG or KEGG entry, provided the score exceeds any other by 20%; otherwise, the gene remains unannotated (Tamames and Puente-Sanchez 2019). The MetaTrans pipeline maps detected genes against the MetaHIT-2014 protein database, which is annotated using both KEGG and egg-NOG (Evolutionary Genealogy of Genes: Non-supervised Orthologous Groups) databases. SAMSA2 uses DIAMOND for gene annotation against the SEED Subsystems hierarchical database and NCBI's RefSeq database. Additionally, other databases such as the CAZy (Carbohydrate-Active Enzyme) database can be incorporated into the pipeline, with these databases first converted into a DIAMOND-searchable binary format for annotation (Westreich et al. 2018). For gene abundance quantification, read counts are normalized, often converted to RPKM (Reads per Kilobase per Million) values, which adjust for both gene length and sequencing depth. The FMAP pipeline, for instance, assumes that the gene length used for normalization corresponds to the minimum gene length for the best hit(s) (Kim et al. 2016). The ASaiM-MT pipeline uses several tools in its workflow, such as Renormalize, Unpack Pathway Abundance to Show Gene Families, and Create Gene-Level Families File. This enables the generation of normalized abundances for gene families, gene pathways, and GO (Gene Ontology) slim terms in each sample (Mehta et al. 2021).

Normalization of read counts is crucial for performing comparative and statistical analyses, such as differential expression analysis (DEA), across samples. It allows for the comparison of different conditions and environmental factors, providing insights into their effects on community functions or monitoring community dynamics over time (Shakya et al. 2019). This approach helps to identify which genes or gene families are expressed at varying levels under different sample conditions. Several analysis pipelines, including SAMSA2, MetaTrans, MOSCA, and CoMW, utilize the DESeq2 R package for DEA. DESeq2 is designed for count data and uses Bayes shrinkage estimators to assess dispersion and fold changes for differential expression analysis. For visualization of differential expression across samples, DESeq2 employs rLog transformation, which performs fold-change shrinkage on a per-sample basis (Love et al. 2014). The COMAN pipeline incorporates GAGE software to infer significantly enriched or depleted pathways when comparing microbiomes under different conditions. A user-defined threshold for FDR-corrected p-values is used to identify these pathways, highlighting coordinated differential expression across entire pathways that may correlate with biological phenotypes (Ni et al. 2016). The FMAP pipeline offers three statistical testing methods for identifying differentially abundant genes and performing pathway and operon enrichment analyses. The default method, the Kruskal-Wallis Rank Sum test, uses RPKM values for DEA. Additional methods include the Quasi-Poisson test (based on RPKM) and MetagenomeSeq, which uses raw count data to detect differentially expressed genes and pathways (Kim et al. 2016).

4 Functional Annotation of Meta-Transcriptomics: Tools and Database

Functional annotation of meta-transcriptomics involves the identification and classification of gene functions from large-scale transcriptomic data obtained from environmental samples. This process helps in understanding the functional roles of microbial communities in various ecosystems. The tabulation of databases used in functional annotation of metatranscriptomic sequence data has been listed in Table 2.

4.1 NCBI RefSeq Database

The RefSeq database is a comprehensive, taxonomically diverse resource that includes sequences from archaea, bacteria, eukaryotes, and viruses. It offers well-annotated, non-redundant genomic, transcriptomic, and protein sequences, making it a reliable reference for gene identification, genome annotation, gene expression studies, and comparative analyses. The database also includes specialized projects, such as fungalITS, RefSeqGene, and rRNA loci, which focus on targeted genetic regions (O'Leary et al. 2016).

4.2 SEED Subsystems Hierarchical Database

The SEED resources organize biological systems into modular subsystems, each representing different biological functions. The aim is to enhance genome interpretation by annotating hundreds of genomes within a single subsystem. The SEED contains nearly 5000 complete genomes, a growing collection of encoded subsystems, a curated set of protein families, and hundreds of genome-scale metabolic models. To facilitate high-performance remote access, four network-based SEED servers have been developed: the Sapling Server, Annotation Support Server, RAST Server, and Model Server. These servers provide access to the underlying relational database, support core annotation services, enable programmatic access to RAST annotation features, and offer an expanding library of metabolic models (Aziz et al. 2012).

4.3 Carbohydrate Active Enzyme (CAZy) Database

The CAZy database is a specialized resource for the study of carbohydrate-active enzymes (CAZymes), which play key roles in the breakdown, modification, or formation of glycosidic linkages. The database organizes enzymes into families based on their structural, catalytic, and carbohydrate-binding modules. Its primary goal is to catalog all carbohydrate-active enzymes across different organisms and glycoscience subfields. The CAZy database includes sequence annotations, family

Table 2 Databases used in functional annotation of metatranscriptomic sequence data

Database name	Pipelines	Latest update/ version	Website link	Current statistics	Reference
NCBI RefSeq database	SAMSA2	July 15, 2022 (v213)	https://www.ncbi.nlm.nih.gov/refseq/	Proteins: 234,520,053 Transcripts: 45,781,716 Organisms: 121,461	O'Leary et al. (2016)
SEED subsystems hierarchical database	SAMSA2	June 2010	https://theseed.org/wiki/Home_of_the_SEED	Complete prokaryotic genomes: > 5000 Annotations: 30,000, metabolic models: 11,000, protein families: 178,000, functional roles: 10,250, subsystems: 1060	Aziz et al. (2012)
CAZy	SAMSA2, CoMW	Nov, 2021	http://www.cazy.org/	Genomes by Kingdom: Bacteria: 24213 Viruses: 460 Archaea: 60 Eukaryota: 377	Cantarel et al. (2009)
COG	SqueezeMeta, COMAN	2014	https://www.ncbi.nlm.nih.gov/research/cog	COGs: 4877 Genomic loci: 3,456,089 Taxonomic categories: 37 Organisms: 1309 Protein IDs: 3,213,255 Complete genome: 1187 bacteria and 122 archaea	Galperin et al. (2021)
KEGG	FMAP, IMP, Leimena et al., SqueezeMeta, COMAN	August 9, 2022	https://www.genome.jp/kegg/pathway.html	KEGG pathways: 552 KEGG organisms: 8333 (785 eukaryotes, 7152 bacteria, 396 archaea) KEGG selected viruses (T4 category): 359 KEGG viruses (Vtax category): 11487	Kanehisa and Goto (2000)
EggNOG	CoMW	v5.0, 2019	http://eggnog5.embl.de/#/app/home	Organisms: 5090 Viruses: 2502 Orthologous groups: 4.4 M	Huerta-Cepas et al. (2019)

Database name	Pipelines	Latest update/ version	Website link	Current statistics	Reference
PFAM	SqueezeMeta	v35.0 nov, 2021	https://pfam.xfam.org/	19,632 families and clans. 61 million sequences	Mistry et al. (2021)
M5nr	Metatrans, CoMW	2012	ftp://ftp.metagenomics.anl.gov/data/MD5nr/code	15,945,780 unique proteins and 5,793,086 protein functional annotations from eight sources comprising 14 databases	Wilke et al. (2012)

classifications, and functional data on these enzymes. As of now, it contains information from 24,213 bacterial, 460 viral and archaeal, and 377 eukaryotic genomes (Cantarel et al. 2009; Terrapon et al. 2017).

4.4 COG Database (Clusters of Orthologous Groups)

The COG database is an essential tool for the phylogenetic classification of proteins across complete genomes. It contains orthologs—proteins that share a direct evolutionary relationship—across a wide range of organisms, particularly bacteria and archaea. The database identifies orthologs through cross-genome comparisons and organizes them into COGs using the COG-making algorithm. New genome sequences are added to the database using the COGMaker method, which helps functionally annotate entire protein families. The COGNITOR tool is utilized to classify novel proteins into COGs and to interpret the functional and phylogenetic data of newly sequenced genomes (Tatusov et al. 2000; Galperin et al. 2021).

4.5 Kyoto Encyclopedia of Genes and Genomes (KEGG)

KEGG is a comprehensive database that provides a reference for understanding the high-level functions of biological systems through genome sequence data. It represents biological systems as molecular networks, with a particular focus on KEGG pathways that are manually curated from the scientific literature. KEGG's mapping procedure is instrumental in revealing systemic functional information embedded within an organism's genome. As of the latest update, KEGG includes 25,297 KEGG orthologous groups, 42,389,832 genes across 8318 KEGG organisms (including 782 eukaryotes, 7141 bacteria, and 395 archaea) (Kanehisa and Goto 2000).

4.6 Evolutionary Genealogy of Genes Non-supervised Orthologous Group (eggNOG)

eggNOG is a comprehensive database designed to elucidate the evolutionary history and orthology relationships among genes, offering functional annotations and predictions across a wide range of genomes. It identifies Orthologous Groups (OGs) through a species-aware clustering approach. eggNOG generates orthology predictions at multiple taxonomic levels, accounting for variations in orthology associations based on the presumed reference speciation event and increasing resolution toward the terminal nodes of the phylogenetic tree. The database encompasses 4.4 million orthologous groups (OGs), complete with associated sequence alignments, phylogenetic trees, hidden Markov models (HMMs), and functional annotations. It includes data from 477 eukaryotic species, 2502 viral proteomes, 4445 bacterial representatives, and 168 archaea, sourced from a total of 25,038 genomes. These genomes were selected for their diversity and filtered for quality (Huerta-Cepas et al. 2019).

4.7 PFAM Database

The Pfam database is a comprehensive resource of protein families, represented by multiple sequence alignments and hidden Markov models (HMMs). In addition to individual families, Pfam organizes related entries into clans—higher-level groups based on sequence, structural, or profile-HMM similarities. The database offers various accessions and redundancy levels across different datasets. Pfam version 35.0 comprises 19,632 families and clans, encompassing 61 million sequences derived from the UniProt Reference Proteomes. To expand its coverage of metagenomic sequence diversity, the database has also established 250 metagenomic protein families. Protein sequences from the MGnify and UniProt databases were clustered, aligned, and subsequently used to generate new SEED alignments, which formed the basis for the creation of additional protein families (Mistry et al. 2021).

4.8 M5NR Database

The M5NR is a non-redundant protein database that integrates multiple sequencing databases along with an indexing system. It utilizes MD5 checksums to differentiate sequence data from metadata across various databases. The database includes sequence identifiers, species annotations, and additional metadata. It maintains mappings to functional hierarchies and facilitates rapid searching for annotations from public sources such as Gene Ontology (GO), KEGG orthologs, SEED Subsystems, eggNOG, and NCBI COGs. Similarity searches are performed using BLAST or BLAT algorithms to identify homologous sequences. The framework is designed to accommodate the importation of additional protein databases with third-party annotations, enabling the creation of new "namespaces." This strategy ensures scalability by appending metadata to annotation sets (Wilke et al. 2012).

5 Application of Metatranscriptomic in Agriculture

As sustainable agriculture increasingly prioritizes not only crop productivity but also long-term environmental sustainability, subsequently, metatranscriptomics has emerged as a key tool for understanding the complex functional profiles of the microbiome. This knowledge is essential for unlocking the vast potential of microbiome-based agricultural practices and various recent application of metatranscriptomic analysis in agriculture have been summarized in Table 3 and Fig. 2. Shotgun RNA sequencing has revealed that most transcripts related to reductive nitrogen transformations in waterlogged paddy soils are primarily derived from Deltaproteobacteria, particularly Anaeromyxobacter and Geobacter (Masuda et al. 2017). Additionally, comprehensive studies have provided valuable insights into the structure and dynamics of microbial consortia responsible for methane emissions in paddy fields (Masuda et al. 2018). Direct sequencing and metatranscriptome analysis also offer structural and functional perspectives on microbial communities, with

Table 3 Various application of metatranscriptomic analysis in agriculture especially in a broad context of crop and their assisted improvements

Sample/plants/ microbes	Applied methods/ algorithms	Inferences form study	References
Paddy soil	Shotgun RNA sequencing	Deltaproteobacteria is predominating microbes	Masuda et al. (2017)
Rice straw	Illumina sequencing	Glycoside hydrolase overexpressed in the thermophiles	Simmons et al. (2014)
Soil comparative analysis	Illumina sequencing	Proteobacteria are dominant in diverse soil	Sharma et al. (2019)
Agriculture soil	Illumina sequencing	Revealed microbial community structure and function	Sharma and Sharma (2019)
Wheat rhizosphere	MG-RAST pipeline	Identified genes linked with the degradation of pesticides	Singh et al. (2018)
	Illumina sequencing	Identified microbiomes in disease suppressive and non-suppressive habitat	Hayden et al. (2018)
Soil biome	MG-RAST pipeline	Seasonal variation in microbial profiles	Barboza et al. (2018)
Lettuce	MinION sequencing	Elucidate microbial transcriptional activity	Haveman et al. (2021)
Grapevine trunk disease	Illumina sequencing	Dissected potential virulence factors	Morales-Cruz et al. (2018)
Tomato and lettuce	Illumina sequencing	Elucidating hidden stress via alternate microbiome	Zolti et al. (2020)
Rice fungal pathogen	Illumina sequencing	Reveals mycoviral diversity	He et al. (2022)
Basil	Illumina sequencing	Identified host defense genes and pathogen virulence genes unitedly	Guo et al. (2016)
Rice field soil	Illumina HiSeq	Functional continuum of methanogenic food web	Peng et al. (2018)
Paddy soil	Shotgun sequencing	Identified consortia driving methane metabolism	Masuda et al. (2018)
	Illumina sequencing	Reveals direct interspecies electron transfer	Holmes et al. (2017)
Paddy soil microcosms	Pyrosequencing	Reveals diverse microbial communities	Kim et al. (2013)
Rice plants	Illumina sequencing	Presence of novel RNA viruses	Chao et al. (2021)
Rice field soil	Illumina sequencing	Functional role of the Ignavibacteria	Bei et al. (2021)

Proteobacteria being the predominant phylum observed in the soils studied (Sharma et al. 2019).

Agricultural soils are increasingly contaminated by synthetic chemicals, such as polyaromatic compounds and metatranscriptome analysis has identified multifunctional genes and enzymes involved in the degradation of aromatic compounds and pesticides in the wheat rhizosphere. Dominant microbial communities associated

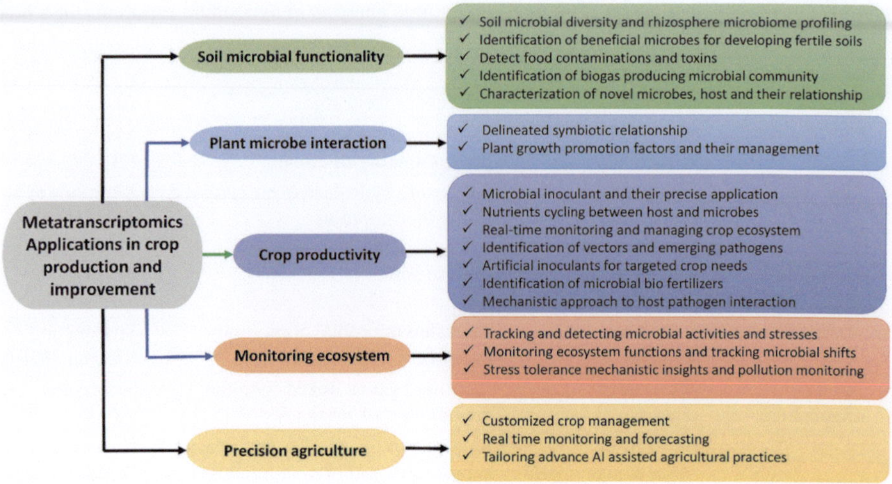

Fig. 2 The figure illustrates various applications of metatranscriptomic studies in agricultural research, with a particular focus on crop performance enhancement and improvements aimed at maintaining eco-friendly solutions for sustainability

with these processes include Proteobacteria, Actinobacteria, Firmicutes, Bacteroidetes, and Cyanobacteria (Singh et al. 2018). Furthermore, metatranscriptomic approaches indicate that the assembly of soil microbial communities is significantly influenced by fluctuations in temperature and moisture, leading to seasonal dynamics that affect both taxonomic and functional microbial profiles in Pampa biome soils (Barboza et al. 2018). Notably, metatranscriptomics of enriched microbial RNA can be utilized to monitor the transcriptional activity of crop microbiomes, facilitating the maintenance of plant health for space-based food production—an essential consideration for long-duration space missions (Haveman et al. 2021).

Grapevines are highly susceptible to trunk diseases, which damage their woody tissues. Metatranscriptomic assessments provide quantitative evaluations of species composition and genome-wide transcriptional profiles of potential virulence factors involved in cell wall degradation, secondary metabolism, and nutrient uptake, as well as insights into the co-infection status (Morales-Cruz et al. 2018). The soil borne fungus *Rhizoctonia solani* is a major pathogen in grain crops, leading to significant production losses. In the absence of resistant cultivars, enhancing biological disease suppression may offer a sustainable approach for disease control. Metatranscriptomic analysis revealed that the dominant taxa in the rhizosphere includes Arthrobacter and Pseudomonas species in non-suppressive samples, and Stenotrophomonas and Buttiauxella in suppressive samples, suggesting the potential of promoting beneficial microflora to reduce infection (Hayden et al. 2018). Additionally, metatranscriptomic analyses of root systems from tomato and lettuce subjected to two different water treatments revealed taxonomic shifts and functional responses within host-associated microbial communities (Zolti et al. 2020).

The thermophilic community of Actinobacteria is overrepresented, along with enzymes such as glycoside hydrolases and carbohydrate-binding module family proteins, during high-solids decomposition of rice straw, as revealed through meta-transcriptomic studies (Simmons et al. 2014). Rice blast, false smut, and rice sheath blight are major fungal diseases that result in significant global reductions in rice yield. Over half of the viral sequences identified in these pathogens are predicted to belong to the Narnaviridae and Botourmiaviridae families, while Polymycoviridae members were identified for the first time in these pathogens (He et al. 2022). Metatranscriptomics enables rapid gene discovery and expression analysis of both plant hosts and parasites through next-generation sequencing of mixed host and pathogen RNA. This approach has identified candidate host defence genes and pathogen virulence factors that are highly expressed during the infection process (Guo et al. 2016). Additionally, metatranscriptomic analyses reveal how differential temperature impacts the food web continuum of methanogenic microbes, which could have implications for rice paddies and natural wetlands in tropical and sub-tropical regions (Peng et al. 2018).

Strikingly, evidence suggests potential cooperation between Methanothrix and Geobacter species in terrestrial methanogenic environments, particularly in paddy soils, where Geobacter species provide electrons to Methanothrix species for methane production via e-pili electrical connections (Holmes et al. 2017). The metatranscriptomic approach enables the analysis of the composition and gene expression of complex soil microbial communities, overcoming the limitations associated with PCR-based methods (Kim et al. 2013). This method has also led to the identification of eight novel viral species and five known ones, including a highly divergent virus from an unclassified family in rice transmitted by the Brown Planthopper (Chao et al. 2021). Additionally, metatranscriptomic analysis of soil microcosms has revealed that denitrification in paddy soils is a collaborative process involving various nitrogen oxide reducers (Masuda et al. 2019). Further, metatranscriptomic analysis of uncultured Ignavibacteria in rice field soil highlighted their metabolic versatility, with primary fermentation being the dominant activity under anoxic conditions, and phenylacetate degradation occurring under micro-oxic conditions, contributing to polymer hydrolysis independent of oxygen availability (Bei et al. 2021). Lastly, metatranscriptomics has revealed the gene expression involved in the production of melanogenesis inhibitors by specific microbial species in fermented unpolished black rice (Sangkaew et al. 2023).

6 Conclusion

The underlying pipelines and algorithms for metatranscriptomic analysis are pivotal in advancing crop improvement by providing a comprehensive understanding of plant-microbe interactions, gene expression, and the functional pathways involved in plant health and productivity. By simultaneously profiling the transcripts of both the host plant and its associated microbial community, metatranscriptomics enables the identification of key genes and metabolic processes that regulate stress responses, nutrient uptake, and disease resistance. The integration of advanced bioinformatics pipelines and machine learning algorithms facilitates the interpretation of complex datasets, enabling the discovery of novel biomarkers, the functional annotation of uncharacterized genes, and the identification of potential targets for genetic modification or breeding programs. With the increasing emphasis on sustainability and resilience in agriculture, metatranscriptomics offers promising prospects for developing crops with enhanced stress tolerance, improved yield, and optimized symbiotic relationships with beneficial microbes. Moreover, these approaches can be seamlessly utilised for characterization of novel pathogenic and non-pathogenic microbes associated with crop performance. Advances in sequencing technologies, coupled with refined computational tools, will further enhance the precision and scalability of these analyses, driving more targeted and efficient crop improvement strategies.

Conflict of Interest The authors declare no conflict of interest.

References

Andrews S (2010) A quality control tool for high throughput sequence data

Anwar MZ, Lanzen A, Bang-Andreasen T, Jacobsen CS (2019) To assemble or not to resemble—a validated comparative metatranscriptomics workflow (CoMW). GigaScience 8(8):giz096

Aziz RK, Devoid S, Disz T, Edwards RA, Henry CS, Olsen GJ, Olson R, Overbeek R, Parrello B, Pusch GD, Stevens RL (2012) SEED servers: high-performance access to the SEED genomes, annotations, and metabolic models. PLoS One 7(10):e48053

Barboza ADM, Pylro VS, Jacques RJS, Gubiani PI, de Quadros FLF, da Trindade JK, Triplett EW, Roesch L (2018) Seasonal dynamics alter taxonomical and functional microbial profiles in Pampa biome soils under natural grasslands. PeerJ 6:e4991

Bei Q, Moser G, Wu X, Müller C, Liesack W (2019) Metatranscriptomics reveals climate change effects on the rhizosphere microbiomes in European grassland. Soil Biol Biochem 138:107604

Bei Q, Peng J, Liesack W (2021) Shedding light on the functional role of the Ignavibacteria in Italian rice field soil: a meta-genomic/transcriptomic analysis. Soil Biol Biochem 163:108444

Bolger AM, Lohse M, Usadel B (2014) Trimmomatic: a flexible trimmer for Illumina sequence data. Bioinformatics 30(15):2114–2120

Braga RM, Dourado MN, Araújo WL (2016) Microbial interactions: ecology in a molecular perspective. Braz J Microbiol 47:86–98

Cantarel BL, Coutinho PM, Rancurel C, Bernard T, Lombard V, Henrissat B (2009) The carbohydrate-active EnZymes database (CAZy): an expert resource for glycogenomics. Nucleic Acids Res 37(suppl_1):D233–D238

Celaj A, Markle J, Danska J, Parkinson J (2014) Comparison of assembly algorithms for improving rate of metatranscriptomic functional annotation. Microbiome 2:1–13

Chao S, Wang H, Yan Q, Chen L, Chen G, Wu Y, Meng B, Jin L, Zhu X, Feng G (2021) Metatranscriptomic sequencing suggests the presence of novel RNA viruses in rice transmitted by brown planthopper. Viruses 13(12):2464

Davis MP, van Dongen S, Abreu-Goodger C, Bartonicek N, Enright AJ (2013) Kraken: a set of tools for quality control and analysis of high-throughput sequence data. Methods 63(1):41–49

Galperin MY, Wolf YI, Makarova KS, Vera Alvarez R, Landsman D, Koonin EV (2021) COG database update: focus on microbial diversity, model organisms, and widespread pathogens. Nucleic Acids Res 49(D1):D274–D281

Garza DR, Dutilh BE (2015) From cultured to uncultured genome sequences: metagenomics and modeling microbial ecosystems. Cell Mol Life Sci 72:4287–4308

Grabherr MG, Haas BJ, Yassour M, Levin JZ, Thompson DA, Amit I, Adiconis X, Fan L, Raychowdhury R, Zeng Q, Chen Z (2011) Trinity: reconstructing a full-length transcriptome without a genome from RNA-Seq data. Nat Biotechnol 29(7):644

Guo L, Allen KS, Deiulio G, Zhang Y, Madeiras AM, Wick RL, Ma LJ (2016) A de novo-assembly based data analysis pipeline for plant obligate parasite metatranscriptomic studies. Front Plant Sci 7:925

Haveman NJ, Khodadad CL, Dixit AR, Louyakis AS, Massa GD, Venkateswaran K, Foster JS (2021) Evaluating the lettuce metatranscriptome with MinION sequencing for future spaceflight food production applications. NPJ Microgravity 7(1):22

Hayden HL, Savin KW, Wadeson J, Gupta VV, Mele PM (2018) Comparative metatranscriptomics of wheat rhizosphere microbiomes in disease suppressive and non-suppressive soils for Rhizoctonia solani AG8. Front Microbiol 9:859

He Z, Huang X, Fan Y, Yang M, Zhou E (2022) Metatranscriptomic analysis reveals rich mycoviral diversity in three major fungal pathogens of rice. Int J Mol Sci 23(16):9192

Holmes DE, Shrestha PM, Walker DJ, Dang Y, Nevin KP, Woodard TL, Lovley DR (2017) Metatranscriptomic evidence for direct interspecies electron transfer between Geobacter and Methanothrix species in methanogenic rice paddy soils. Appl Environ Microbiol 83(9):e00223–e00217

Huerta-Cepas J, Szklarczyk D, Heller D, Hernández-Plaza A, Forslund SK, Cook H, Mende DR, Letunic I, Rattei T, Jensen LJ, von Mering C (2019) eggNOG 5.0: a hierarchical, functionally and phylogenetically annotated orthology resource based on 5090 organisms and 2502 viruses. Nucleic Acids Res 47(D1):D309–D314

Kanehisa M, Goto S (2000) KEGG: Kyoto encyclopedia of genes and genomes. Nucleic Acids Res 28(1):27–30

Kaul S, Choudhary M, Gupta S, Dhar MK (2021) Engineering host microbiome for crop improvement and sustainable agriculture. Front Microbiol 12:635917

Kim Y (2013) Metatranscriptome analysis of microbial communities in rice microcosms

Kim J, Kim MS, Koh AY, Xie Y, Zhan X (2016) FMAP: functional mapping and analysis pipeline for metagenomics and metatranscriptomics studies. BMC Bioinform 17:1–8

Kopylova E, Noé L, Touzet H (2012) SortMeRNA: fast and accurate filtering of ribosomal RNAs in metatranscriptomic data. Bioinformatics 28(24):3211–3217

Koseoglu A (2024) Metatranscriptomics analysis using microbiome RNA-seq data

Kumar A, Yadav A (2024) Next generation sequencing in metagenomics and metatranscriptomics. In: Multi-omics analysis of the human microbiome: from technology to clinical applications. Springer, Singapore, pp 49–75

Leimena MM, Ramiro-Garcia J, Davids M, van den Bogert B, Smidt H, Smid EJ, Boekhorst J, Zoetendal EG, Schaap PJ, Kleerebezem M (2013) A comprehensive metatranscriptome analysis pipeline and its validation using human small intestine microbiota datasets. BMC Genomics 14:1–14

Love MI, Huber W, Anders S (2014) Moderated estimation of fold change and dispersion for RNA-seq data with DESeq2. Genome Biol 15:1–21

Martin M (2011) Cutadapt removes adapter sequences from high-throughput sequencing reads. EMBnet J 17(1):10–12

Martinez X, Pozuelo M, Pascal V, Campos D, Gut I, Gut M, Azpiroz F, Guarner F, Manichanh C (2016) MetaTrans: an open-source pipeline for metatranscriptomics. Sci Rep 6(1):26447

Masuda Y, Itoh H, Shiratori Y, Isobe K, Otsuka S, Senoo K (2017) Predominant but previously-overlooked prokaryotic drivers of reductive nitrogen transformation in paddy soils, revealed by metatranscriptomics. Microbes Environ 32(2):180–183

Masuda Y, Itoh H, Shiratori Y, Senoo K (2018) Metatranscriptomic insights into microbial consortia driving methane metabolism in paddy soils. Soil Sci Plant Nutr 64(4):455–464

Masuda Y, Matsumoto T, Isobe K, Senoo K (2019) Denitrification in paddy soil as a cooperative process of different nitrogen oxide reducers, revealed by metatranscriptomic analysis of denitrification-induced soil microcosm. Soil Sci Plant Nutr 65(4):342–345

Mehta S, Crane M, Leith E, Batut B, Hiltemann S, Arntzen MØ, Kunath BJ, Pope PB, Delogu F, Sajulga R, Kumar P (2021) ASaiM-MT: a validated and optimized ASaiM workflow for metatranscriptomics analysis within Galaxy framework. F1000Research 10:103

Mistry J, Chuguransky S, Williams L, Qureshi M, Salazar GA, Sonnhammer EL, Tosatto SC, Paladin L, Raj S, Richardson LJ, Finn RD (2021) Pfam: the protein families database in 2021. Nucleic Acids Res 49(D1):D412–D419

Morales-Cruz A, Allenbeck G, Figueroa-Balderas R, Ashworth VE, Lawrence DP, Travadon R, Smith RJ, Baumgartner K, Rolshausen PE, Cantu D (2018) Closed-reference metatranscriptomics enables in planta profiling of putative virulence activities in the grapevine trunk disease complex. Mol Plant Pathol 19(2):490–503

Mukherjee A, Reddy MS (2020) Metatranscriptomics: an approach for retrieving novel eukaryotic genes from polluted and related environments. 3 Biotech 10(2):71

Narayanasamy S, Jarosz Y, Muller EE, Heintz-Buschart A, Herold M, Kaysen A, Laczny CC, Pinel N, May P, Wilmes P (2016) IMP: a pipeline for reproducible reference-independent integrated metagenomic and metatranscriptomic analyses. Genome Biol 17:1–21

Nath D, Das S, Ghangrekar MM (2021) High throughput techniques for the rapid identification of electroactive microorganisms. Chemosphere 285:131489

Ni Y, Li J, Panagiotou G (2016) COMAN: a web server for comprehensive metatranscriptomics analysis. BMC Genomics 17:1–7

O'Leary NA, Wright MW, Brister JR, Ciufo S, Haddad D, McVeigh R, Rajput B, Robbertse B, Smith-White B, Ako-Adjei D, Astashyn A (2016) Reference sequence (RefSeq) database at NCBI: current status, taxonomic expansion, and functional annotation. Nucleic Acids Res 44(D1):D733–D745

Ortiz R, Gera P, Rivera C, Santos JC (2021) Pincho: a modular approach to high quality de novo transcriptomics. Genes 12(7):953

Peng J, Wegner CE, Bei Q, Liu P, Liesack W (2018) Metatranscriptomics reveals a differential temperature effect on the structural and functional organization of the anaerobic food web in rice field soil. Microbiome 6:1–16

Rho M, Tang H, Ye Y (2010) FragGeneScan: predicting genes in short and error-prone reads. Nucleic Acids Res 38(20):e191–e191

Sangkaew O, Prombutara P, Roytrakul S, Yompakdee C (2023) Metatranscriptomics reveals sequential expression of genes involved in the production of melanogenesis inhibitors by the defined microbial species in fermented unpolished black rice. Microbiol Spectr 11(2):e03139–e03122

Sequeira JC, Rocha M, Madalena Alves M, Salvador AF (2019) MOSCA: an automated pipeline for integrated metagenomics and metatranscriptomics data analysis. In: Practical applications of computational biology and bioinformatics, 12th International Conference. Springer International Publishing, pp 183–191

Shakya M, Lo CC, Chain PS (2019) Advances and challenges in metatranscriptomic analysis. Front Genet 10:904

Sharma PK, Sharma V, Sharma S, Bhatia G, Singh K, Sharma R (2019) Comparative metatranscriptome analysis revealed broad response of microbial communities in two soil types, agriculture versus organic soil. J Genetic Eng Biotechnol 17(1):6

Sharuddin SS, Ramli N, Yusoff MZM, Muhammad NAN, Ho LS, Maeda T (2022) Advancement of metatranscriptomics towards productive agriculture and sustainable environment: a review. Int J Mol Sci 23(7):3737

Simmons CW, Reddy AP, D'haeseleer P, Khudyakov J, Billis K, Pati A, Simmons BA, Singer SW, Thelen MP, VanderGheynst JS (2014) Metatranscriptomic analysis of lignocellulolytic microbial communities involved in high-solids decomposition of rice straw. Biotechnol Biofuels 7:1–13

Singh DP, Prabha R, Gupta VK, Verma MK (2018) Metatranscriptome analysis deciphers multifunctional genes and enzymes linked with the degradation of aromatic compounds and pesticides in the wheat rhizosphere. Front Microbiol 9:1331

Taj B, Adeolu M, Xiong X, Ang J, Nursimulu N, Parkinson J (2023) MetaPro: a scalable and reproducible data processing and analysis pipeline for metatranscriptomic investigation of microbial communities. Microbiome 11(1):143

Tamames J, Puente-Sanchez F (2019) SqueezeMeta, a highly portable, fully automatic metagenomic analysis pipeline. Front Microbiol 9:3349

Tatusov RL, Galperin MY, Natale DA, Koonin EV (2000) The COG database: a tool for genome-scale analysis of protein functions and evolution. Nucleic Acids Res 28(1):33–36

Terrapon N, Lombard V, Drula E, Coutinho PM, Henrissat B (2017) The CAZy database/the carbohydrate-active enzyme (CAZy) database: principles and usage guidelines. A practical guide to using glycomics databases:117–131

Terrón-Camero LC, Gordillo-González F, Salas-Espejo E, Andrés-León E (2022) Comparison of metagenomics and metatranscriptomics tools: a guide to making the right choice. Genes 13(12):2280

Thingujam D, Liu J, Majeed A, Mukhtar MS (2024) Plant–microbiome dynamics through spatial metatranscriptomics and network biology. Trends Plant Sci 29:1176

Truong DT, Franzosa EA, Tickle TL, Scholz M, Weingart G, Edoardo P, Tett A, Huttenhower C, Segata N (2015) MetaPhlAn2 for enhanced metagenomic taxonomic profiling. Nat Methods 12(10):902–903. https://doi.org/10.1038/nmeth.3589

Westreich ST, Treiber ML, Mills DA, Korf I, Lemay DG (2018) SAMSA2: a standalone metatranscriptome analysis pipeline. BMC bioinformatics 19:1–11

Wilke A, Harrison T, Wilkening J, Field D, Glass EM, Kyrpides N, Mavrommatis K, Meyer F (2012) The M5nr: a novel non-redundant database containing protein sequences and annotations from multiple sources and associated tools. BMC Bioinform 13:1–5

Wolf JB (2013) Principles of transcriptome analysis and gene expression quantification: an RNA-seq tutorial. Mol Ecol Resour 13(4):559–572

Zolti A, Green SJ, Sela N, Hadar Y, Minz D (2020) The microbiome as a biosensor: functional profiles elucidate hidden stress in hosts. Microbiome 8(1):71

Metaproteomics: Recent Advances and Applications

Preeti Chaurasia, Ashutosh Kumar, and Khan Mohd. Sarim

1 Introduction

Microbial association with plant, animal, and human systems is essential to their growth and development. Such interactions are generally not well studied and are required to uncover to understand the overall molecular network functioning behind association. These are multidirectional interactions, and proteins in biological systems are key components in every metabolic function, including cell division, construction of cellular architecture, cellular communication, nutrient transport, etc. The overall component of such proteins refers to the proteome of an organism. Similarly, multiple proteomes from different organisms in a sample collected from an environment are considered a metaproteome, and the study corresponds to metaproteomics. In the OMICS era, metagenomic data have uncovered a lot of hidden species, some are possibly key players in various ecological functions. However, in the last two decades, researchers have focused on understanding gene expression profiling and proteins involved in performing different biological functions in challenging environments. Initially, Wilmes and Bond (2004) put forward the concept of metaproteomics using total proteome extraction from an activated sludge system, followed by 2D-PAGE and protein spot identification. Since then, a large number of studies in the medical, environmental, and agricultural fields have been conducted using the metaproteomics approach. The general workflow of metaproteomics is to extract total protein from environmental samples, including contaminated

P. Chaurasia
Department of Microbiology, SS Jain Suboth PG College, Jaipur, Rajasthan, India

A. Kumar
G Surgiwear Limited, Shahjahanpur, Uttar Pradesh, India

K. M. Sarim (✉)
Institut Ruđer Bošković, Zagreb, Croatia

Renu et al. (eds.), *Meta-omics in Crop Improvement*,
https://doi.org/10.1007/978-981-95-2294-1_5

agricultural soil, rhizospheric soil, or medical samples, digest it into peptides, and analyze using Matrix-assisted laser desorption ionization-Time of flight (MALDI-TOF) and now in Liquid Chromatography-Mass spectrometer (LC-MS/MS). Advanced data analysis tools are available, coupled with improved algorithms to sort and identify mixed peptides and distinguish closely related proteins.

As the soil environment is a highly variable system that encounters different challenges, including deposited salt in the subsurface, high temperature, flood, drought, various pesticide exposure, and other environmental contaminants. These stresses have the potential to alter the microbial community, and such changes in microbial composition in a set of environments are also a key indicator of certain types of environmental conditions/stress, vice versa.

Protein extraction from isolated bacterial cultures is a comparatively labile process than from complex environmental samples, as not only is microbial diversity in such samples rich, but also affected by several factors, including the physicochemical properties of the samples.

OMICs data integration is indeed a powerful approach to extract maximum and nearly accurate information at the genomic level and its associated proteome response. Metagenomics at an initial stage delivers the genomic sequence of entire genomes present in an environmental sample.

2 Current Methods and Technological Advancements

Sampling and extraction: Extraction of metaproteome from different samples is quite challenging, specifically from the soil environment, either in bulk soil or rhizospheric soil, both containing a lot of humic substances that hinder proteome extraction. Besides this, soil may vary in pH, electrical conductivity, or even C, N, and P dynamics (Keiblinger and Riedel 2018). Humic acids are large organic molecules that play an important role in soil properties, agronomy, and plant growth as well. Major humic acid substances in soil are humic acid, fulvic acid, and humin. All the above-mentioned components in soil bind protein, making further protein extraction and purification troublesome. Besides this, another key challenge is the diversity of microbial species associated with the environment. To obtain meaningful results, essential proteins from the sample must represent corresponding microbial species of a given environment. The complexity of proteins and microbial composition makes this objective cumbersome. Benndorf et al. (2007) optimized a method for metaproteome extraction and analysis of soil and groundwater samples by using 0.1 M NaOH alkaline method, from which solubilized proteins were subjected to centrifugation followed by phenol extraction, and final precipitation in methanol. In continuation, during the extraction of metaproteome from an anaerobic benzene microbial community in sediment to disrupt cells, researchers utilized ultrasonic treatment before high centrifugation at 4 °C (Benndorf et al. 2009). French press is a traditional method for protein extraction from microbial cells. It is an equipment for disrupting the cell wall and membrane using hydraulic pressure. Abram et al. (2009) used the French press method along with another sonication method. They found that protein from the second sonication method was comparatively higher than in French press methods. Further, a direct method was developed for soil metaproteome isolation which follows lysis by detergent and subsequently TCA treatment for protein precipitation and clean-up. Comparatively, in the SDS-TCA method, researchers find nearly 500 unique proteins, and thus without any biases of size of protein or localization (Chourey et al. 2010).

In addition to this, a study of plant-associated microbes requires a large input of plant material which limits the potential of metaproteomics capability to explore useful insight. Salvato et al. (2022) evaluated eight different protein extraction methods, although they were able to get the maximum representable protein with less input plant material nearly 40–70 mg, but also concluded that no single method applies to all types of plant sample protein extraction, and individual optimized methods may require for different plant samples. In a similar effort, researchers have evaluated different lysis agents including SDS, Urea, B-PER, RIPA, PreOmics, and AllPrep methods. It was observed that for protein yield RIPA was best, while the highest protein was extracted by SDS followed by PreOmics tested against chlorinated and non-chlorinated wastewater biofilm (Almulhim and Hong 2023). In addition to this, the efficiency of SDS optimization was also found significant when Mikolitis et al. (2024) tested different protein extractions, including SDS with bead beating, SDS with boiling, and NaOH, and the SDS with bead beating method was found effective in protein lysis and protein extraction. Interestingly, similar evidence of the SDS bead beating method with improved protein extraction was also put forward, where the already developed SDS method, along with NaOH method, researcher used SDS-Phenol protocol with bead beating, and bead beating method were found effective for maximum protein digestion (Tartaglia et al. 2020). With a different theme, researchers during grass metaproteome isolation evaluated two protein extraction methods as a complementary modified method. First, to remove humic acid by using the established HTPC method commonly known as heat, thaw, phenol, chloroform, and second surfactant method that co-extracts humic acid. Combined methods provided a wide range of proteins that was not possible with either of single methods (Quinn et al. 2022). Optimization of lysis agents and other reagents is key to obtaining a fine-tuned method for efficient and high-yield protein. Besides soil or plant samples, clinical researchers have also tested a variety of methods in protein extraction, and their insight in terms of microbial lysis can be beneficial in all types of protein extraction. In this context, Zhang and associates evaluated four different protein extraction methods and found that SDS optimization yielded a high quantity of protein. In the first protocol, microbial cells were resuspended in 4% SDS, followed by incubation at 95 °C for 10 min, and performed bead beating. In protocol 2, cells were subjected to sonication instead of bead beating. Protocol 3 was performed with 8 M urea in SDS, but without 95 °C treatments, followed by bead beating, and protocol 4 was similar to protocol 2 without 95 °C treatments, with 8M Urea, followed by sonication. It was concluded that SDS with ultrasonic lysis was a comparatively better method for efficient protein extraction (Zhang et al. 2018).

Collecting a soil sample, a water sample, or animal or human samples requires different strategies to achieve the objective. In case research is intended to study microbial diversity in salt-affected agricultural soil, samples may be collected from the topsoil, nearly 6–25 cm, as it is the region where salt accumulates due to evaporation, and microbes have to deal with it. Similarly, in petroleum-contaminated soil or heavy metal-contaminated soil, a little deeper digging is required, considering petroleum hydrocarbons might be present in different zones, or heavy metals bind with soil particles but, due to soil chemistry, may leach out.

Consistently increasing metaproteomics-based research demands a large database for the validation of peptides generated during LC-MS analysis of sample proteins. Although a huge amount of data generation also strengthens such databases, there is still a dire need to organize and manage them using advanced algorithms and database management tools. The mass spectra from LC-MS/MS must match the available data to identify the correct protein. To strengthen the database and retrieval algorithm, various researchers have improved it.

Database search tools: In the early 1990s, researchers started developing and using databases to identify uninterrupted peptide sequences generated from mass spectrometry, which revolutionized the proteomic field. Several advanced search tools and databases came into existence, and common and frequently used database tools (Table 1).

Table 1 Tools and databases for metaproteomic studies

S. No.	Database search tool	Type	Year	References
1.	SEQUEST	First database search program based on a two-step program where the first step was to pass through the peptide database and search for a linear peptide with a score, and further second step was to re-score the best-scoring peptide using a cross-correlation algorithm.	1993	Eng et al. (1994)
2.	Mascot	Peptide search engine which uses a primary protein sequence database. The digested protein data from a mass spectrometer generates mass values, which are further compared with calculated values before digestion. Protein sequences are digested virtually, which generates theoretical peptides, and the closest match of a scoring algorithm then identifies the closest match.	1993	Perkins et al. (1999)
3.	X! Tandem	A search database that identifies proteins in tandem mass spectra (MS/MS data)	2004	Fenyö and Beavis (2003)

(continued)

Table 1 (continued)

S. No.	Database search tool	Type	Year	References
4.	OMSSA (Open Mass Spectrometry Search Algorithm)	A fast algorithm written primarily in C++ and NCBI toolkit. It takes data files and generic file format, extracts ms spectra, and compares with the in silico digested protein samples.	2004	Geer et al. (2004)
5.	Myrimatch	Database implying a scorer based on multivariate hypogeomatric distribution with improved ability to distinguish an intense and a minor peak. This scorer algorithm is composed of two elements, first retains nearly all spectra's intensity, and second peak stratification into configurable number of classes	2007	Tabb et al. (2007)
6.	Andromeda	A search engine that can be used alone or with the MaxQuant platform and works on a probabilistic score-based method that compares theoretical and experimental fragment ions.	2011	Cox et al. (2011)
7.	Comet	A database tool of protein search, an upgraded version of Sequest, written in C++ with a Linux and window compatible program. It uses MSToolkit to read various types of spectra.	2012	Eng et al. (2013)
8.	MS-GF+	A database search tool sensitive in terms of a greater number of peptide identifications than any available resource It is universal and can be used with various protocols, MS spectra, and mass spectrometry instruments.	2014	Kim and Pevzner (2014)
9.	MetaProteomeAnalyzer (MPA)	A powerful standalone tool designed for high throughput metaproteomic analysis coupled with multiple search engines and improved peptide identification with interactive graphical visualization. It supports Comet, MS-GF+ and X!Tandem datbases.	2015	Muth et al. (2015)
10.	MetaLab	An automated pipeline to address protein complexity and species diversity supporting data dependent as well as independent modes across multiple MS platforms	2017	Cheng et al. (2017)

(continued)

Table 1 (continued)

S. No.	Database search tool	Type	Year	References
11.	MSFragger	A tool for ultrafast and comprehensive peptide identification from tandem mass spectra. It works on the concept of open database search and is able to identify modified and unmodified peptides, even novel ones. It accesses all three forms of file format mzXML/mzML/MGF, using MSFTBX, and data libraries, rescales peaks and top peaks, restrains, and filters based on m/z and minimum intensity ratio.	2017	Kong et al. (2017)
12.	pDeep	A deep learning-based database search tool addresses a key important parameter in metaproteomics, which is an accurate identification of peptides by predicting mass spectra from peptide sequences.	2017	Zhou et al. (2017)

3 Data Collection and Bioinformatics Analysis

Data collection and bioinformatics analysis form the twin pillars of modern proteomics workflows. Together, they ensure that the biological questions are addressed by accurately capturing the complex protein landscape of a sample and translating this raw data into meaningful biological insights (Mohanasundaram et al. 2023).

A successful proteomics experiment really starts with gathering top-notch samples. The entire analysis depends on how accurately these samples reflect the biological process under study. That's why strict protocols are essential as ensuring that samples from both control and experimental groups are collected under the same conditions to avoid any bias. It's also crucial to process the samples quickly and keep them cold; this helps prevent enzymes from breaking down the proteins. Additionally, properly dividing the samples into smaller portions minimizes damage from repeated freeze–thaw cycles. All these careful steps work together to preserve the natural state of the proteins and maintain the overall integrity of the complex mixtures being examined (Goh and Wong 2019).

Extracting proteins from various sources such as tissues, fluids, or environmental samples can be quite challenging. Researchers often use a blend of chemical, physical, and enzymatic methods to break open cell walls and dissolve the proteins effectively (Chen et al. 2020). After this step, proteins are typically chopped into smaller pieces using enzymes like trypsin, which makes them easier to analyze via mass spectrometry. The primary goals here are to get as much protein as possible, avoid any contamination, and ensure that the final data truly mirrors the actual biological state of the sample.

Modern instruments, including high-resolution liquid chromatography, tandem mass spectrometry (LC–MS/MS), allow researchers to capture a broad spectrum of peptides, even those at low abundance. Advances in these technologies have not only improved sensitivity and resolution but have also enabled methods like data-dependent (DDA) and data-independent acquisition (DIA), which capture detailed and unbiased snapshots of complex proteomes. Each of these steps is carefully optimized to ensure that the "raw material" entering the bioinformatics pathway (Chen et al. 2020).

4 Bioinformatics Analysis in Proteomics

Once the data is collected, transforming raw MS spectra into biologically meaningful information is paramount (Chen et al. 2020).

4.1 Data Preprocessing and Quality Control

Before interpreting the biological meaning of mass spectrometry data, it first needs a good clean-up. This involves reducing any unwanted noise, finding the key signal peaks, aligning the data correctly, and normalizing it to remove technical inconsistencies. Techniques like log transformation and filling in missing data help stabilize everything, ensuring that the dataset is well-prepared for quantitative analysis. Software tools which are used for quality control (QC) and trimming in mass spectrometry data analysis, such as Spectroswiss Software Suite, OpenMS, etc. (Gihawi et al. 2023). In short, every step is carefully managed so that the processed data is clean, high quality, and ready for deeper exploration (Kumar and Mann 2009).

4.2 Peptide and Protein Identification

In MS/ MS experiments, only peptide information is obtained, not protein, which leads to ambiguities. ProteinExtractor utilizes spectra of peptides as input and generates protein lists with the same algorithm. It also combines the peptide search results of several search engines allowing to combine the sensitivity and selectivity of each search engine (Thiele et al. 2010). Once the data is polished, the next step is to match the observed signals to theoretical predictions from protein databases. Tools like Mascot, SEQUEST, and even more modern de novo sequencing algorithms help in figuring out which peptides are present and, by extension, which proteins they come from. Since proteomic datasets are very complex and often feature many protein modifications, advanced algorithms (including those that use artificial intelligence) are now crucial for speeding up the process and improving accuracy (Calderón-González et al. 2016). However, there are some decoy filters used in proteomics for the elimination of falsified results (Lee et al. 2021) such as:

(a) *Reverse Decoy Filter*: Creates a decoy database by reversing protein sequences. This method is simple and widely used for estimating false discovery rates (FDR).

(b) *Shuffle Decoy Filter*: Randomly shuffles amino acids in protein sequences to generate a decoy database. It ensures randomness but may overestimate FDR in certain cases.
(c) *Pseudo-Reverse Decoy Filter*: Similar to the reverse method but introduces slight modifications to avoid redundancy in peptide sequences.
(d) *De Bruijn Decoy Filter*: Uses a repeat-preserving method based on de Bruijn graphs to construct decoy databases. This approach addresses issues of randomness and redundancy.
(e) *Target-Decoy Competition (TDC)*: Combines target and decoy results, selecting the best match for a spectrum based on the highest score.

4.3 Quantitative Analysis and Functional Annotation

After identifying the proteins, researchers measure how much of each one is present. This is accomplished by using methods such as spectral counting, label-free quantification, or isotope labeling techniques like stable isotope labeling by amino acids in cell culture (SILAC) or tandem mass tags (TMT). Statistical tests such as t-tests or ANOVA are then used to compare protein levels under different conditions, highlighting any significant changes. Beyond the numbers, tools for functional annotation help convert lists of proteins into understandable insights about cellular processes and signalling pathways by referencing frameworks like Gene Ontology (GO) and KEGG pathways (Goeminne 2019). User-friendly resources and tutorials make these complex workflows more accessible to researchers. Some advanced tools and automated pipelines are making waves in proteomics for functional analysis:

4.3.1 ProtPipe

This multifunctional pipeline is designed for high-throughput proteomics and peptidomics. It automates processes like differential abundance analysis, pathway enrichment, and protein-protein interaction analysis. It also provides annotated tables and visualizations for downstream analysis (Li et al. 2024).

4.3.2 APP (Automated Proteomics Pipeline)

APP integrates multiple open-access tools for mass spectrometry data analysis. It simplifies tasks like peptide identification, validation, and quantitation, while supporting distributed computing for handling large datasets (Malm et al. 2014).

4.3.3 Prophane

Proteomics result Pruning & Homology group ANnotation Engine (Prophane) in general complements the metaproteome analyzer by adding various databases and further improve protein spectrum data search. Prophane is a Perl script-based program that automates data search and gives a broader range of peptide matches (Schneider et al. 2011).

4.3.4 iMetalab

It is a complete automated pipeline and suite that integrates multi-database search, including a iMetaReport allow to have a quick look at data along with interactive iMetaShiny toolkit for taxonomic, functional classification visualization (Li et al. 2020).

4.3.5 MetaNovo

Another advanced tool that can extract more protein from the microbiome data along with is comparatively more confident in not ignoring host protein. It uses a novel algorithm for probabilistic optimization of the entire available Uniprot database and creates a tailored database for decoy search at the proteome level (Potgieter et al. 2023).

4.3.6 Meta4P

MetaProteins-Peptides-PSMs Parser (Meta4P) is a user-friendly tool that can retrieve and provide data from three different formats: protein, peptides, and PSM, and can perform label free data and quantitative metaproteoeme data analysis. It is a Python-based program uses pandas, numpy and other (Porcheddu et al. 2023).

4.4 Integration of Multi-omics and Visualization

Scientists are increasingly blending different "omics" approaches, such as metaproteomics, metagenomics, metatranscriptomics, and metabolomics, so as to capture a more complete picture of both environmental and agricultural systems. By linking the genetic blueprints revealed through metagenomics with the actual proteins detected via metaproteomics, researchers can gain deeper insights into gene functions, microbial interactions, and how these processes influence entire ecosystems. This integrated, holistic approach is especially important for practical applications such as assessing soil health, breaking down pollutants, and enhancing precision agriculture. Modern techniques are increasingly blending proteomics data with other "omics" information, such as transcriptomics or metabolomics, to build a comprehensive picture of biological systems. State-of-the-art visualization and network analysis tools then work their magic by transforming complex, multi-layered data into clear visuals that shed light on key proteins, interactions, and pathways (Wöhlbrand et al. 2017; Schmidt et al. 2014; Kumar and Yadav 2024).

4.5 Bridging Collection and Computational Analysis

In essence, the path from collecting high-quality samples in the lab to extracting actionable insights via bioinformatics is a carefully orchestrated transition. Robust collection protocols ensure that samples genuinely represent the biological system, while advanced computational tools turn raw spectral data into meaningful insights about protein functions, disease mechanisms, or environmental responses. This seamless integration for analysis at every stage is what makes proteomics so powerful, reproducible, and reliable (Abdi et al. 2024).

5 Contribution of Metaproteomics in Strengthening Agricultural and Environmental Studies

Metaproteomics plays a pivotal role in advancing agricultural and environmental studies by providing insights into microbial communities and their functional dynamics. In agricultural studies by enhancing soil health, management of pests and diseases, and by boosting crop productivity. Additionally, in environmental studies, it contributes in pollution remediation, climate change migration, ecosystem monitoring, etc. (Bastida et al. 2009). In agriculture, metaproteomics reveals critical insights into soil health and crop management. Through this approach, scientists can monitor microbial activity, identifying the proteins responsible for breaking down organic matter, cycling nutrients, and defending against pests and pathogens. For example, understanding how microbes respond to nutrient shortages or environmental stress can lead to more precise farming techniques, such as targeted fertilization or eco-friendly pest control strategies. This not only boosts crop yields but also promotes long-term soil fertility and sustainability.

Metaproteomics is making significant strides in unraveling the complexities of microbial communities, bridging the gap between microscopic processes and their real-world impacts on agriculture and the environment. By examining the full range of proteins expressed by microorganisms, metaproteomics offers a dynamic understanding of how microbes function, interact, and adapt an essentially uncovering their role in sustaining life and tackling environmental challenges (Bastida et al. 2009). From an environmental perspective, the field contributes immensely to pollution remediation, ecosystem monitoring, and understanding the effects of climate change. By analyzing proteins involved in degrading pollutants or adapting to changing conditions, researchers can identify natural processes that mitigate damage. For example, metaproteomics can highlight how microbial enzymes break down oil spills or detoxify heavy metals in contaminated sites, paving the way for innovative bioremediation techniques. Furthermore, shifts in microbial protein expression can serve as sensitive indicators of ecosystem health, enabling proactive measures to restore balance (Goh and Wong 2019).

What sets metaproteomics apart is its ability to integrate seamlessly with other "omics" technologies, such as metagenomics and metabolomics (Chaurasia et al. 2022). While metagenomics maps the genetic potential of microbial communities, metaproteomics shows which genes are actively expressed, and metabolomics sheds light on the resulting biochemical pathways. Together, these approaches offer a holistic view of microbial ecosystems, enhancing our understanding of their structure, function, and influence on larger environmental and agricultural systems. In essence, metaproteomics is not just a tool but a lens through which we can observe and optimize nature's hidden mechanisms. Its application is to make agriculture more productive and sustainable, or in protecting delicate ecosystems, which has underscored its critical role in addressing some of the most pressing challenges of our time. As the field continues to evolve, it holds promise for even deeper insights and more groundbreaking solutions.

6 Future Trends

The future of metaproteomics is set to revolutionize our understanding of microbiomes, disease mechanisms, and environmental sustainability. With advancements in technology, particularly in mass spectrometry (MS) resolution, researchers will be able to generate more precise and comprehensive datasets, leading to unprecedented insights into microbial protein expression. As metaproteomics continues to evolve, its applications will expand across various scientific domains, including medicine, agriculture, and biotechnology (Heyer et al. 2017). Despite many advancements, challenges in bioinformatics remain a significant hurdle in metaproteomics. Protein identification is especially difficult when the taxonomic composition is unknown or when databases lack relevant protein entries. For instance, although databases like UniProt/TrEMBL contain protein sequences from over a million species, this number represents only a tiny fraction of the estimated trillion microbial species on Earth. This gap in cataloged proteins presents a major challenge for accurate identification, as even small mutations in protein sequences can significantly disrupt detection. A minor change in every tenth amino acid can result in drastically different tryptic peptides, making it nearly impossible to identify proteins through conventional methods. Overcoming this limitation will require expanded databases and advanced algorithms that can predict protein structures and functions in previously uncharacterized microorganisms.

Another pressing issue is reproducibility in metaproteomics research. Different computational workflows and software tools often yield varying results, making it difficult to compare studies or establish standardized methodologies. Currently, only a few studies have rigorously tested complete metaproteomics workflows using mixed microbial cultures. A broader, collaborative approach among research groups to evaluate identical samples with different tools would be invaluable in improving the consistency and reliability of results. Establishing universal protocols for sample preparation, data analysis, and interpretation will be crucial for the future success of metaproteomics.

Understanding many interactions requires integrated, multi-omics approaches that combine metaproteomics with metagenomics, metabolomics, and transcriptomics. By synthesizing data from various omics technologies, researchers can construct a more holistic picture of microbial function and interactions within the host environment. This comprehensive approach will enable scientists to differentiate between transient microbial fluctuations and long-term shifts in microbiome behavior.

Looking ahead, metaproteomics will continue to expand its role in agricultural, medical, and environmental research. In medicine, it will aid in biomarker discovery, disease diagnostics, and therapeutic development. In agriculture and environmental sciences, it will enhance our knowledge of microbial adaptation to climate change, plant growth and function, pollution mitigation, and bioremediation. As research advances, metaproteomics will transition from simply generating vast amounts of data to deriving meaningful biological insights that influence real-world applications.

Overall, metaproteomics is poised for a transformative impact across scientific disciplines. By overcoming challenges in protein identification, bioinformatics evaluation, and standardization, the field will unlock new possibilities in microbiome research, personalized medicine, and ecological conservation. The journey of metaproteomics is only beginning, and its future promises revolutionary discoveries that will shape the way we understand microbial life and its profound influence on health and the environment.

7 Conclusion

The evolution of metaproteomics has significantly enhanced our ability to understand the functional roles of microbial communities. Extensive metagenomic studies have uncovered a vast repository of uncultured microorganisms, highlighting their critical contributions to ecological processes. However, while metagenomics provides genetic insights, it is metaproteomics that offers a deeper understanding of actual protein expression, linking genotype to phenotype. The advancements in analytical techniques, particularly the transition from 2D gel electrophoresis to high-throughput mass spectrometry and LC-MS/MS, have revolutionized microbial protein identification. These developments have facilitated the study of host-pathogen interactions, disease mechanisms, and plant-microbe relationships, providing critical insights into environmental adaptation and stress tolerance. By refining methodological workflows and integrating bioinformatics tools, metaproteomics has emerged as a powerful approach for deciphering microbial functionality. Its applications span medicine, agriculture, and environmental sciences, addressing key biological questions previously beyond reach. As technology continues to improve, metaproteomics will further expand its impact, enabling a comprehensive understanding of microbiomes and their contributions to global ecosystems. The integration of multi-omics approaches will strengthen data accuracy, fostering discoveries and applications. Metaproteomics is poised to play a crucial role in future biological and biomedical research, driving innovation across diverse scientific domains.

Conflicts of Interest The authors declare no conflict of interest.

References

Abdi G, Jain M, Barwant M, Tendulkar R, Tendulkar M, Tariq M, Amir A (2024) Unveiling the dynamic role of bioinformatics in automation for efficient and accurate data processing and interpretation. In: Advances in bioinformatics. Springer, Singapore, pp 279–319

Abram F, Gunnigle E, O'Flaherty V (2009) Optimisation of protein extraction and 2-DE for metaproteomics of microbial communities from anaerobic wastewater treatment biofilms. Electrophoresis 30(23):4149–4151

Almulhim F, Hong PY (2023) Evaluation of protein extraction methods to improve meta-proteomics analysis of treated wastewater biofilms. Proteomics 23(20):2300191

Bastida F, Moreno JL, Nicolas C, Hernandez T, Garcia C (2009) Soil metaproteomics: a review of an emerging environmental science. Significance, methodology and perspectives. Eur J Soil Sci 60(6):845–859

Benndorf D, Balcke GU, Harms H, Von Bergen M (2007) Functional metaproteome analysis of protein extracts from contaminated soil and groundwater. ISME J 1(3):224–234

Benndorf D, Vogt C, Jehmlich N, Schmidt Y, Thomas H, Woffendin G, Shevchenko A, Richnow HH, von Bergen M (2009) Improving protein extraction and separation methods for investigating the metaproteome of anaerobic benzene communities within sediments. Biodegradation 20:737–750

Calderón-González KG, Hernández-Monge J, Herrera-Aguirre ME, Luna-Arias JP (2016) Bioinformatics tools for proteomics data interpretation. In: Modern proteomics–sample preparation, analysis and practical applications, pp 281–341

Chaurasia P, Jasuja ND, Kumar S (2022) Bioremediation assessment in industrial wastewater treatment: the omics approach. In: Omics insights in environmental bioremediation. Singapore, Springer, pp 455–485

Chen C, Hou J, Tanner JJ, Cheng J (2020) Bioinformatics methods for mass spectrometry-based proteomics data analysis. Int J Mol Sci 21(8):2873

Cheng K, Ning Z, Zhang X, Li L, Liao B, Mayne J, Stintzi A, Figeys D (2017) MetaLab: an automated pipeline for metaproteomic data analysis. Microbiome 5:157

Chourey K, Jansson J, VerBerkmoes N, Shah M, Chavarria KL, Tom LM, Brodie EL, Hettich RL (2010) Direct cellular lysis/protein extraction protocol for soil metaproteomics. J Proteome Res 9(12):6615–6622

Cordier T, Lanzén A, Apothéloz-Perret-Gentil L, Stoeck T, Pawlowski J (2019) Embracing environmental genomics and machine learning for routine biomonitoring. Trends Microbiol 27:387–397

Cox J, Neuhauser N, Michalski A, Scheltema RA, Olsen JV, Mann M (2011) Andromeda: a peptide search engine integrated into the MaxQuant environment. J Proteome Res 10(4):1794–1805

Eng JK, McCormack AL, Yates JR (1994) An approach to correlate tandem mass spectral data of peptides with amino acid sequences in a protein database. J Am Soc Mass Spectrom 5(11):976–989

Eng JK, Jahan TA, Hoopmann MR (2013) Comet: an open-source MS/MS sequence database search tool. Proteomics 13(1):22–24. https://doi.org/10.1002/pmic.201200439

Fenyö D, Beavis RC (2003) A method for assessing the statistical significance of mass spectrometry-based protein identifications using general scoring schemes. Anal Chem 75(4):768–774

Geer LY, Markey SP, Kowalak JA, Wagner L, Xu M, Maynard DM, Yang X, Shi W, Bryant SH (2004) Open mass spectrometry search algorithm. J Proteome Res 3(5):958–964

Gihawi A, Cardenas R, Hurst R, Brewer DS (2023) Quality control in metagenomics data. In: Metagenomic data analysis. Springer, New York, NY, pp 21–54

Goeminne L (2019) Statistical methods for differential proteomics at peptide and protein level (Doctoral dissertation, Ghent University)

Goh WWB, Wong L (2019) Advanced bioinformatics methods for practical applications in proteomics. Brief Bioinform 20(1):347–355

Gómez-Godínez LJ, Fernandez-Valverde SL, Martinez Romero JC, Martínez-Romero E (2019) Metatranscriptomics and nitrogen fixation from the rhizoplane of maize plantlets inoculated with a group of PGPRs. Syst Appl Microbiol 42:517–525

Heyer R, Schallert K, Zoun R, Becher B, Saake G, Benndorf D (2017) Challenges and perspectives of metaproteomic data analysis. J Biotechnol 261:24–36

Keiblinger KM, Riedel K (2018) Sample preparation for metaproteome analyses of soil and leaf litter. In: Microbial proteomics: methods and protocols, pp 303–318

Kim S, Pevzner PA (2014) MS-GF+ makes progress towards a universal database search tool for proteomics. Nat Commun 5(1):5277

Kong AT, Leprevost FV, Avtonomov DM, Mellacheruvu D, Nesvizhskii AI (2017) MSFragger: ultrafast and comprehensive peptide identification in mass spectrometry–based proteomics. Nat Methods 14(5):513–520

Kumar C, Mann M (2009) Bioinformatics analysis of mass spectrometry-based proteomics data sets. FEBS Lett 583(11):1703–1712

Kumar A, Yadav A (2024) Next generation sequencing in metagenomics and Metatranscriptomics. In: Multi-omics analysis of the human microbiome: from technology to clinical applications. Springer, Singapore, pp 49–75

Lee S, Park H, Kim H (2021) Comparison of false-discovery rates of various decoy databases. Proteome Sci 19:11. https://doi.org/10.1186/s12953-021-00179-7

Li C, Quan Q, Gan Y, Dong J, Fang J, Wang L, Liu J (2020) Effects of heavy metals on microbial communities in sediments and establishment of bioindicators based on microbial taxa and function for environmental monitoring and management. Sci Total Environ 749:141555. [CrossRef]

Li Z, Weller CA, Shah S, Johnson NL et al (2024) ProtPipe: a multifunctional data analysis pipeline for proteomics and peptidomics. Genom Proteom Bioinform 22(6):qzae083. https://doi.org/10.1093/gpbjnl/qzae083

Malm EK, Srivastava V, Sundqvist G, Bulone V (2014) APP: an automated proteomics pipeline for the analysis of mass spectrometry data based on multiple open access tools. BMC Bioinform 15:1–8

Mikolitis AS, Mach PM, Kroeger ME, McBride EM, Glaros TG (2024) Comparison of protein extraction methods and data analysis strategies for complete metaproteomic soil analysis. bioRxiv 13:2024–2006

Mohanasundaram S, Dhatwalia D, Vijayaraghavan P, Alzubaidi LH, Makhzuna K (2023) Bioinformatics: computational approaches for genomics and proteomics. In: E3S web of conferences, vol 399. EDP Sciences, p 04042

Muth T, Behne A, Heyer R, Kohrs F, Benndorf D, Hoffmann M, Lehteva M, Reichl U, Martens L, Rapp E (2015) The MetaProteomeAnalyzer: a powerful open-source software suite for metaproteomics data analysis and interpretation. J Proteome Res 14(3):1557–1565

Perkins DN, Pappin DJ, Creasy DM, Cottrell JS (1999) Probability-based protein identification by searching sequence databases using mass spectrometry data. Electrophoresis 20(18):3551–3567

Potgieter MG, Nel AJ, Fortuin S, Garnett S, Wendoh JM, Tabb DL, Mulder NJ, Blackburn JM (2023) MetaNovo: an open-source pipeline for probabilistic peptide discovery in complex metaproteomic datasets. PLoS Comput Biol 19(6):e1011163

Porcheddu M, Abbondio M, De Diego L, Uzzau S, Tanca A (2023) Meta4P: a user-friendly tool to parse label-free quantitative metaproteomic data and taxonomic/functional annotations. J Proteome Res 22(6):2109–2113

Quinn GA, Abdelhameed A, Banat IM, Berrar D, Doerr SH, Dudley E, Francis LW, Gazze SA, Hallin I, Matthews GP, Swain MT (2022) Complementary protein extraction methods increase the identification of the Park grass experiment metaproteome. Appl Soil Ecol 173:104388

Salvato F, Vintila S, Finkel OM, Dangl JL, Kleiner M (2022) Evaluation of protein extraction methods for metaproteomic analyses of root-associated microbes. Mol Plant-Microbe Interact 35(11):977–988

Schmidt A, Forne I, Imhof A (2014) Bioinformatic analysis of proteomics data. BMC Syst Biol 8:1–7

Schneider T, Schmid E, de Castro Jr JV, Cardinale M, Eberl L, Grube M, Berg G, Riedel K (2011) Structure and function of the symbiosis partners of the lung lichen (Lobaria pulmonaria L. Hoffm.) analyzed by metaproteomics. Proteomics 11(13):2752–2756

Tabb DL, Fernando CG, Chambers MC (2007) MyriMatch: highly accurate tandem mass spectral peptide identification by multivariate hypergeometric analysis. J Proteome Res 6(2):654–661

Tartaglia M, Bastida F, Sciarrillo R, Guarino C (2020) Soil metaproteomics for the study of the relationshipsbetween microorganisms and plants: a review of extraction protocols and ecological insights. Int J Mol Sci 21(22):8455

Thiele H, Glandorf J, Hufnagel P (2010) Bioinformatics strategies in life sciences: from data processing and data warehousing to biological knowledge extraction. J Integr Bioinform 7(1):32–40

Wilmes P, Bond PL (2004) The application of two-dimensional polyacrylamide gel electrophoresis and downstream analyses to a mixed community of prokaryotic microorganisms. Environ Microbiol 6(9):911–920

Wöhlbrand L, Wemheuer B, Feenders C, Ruppersberg HS, Hinrichs C, Blasius B, Daniel R, Rabus R (2017) Complementary metaproteomic approaches to assess the bacterioplankton response toward a phytoplankton spring bloom in the Southern North Sea. Front Microbiol 8:442

Zhang X, Li L, Mayne J, Ning Z, Stintzi A, Figeys D (2018) Assessing the impact of protein extraction methods for human gut metaproteomics. J Proteome 180:120–127

Zhou XX, Zeng WF, Chi H, Luo C, Liu C, Zhan J, He SM, Zhang Z (2017) pDeep: predicting MS/MS spectra of peptides with deep learning. Anal Chem 89(23):12690–12697. https://doi.org/10.1021/acs.analchem.7b02566

Exploring Plant Disease Ecology through Meta-omics: Current Insights and Future Perspectives

Soumyayan Roy, Mousumi Murmu, Shayree Chatterjee, Mainak Barman, Arpita Das, Shiv Kumar, and Sanjeev Gupta

1 Introduction

The disease development in plants concerning soil-borne pathogen is multi-faceted and arises from the crosstalk between the plant, the microbiome within the rhizosphere region, and the microenvironment surrounding the root zone. Earlier studies concentrated on a simplistic approach of focussing on individual components (e.g., pathogen and plant interaction) of disease development rather than the study of the community for the understanding of the disease in a holistic manner (Knief 2014). Recent magnificent progression in molecular techniques, including Next Generation Sequencing (NGS) and advanced chromatographic and spectrophotometric methods for molecule identification and characterization, has revolutionized the study of disease development. These innovations have shifted the focus from individualistic approaches to community-based "meta" studies. This includes investigating disease ecology through an in-depth understanding of the plant microbiome, encompassing microorganisms that interact with both the host and the pathogen, influencing disease development either positively or negatively (Singh and Kothari 2017). These studies offer valuable insights into the beneficial soil microbes that suppress and

S. Roy · M. Murmu · M. Barman · A. Das (✉)
Department of Genetics and Plant Breeding, Bidhan Chandra Krishi Viswavidyalaya, Mohanpur, Nadia, West Bengal, India

S. Chatterjee
ICAR-National Institute for Plant Biotechnology, New Delhi, India

S. Kumar
International Centre for Agricultural Research in the Dry Areas (ICARDA), South Asia and China Regional Programme, New Delhi, India

S. Gupta
Crop Science Division, Indian Council of Agricultural Research, New Delhi, India

Renu et al. (eds.), *Meta-omics in Crop Improvement*,
https://doi.org/10.1007/978-981-95-2294-1_6

mitigate plant pathogens. This knowledge forms the foundation for developing environment-friendly biological control strategies for effective disease management. The disease ecology exploration using meta-omics approaches involves initial studies to understand the diverse microbial communities present above and below the ground, with a profound insight into the core and satellite microbiota in the microbiome of the plants. Upon identification, the next step is to characterize and understand the interaction among the microbes and between the microbes and the plant. This step is crucial for adequately identifying pathogens associated with specific plant diseases. The next phase of the disease ecology study concerns the role of the interacting non-pathogenic microbiota from the host's microbiome in aggravating or attenuating the disease. Metagenomic techniques reveal the microbial composition, metatranscriptomic approaches elucidate disease development pathways and pathogen identification, while metaproteomic and meta-metabolomic methods uncover the interplays between the microbiome and the host.

This chapter explores microbial interactions in plant disease ecology, providing insights into the role of various factors and the application of advanced technologies in understanding these dynamics.

2 Diversity of Plant Microbiome

Plant microbiome is predominantly occupied by communities like bacteria, fungi, nematodes, oomycetes, viruses, and algae (Nadarajah et al. 2021). There different microbiome diversity found functional in important agricultural crop plants. The diversity of microbiota depends on environment, vegetation, ecosystem, geographical location, plant genotypes, developmental stages of plants, secondary metabolites produced by the plants, effects of other microbiota, agricultural practices etc. (Gupta et al. 2022).

2.1 Below-Ground Plant Microbiota

The rhizosphere serves as a dynamic hotspot of microbial activity and represents one of the most complex and interactive ecosystems on Earth. The plant root system creates unique ecological niches for soil microbiota, supporting colonization in the rhizosphere and roots. The rhizosphere is enriched by bacterial phyla consisting of Actinobacteria, Bacteroidetes, Firmicutes, Proteobacteria and Acidobacteria, also with fungal phyla like Ascomycota and Basidisomycota (Saritha 2023).

Generally, bacteria are transmitted vertically by seeds. The diversity of root microbiota depends on the secondary metabolites and root exudates, thus, genotype dependent. Root exudates including various organic acids, alkaloids, fatty acids, flavonoids, amino acids, plant growth promotors, putrescine, nucleotides, phenolics, sterols, sugars as well as vitamins, influence microbial composition surrounding roots. Secondary metabolites change the composition of the organisms, hence, alter the diversity of the microbiota. The pattern of community assembly is primarily driven by the chemistry of root exudates and the substrate preferences of the microbiota. As an example, micro-organisms from Actinobacteria and Proteobacteria are influenced by the secondary metabolite benzoxazinoids (BXs) secreted by maize roots (Hu et al. 2018). Furthermore, the bacterial community around the wheat root system impacts the abundance of Pseudomonads, Copiotrophs and Actinobacteria in the rhizosphere as compared to bulk soil (Donn et al. 2015). It is observed that, endophytic bacteria are abundant and enter plants either passively through cracks, wounds, and lateral root emergence points, or actively by colonizing root tissues. The diversity of endophytic bacterial community present in the rhizosphere varies with the crop plant. As an example, rice plants in its root zone, permits the endophytic bacteria from the families of Bradyrhizobiaceae, Rhizobiaceae, Streptomycetaceae, and Comamonadaceae (Edward et al. 2015). Another report of Correa-Galeote et al. (2018) claimed that, phyla of Firmicutes, Proteobacteria and Bacteroidetes were abundant endophytes in root of maize. Earlier report confirms less diversity in endosphere microbiota than rhizosphere microbiota (Fitzpatrick et al. 2018). The presence of various bacterial communities is highly influenced by many parameters like soil pH, mycorrhizal type, and environmental conditions. In contrast, fungal community composition is directly and indirectly shaped by ectomycorrhizal type, habitat, and soil pH. Notably, except for Umbelopsidomycetes and Agaricomycetes, the abundance of higher-level fungi increases in arbuscular mycorrhizal habitats. These findings highlight that microbiome composition is ecosystem-dependent, driven by plant types with distinct nutrient acquisition strategies and associated mycorrhizal types (Bahram et al. 2020). Sometimes presence of one or more genera invites other genera to colonise and hence it increases the diversity. The bacterial community structure of rice is enhanced when there is a presence of exopolysaccharide-producing microorganisms like *Bacillus* and *Azozpirillum* (Yudina et al. 2023). The composition as well as diversity of fungal community present within the soil depends on biotic factors such as plant derived exudates and feedback to soil (Raaijmakers and Mazzola 2016) as well as abiotic factors such as soil texture, structure (Xia et al. 2020) and nutrients availability (Vasco-palacios and Bahram 2020). In the study of Birt et al. (2023), *Fusarium* spp. was detected as the dominant fungal species in the root endophyte of banana; however, contradictory report was given by Liu et al. (2019).

2.2 Above-Ground Plant Microbiota

Above-ground plant organs create a conducive environment for hosting endophytic and ectophytic phyllospheric microbiota. The endophytic microbiota gets entry through flowers, fruits, wounds, and then systematically spread through xylem to other plant compartments. The diversity of the phyllosphere varies significantly depending on its source of colonization and the specific environmental conditions of its location. Bacterial ecto microbiota of the taxa Sphingomonas, Pantoea,

Pseudomonas, Enterobacter, Curtobacterium, Citrobacter, Bacillus, Acinetobacter, Erwinia, and Methylobacterium are prevalent in the phyllospheric region of grapevine (Zarraonaindia et al. 2015; Kecskeméti et al. 2016). Earlier report identified Sphingomonads and Methylobacteria as core microbiota in maize, with their composition influenced by environmental parameters (Wallace et al. (2018). Microbiota sources include soil, air, and seed, eventually establishing on or within plant tissues. Leaf surfaces, particularly sites like stomata, epidermal junctions, hydathodes, trichomes, veins, and cuticle depressions, are key colonization points for bacteria (Beattie and Lindow 1995). Seed-associated bacteria such as Firmicutes, Proteobacteria, Actinobacteria, and Bacteroidetes (Liu et al. 2012; Barret et al. 2015) are also found in phyllospheric regions like floral structure and fruits (Compant et al. 2010; Glassner et al. 2018). Soil-driven bacteria can act as phyllospheric microbiota, with diversity influenced by environmental factors and plant types (Vorholt 2012; Zarraonaindia et al. 2015). This environment-driven microbial diversity affects global food webs and nutrient cycles. Notably, *Pseudomonas* are common in cultivated plants such as almond, tobacco, pumpkin, grape, and apple (Aleklett et al. 2014), with *Enterobacteriaceae* and *Pseudomonas* being predominant in apple flowers (Steven et al. 2018).

Common fungal genera found on leaves include *Penicillium*, *Cladosporium*, *Acremonium*, *Alternaria*, *Mucor*, and *Aspergillus*. Yeast genera such as *Rhodotorula, Cryptococcus,* and *Sporobolomyces* are also prevalent. The fungus *Aureobasidium pullulans* is known to outcompete pathogens and is predominant in the phyllosphere, including on fruits (Cordier et al. 2012). In rice, *Curvularia lunata* and *Bipolaris ravenelii* were identified in the phyllosphere by Venkatachalam et al. (2016). While reports on cyanobacteria and algae diversity in the phyllosphere are limited, forest vegetation has been noted for such diversity. Venkatachalam et al. (2016) identified Nostocales and Oscillatoriales as the most abundant phyllospheric cyanobacteria in rice.

2.3 Core and Satellite Microbiomes

The core microbiome consists of taxa that congruously exist within a specific niche, showing genotype dependency, regardless of soil and environmental factors. This concept is based on taxon abundance and occupancy across multiple samples. For example, the potato core microbiome includes Sphingobium, Bradyrhizobium, and Microvirga (Pfeiffer et al. 2017), while the core microbiome of grape consists of Micrococcaceae, Pseudomonadaceae, and Hyphomicrobiaceae (Zarraonaindia et al. 2015), and the rice core microbiome includes Alphaproteobacteria, Deltaproteobacteria, and Actinobacteria (Edwards et al. 2015). In cultivated crops, fungal core microbiota is also prevalent, such as *Cladosporium* and *Pleosporales* in sugarcane, although 45 taxa were considered core, but not universally across soils (De Souza et al. 2016). Similarly, *Cladosporium* and *Fusarium* were identified as core microbiota in grapevine (Liu and Howell 2021) and chili pepper (Gao et al. 2021). *Fusarium oxysporum* and related class, including Dothideomycetes,

Sordariomycetes, and Mortierellomycota, are core fungal microbes in all plant compartments of banana (Birt et al. 2023). Chemical compounds, such as defense phytohormones, can alter microbiome diversity. For instance, salicylic acid (SA) can disrupt community structure in the phyllosphere, but core microbiota can overcome these effects and maintain stable activity.

Satellite taxa, though present in low abundance and at fewer sites, are gaining recognition for their crucial role in ecosystems. Their importance is evident based on local abundance, geographical range, and specificity to particular habitat (Jousset et al. 2017). Satellite taxa help to prevent the unwanted entry of microbiota into soil (Mallon et al. 2015) and contribute to plant defense mechanism through secreting antifungal volatile compounds. Despite their low abundance, their role is dependent on several factors. Hol et al. (2015) reported that the loss of rare soil microbes negatively correlates with plant yield. The ratio of satellite to core taxa significantly influences the plant microbiome's ability to reduce environmental stresses, enhancing its function and buffering capacity.

3 Potential Role of Microbiome Towards Sustainable Agriculture

Microbiota plays a pivotal role in sustainable agriculture through supporting plant growth, yield potential, and boosting resilience to both biotic and abiotic stressors. It is found rational to accept microbiome-based approaches over the chemical based agronomic improvements, especially when their mechanisms and interactions with plants are well understood. Environmental factors also shape microbial communities in different plant genotypes. Since microbiome composition is genotype-dependent, incorporating beneficial microbiota into breeding strategies can optimize plant-microbe interactions for improved crop performance.

Studies have shown important role of bacterial microbiota for increasing plant yield through management of soil fertility in a stable manner (Desai et al. 2012; Kumar et al. 2014). Core microbiome is an impactful approach to incorporate all effective tools from microbiota, their metabolites and their combinations to enhance the yield attributes productively (Qiu et al. 2022). Increasing plant yield under stress conditions is a key objective, with microbiota demonstrating significant potential to improve yields in challenging environments, such as desert agriculture (Köberl et al. 2011) and tree farming for reforestation programme (Chanway 1997; Bashan et al. 2012). Desert microbiota also promotes plant growth by defending against phytopathogens (Köberl et al. 2011). The use of biofertilizers to boost yield is trending, with microbiota playing a key role in atmospheric nitrogen fixation, primarily by diazotrophic bacteria and archaea (Kuypers et al. 2018). Crops like corn and wheat have shown increased yields through the expression of nitrogen-related genes (*nifH*, *nirK*, *nirS*, *amoA*, and *nosZ*) after biofertilizer application (Grzyb et al. 2021). Additionally, Chloroflexi and Bacteroidetes reduce nitrogen losses by 54% and improve plant growth by increasing beneficial bacteria like *Bacillus, Rhodoplanes*, and *Nitrospira* when biofertilizers such as *Bacillus subtilis* are applied (Chen et al. 2019; Sun et al. 2020).

Plant growth-promoting bacteria enhance plant defense against an array of biotic and abiotic stressors by producing secondary metabolites, phytohormones, and siderophores (Bhattacharyya et al. 2012; Malusa et al. 2014). These interactions facilitate plants to cope with vagaries such as drought, extreme temperatures, nutrient limitations, flooding, high soil salinity, and contamination (Pramanick et al. 2020; Zuluaga et al. 2021; Chai et al. 2022). Microbiota modulate phytohormone production, including abscisic acid, brassinosteroids, cytokinins, ethylene, gibberellins, and jasmonic acid, enhancing resistance to pathogens and drought (Zhang et al. 2018). Microorganisms, particularly fungi like *Piriformospora indica*, are increasingly used in agricultural pest and disease management, promoting plant growth and stress resistance (Siddiqui 2006; Ait Kaki et al. 2013).

Microbiota have recently garnered significant attention for their critical role in phytoremediation, where they enhance plant growth and facilitate the restoration of polluted soils and water bodies (Afzal et al. 2014). *Bacillus cereus* promotes heavy metal extraction, such as lead, in *Brassica nigra* (Aktar et al. 2022). Direct bioremediation has also been achieved deploying strains along with enzymes from harsh environments to treat polluted soils (Mapelli et al. 2012). Soil microbiomes mediate nutrient cycles, with autotrophic bacteria promoting carbon flow (Cao et al. 2023). Methanotrophic bacteria contribute to methane consumption, reducing the greenhouse effect by absorbing up to 60% of methane produced on Earth (Leng et al. 2015; Dutta and Dutta 2016). Additionally, the halo-alkalitolerant endoglucanase produced by *Paenibacillus tarimensis*, isolated from saline systems, shows functional activity across a wide range of pH, temperature, and salt concentrations (Raddadi et al. 2013). These microbiotas and their enzymes are promising tools for stress breeding in plants.

4 Disease Dynamics and Plant Microbiota Interaction

The dynamics of plant diseases is not a simple note-down occurrence, rather it is a complex crosstalk of plants with their associated pathogens and microbiota within the natural framework. Interestingly, these can either overturn diseases by encouraging growth in plant health or intensify it by aiding pathogen expansion.

4.1 Disease Suppression

Disease suppression takes place when the pathogen(s) activity is hindered or the resilience of plant is encouraged by the components of plant microbiota. Competition for resources, production of compounds which are antimicrobial, and host-immune inflection can arbitrate disease suppression.

4.1.1 Resource Competition

It is often observed that beneficial microbes outpace the pathogens for crucial nutritional sources in the rhizospheric zone, for example, *Pseudomonas fluorescens* and other bacteria that produce siderophores sequester iron, limiting its accessibility to pathogens and so prevent their development. Metagenomic research has linked microbial activity to disease suppression by identifying siderophore biosynthesis pathways in suppressive soils (Lugtenberg and Kamilova 2009; Raaijmakers et al. 2010).

4.1.2 Production of Antimicrobial Compound

Pathogens are directly inhibited by bioactive compounds produced by some microbiota members. *Bacillus subtilis*, for instance, produces cyclic lipopeptides that break down bacterial and fungal cell membranes, such as iturin and surfactin. These antimicrobial substances have been described by metabolomics, and their production and modulation have been mapped by metagenomics and metaproteomics (Sessitsch et al. 2013; Berendsen et al. 2012).

4.1.3 Induction of Host Immune Responses

Plants that are exposed to beneficial microorganisms may develop induced systemic resistance (ISR), which primes the immune system for improved defense. Plant defense is usually mediated through activation of signal transduction including SA and jasmonic acid by rhizobacteria like *Bacillus amyloliquefaciens*. Defense-related genes are upregulated in ISR-activated plants and facilitate defense response against a wide array of pathogens (Pieterse et al. 2014; Mendes and Raaijmakers 2013).

4.2 Disease Induction

Pathogens take advantage of specific interactions within the microbiota or cause changes in microbial composition in order to promote infection and the development of disease. This approach frequently entails host-microbiota manipulation, synergistic pathogen interactions, or microbial imbalances.

4.2.1 Microbial Dysbiosis

Disease susceptibility may result from dysbiosis, or disturbance of the microbial ecosystem. Overuse of agrochemicals might decrease beneficial microorganisms and provide habitat for pathogens such as *Rhizoctonia* or *Fusarium* and *Sclerotium*. Such dysbiosis has been identified by metagenomic research with rising pathogen prevalence and decreased functional diversity (Friesen et al. 2006; Van der Heijden et al. 2008).

4.2.2 Synergistic Pathogen Interactions

The plant pathogens frequently interact with other microorganisms to increase their virulence in a synergistic manner. *Ralstonia solanacearum*, for instance, releases nutrients that are advantageous to both itself and its opportunistic microbial partners

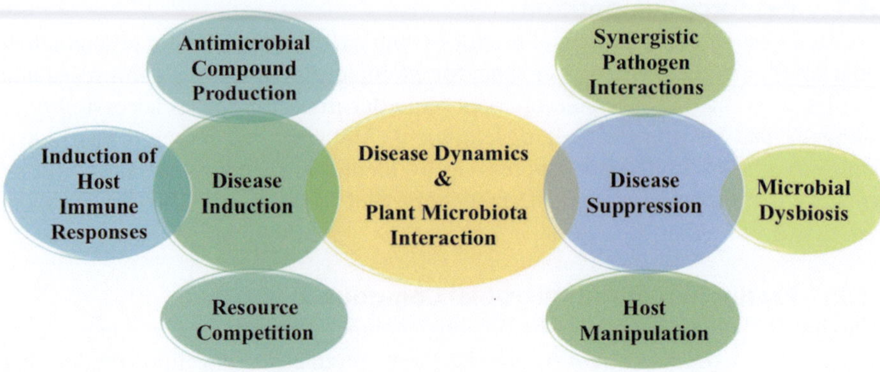

Fig. 1 Disease dynamics and plant microbiota interaction

by breaking down plant cell walls with the help of enzymes. These enzymes and their functions in pathogen synergy have been discovered through metaproteomic research (Lareen et al. 2016; Gould et al. 2018).

4.2.3 Host Manipulation

Various diseases alter the physiology of their hosts by manipulating the microbiota for their benefit. For example, *Xanthomonas campestris* alters plant exudates to attract disease causing microorganisms and inhibit beneficial ones. While metagenomic techniques have documented the resultant changes in microbial community composition and metabolomic investigations described the exudate alterations (Brader et al. 2017).

Meta-omics technologies bridge the gap between microbial composition and function, providing researchers with a holistic understanding of plant disease dynamics (Fig. 1). Community structures are revealed by metagenomics, active genes are clarified by metatranscriptomics, functional proteins are identified by metaproteomics, and biochemical interactions are shown by metabolomics. Researchers can discover the genetic and ecological processes behind diseases induction and suppression by combining these methods.

5 Concept of Meta-omics for Exploring Plant Microbiome

The microbiome study is an essential facet of the study of disease. Traditional approaches concentrate mainly on the pathogen and the host in order to understand the disease. The streamlined approach was further constricted by the availability of a myriad of non-culturable microbes, which cannot be isolated and sequenced through the prevalent method of microbe identification, i.e., culture development and nucleotide sequencing using Sanger sequencing techniques (Hugenholtz and Tyson 2008; Fadiji and Babalola 2020). Over time, it was illustrated that the complex interaction between pathogens and other microorganisms (fungi, bacteria, as well as other protists, actinomycetes, viruses and phage) plays a significant role in the disease development process (Chow et al. 2011; Rogers 2012). The recent advances in the NGS techniques eased the process of understanding the complex interactions as the platform has enabled data retrieval directly from environmental samples (Sogin et al. 2006), which pave the way in the study of metagenomics (Shokralla et al. 2012). The concept further progressed over the years with the advancement of high-throughput techniques to unravel the intricacies of plant-pathogen as well as pathogen-microbiome interactions culminating in a multi-dimensional omics cascade for the study of microbiome through metagenomics, meta-transcriptomics, meta-proteomics, and meta-metabolomics. The holistic research can provide insights into important interactions in the emergence, ecology, epidemiology and occurrence of diseases, plant rhizospheric interactions and co-evolution of microbes and plants with the additional benefit of providing environment-friendly, effective management for the diseases.

5.1 Meta-genomics

The metagenomic approach is one of the most evolving approaches to identify and demarcating organisms in the microbiome. The earlier methodologies identified that culturable microorganisms are just a tiny fraction of the total diversity of microbes available in the soil. So, to understand the complete microbiome and all the holobiont of the plants, integrating modern techniques is of utmost necessity. Present metagenomic approach involving the latest NGS technologies from minuscule DNA samples directly obtained from the environment can contribute significantly for understanding the vast microbial diversities available in the soil environment. This provides an insight about the significance of the microorganisms in the development of various diseases in plants (Garrido-Oter et al. 2018).

The microbiome of the plants involves a great variety of microorganisms like archae, bacteria, viruses, fungi, and actinomycetes (Brader et al. 2017), which either interact as pathogenic or non-pathogenic to plants (Jones et al. 2019). Non-pathogenic microbial communities can interact with plants or other pathogenic microbes to enhance or suppress plant pathogens. The metagenomic techniques have identified pathogens responsible for citrus huanglongbing and boxwood blight (*Calonectria pseudonaviculata*) along with the pathogens in the rice leaf microbiome (Masuda et al. 2024). Metagenomics techniques involving microarray can help to unveil the genes responsible for the disease development, like in maize by Baldwin et al. (1999) for the pathogen *Cochliobolus carbonum.* Microarray technology can also differentiate the different microbial communities present in the rhizosphere using Expressed Sequence Tags (ESTs) to identify their taxonomic and functional roles. Presently, almost 126 plant pathogenic bacteria (Xu and Wang 2019), and around 117 species of plant pathogenic fungus (Aylward et al. 2017) are being characterized and the process is still ongoing as newer pathogens are getting identified and reported. The metagenomic study can also help to identify the evolution of the pathogens upon interaction with antibiotics or fungicides. Emerging diseases due to global changes in the ecosystem can also be gauged through studies of the whole abiotic factors, biotic factors, plant and microbial activities with the ecological interactions through the other -omics approaches in integration with metagenomics (Masuda et al. 2024).

5.2 Meta-transcriptomics

Metatranscriptomics is most significant in soil-borne diseases and their interaction with the microbiota prevalent in the rhizosphere. Hayden et al. (2018) studied wheat to explore the non-supressive role of *Pseudomonas* spp. and *Arthrobacter* spp., and suppressive role of *Buttiauxella* spp. and *Stenotrophomonas* spp. against *Rhizoctonia solani* anastomosis group (AG) 8 pathogensin the rhizosphere. The study also illustrated the role of polyketide cyclase (*pc*), terpenoid biosynthesis backbone gene (*dxs*), and cold shock proteins (*csp*) in the suppressiveness of the soils. Another domain where the meta-transcriptomic approach is of immense value is understanding viromes. Marzano et al. (2016) studied the viromes of various pathogenic fungal species to report the association of mycoviruses in the *Mitovirus* genus of the family *Narnaviridae* associated with the various fungal lineages like *Benyviridae, Ourmiavirus, Barnaviridae, Fusariviridae, Mononegavirales, Chrysoviridae, Tombusviridae, Endornaviridae, Partitiviridae, Hypoviridae, Virgaviridae, Narnaviridae, Totiviridae, Ophioviridae,* and *Tymoviridae*. The infectious nature of imported fruits due to presence of viral pathogens and their interaction through RNAs can be illustrated through metatranscriptomic approaches (Lee et al. 2024; Musweswe et al. 2024). Another domain of implementation for the metatranscriptomic approach is the identification of viral pathogens in plants. Studies in this aspect in bitter apples reported mixed infection of viruses with the identification of a novel virus strain Bitter apple aphid-borne yellows virus (BaABYV) of genus *Polerovirus* (Ghorani et al. 2024). Similar study in citrus grown in Korea uncovered several viruses and viroids along with known viral pathogens of citrus (Kim et al. 2024). Finally, it can also help to unveil the antisense RNA released by viruses for pathogen attenuation or host infection to provide an insight into the viral pathogens and their mode of infection (Saarenpää et al. 2024).

5.3 Meta-proteomics

Metaproteomic analysis of the microbiome is a powerful method for understanding the functional activity of the microbiota in the rhizosphere microbiome. The process of the exo-protein and the protein complement at a specific time for the entire environmental microbiota uncovers valuable information regarding the complex intermicrobial interactions like synergism, antagonism, and resource competition (Herbst et al. 2016). Metaproteomic interventions can predict the behavior of the microbes towards the host as well as the pathogen. The studies can provide accurate details on the microbial biomass, their membrane proteins and the interaction between proteins and other molecules within the host for a robust idea of the specific interaction and mode of action of protein molecules with the host protein along with the essential functions undertaken by the protein in the pathogen. However, the whole community-level meta-proteomic study is hindered by the incorrect understanding of the microbes' growth level estimated through biomass accumulation and the constraints in extracting the extracellular proteins from environmental samples (Bastida et al. 2014). On the other hand, the proteomic study of the interactome of the host-pathogen can only identify the proteins and enzymes in action during the host-pathogen interaction without any profound understanding of the rhizospheric interaction between the microbiome present and the pathogen (Jayaraman et al. 2012).

5.4 Meta-metabolomics

Metabolomics concentrates on the study of small molecules resulting from the various cascade of pathways in a biological system under the influence of a specific condition. When extended to the community level, the study provides a holistic understanding of the plant's rhizospheric microbiome and provide insights into the disease development (van Dam and Bouwmeester 2016). Integrating meta-metabolomics with metagenomic study can provide various deeper insights into multiple metabolites for disease attenuation or suppression, like the disease suppressiveness of composts in economic scenarios through metagenomics study using the common DNA barcode like the 16S RNA and internal transcribed spacer (ITS) sequences, and compound analysis with ultra-high performance liquid chromatography with mass spectrometry (UHPLC-MS-TOF) and Carbon-13 based nuclear magnetic resonance (^{13}C NMR) (Blaya et al. 2016), impact of root exudates on soil microbes in the phyllosphere during the growth of the plant (Chaparro et al. 2013), and the understanding of soil microbe-plant metabolite linkage without the interference of soil matrix effect (Zhalnina et al. 2018). The comprehensive study of the multiple omics techniques in combination with each other can provide a profound understanding of the microbes present in the microbiome, their pathways for interaction, the enzymes and the subsequent metabolites released by the plant into the rhizosphere, and finally, the importance of rhizosphere microbiome during the disease development process in the plant.

6 Cataloguing Plant Microbiota Using High Throughput Sequencing (HTS)

Microbiomes inhabit in a range of niches and, in addition to having a wide range of compositions, they have important functional roles that affect plant diseases, environment, and agriculture. Microbiome research has advanced due to technological and analytical advancements such improved long-read sequencing, more inexpensive high-throughput DNA and RNA sequencing, and creative computational approaches. These advancements allow for a more thorough understanding of microbiomes, particularly those of bacteria and fungi, at the genomic, transcriptional, and translational levels, providing information on their compositions and functions that is more precise than that of individual species.

6.1 Characterization of Fungal Communities Through HTS

By cycling nutrients across trophic levels, the fungal kingdom, which has great diversity ranging from saprotrophy to parasitism, plays an essential function in ecosystems (Tedersoo et al. 2018). Fungus ecology expanded quickly with the introduction of high-throughput sequencing in the late 2000s, revealing enormous unexplored mycobiota, even though genetic investigations have discovered many uncultivable fungus species (Hibbett et al. 2009). Second- and third-generation sequencing technologies are being used in high-throughput fungal community research, and this chapter provides a comprehensive overview of the most advanced molecular techniques for species-level identification. Here, we have covered HTS metabarcoding processes, new methods, and sophisticated fungal community research, before wrapping up with next mycological potential and difficulties.

6.2 Overall Fungal and Bacterial Diversity

Tens of thousands of OTUs are frequently recovered in extensive HTS experiments, indicating a notable variety of fungal species, a large portion of which is yet unknown. 44,563 non-singleton OTUs were discovered in a global soil fungal investigation; over 6% of them were unassigned, even at the class level, indicating the presence of several undescribed fungal groupings (Tedersoo et al. 2014). Although patterns differ by fungal group, climate, soil chemistry, and geographic location, all have an impact on global fungal diversity. For instance, arbuscular mycorrhizal fungi are more prevalent in grasslands, whereas ectomycorrhizal fungi are more varied in temperate biomes (Davison et al. 2015). Environmental factors like soil organic carbon (Maestre et al. 2015), temperature (Newsham et al. 2016), and pH (Djemiel et al. 2023) also shape fungal communities. In the near future, metabarcoding and metaproteomics may help manage plant pathogens biologically and suppress disease dynamics by revealing information about fungal proteins involved in plant-pathogen interactions. Similarly, HTS of bacterial populations is essential for comprehending biodiversity and environmental health. It sheds light on ecological relationships and the practical functions of bacteria in biodegradation, disease prevention, and nutrient cycling. It can help guide sustainable agricultural methods that improve crop productivity and soil health. All things considered, HTS is a potent technique for expanding our understanding of microbiomes and their effects on the environment and human health (Table 1).

A major development in molecular diagnostics is represented by HTS technologies, which are especially useful for identifying and describing plant diseases and pests. With the goal of routine use, their adoption by plant pest diagnosis labs has been gradually growing. However, dependable procedures are required due to the intricacy of HTS and the ensuing bioinformatics analysis. These guidelines, which

Table 1 Characterization of fungal and bacterial communities through High Throughput Sequencing (HTS) techniques

Fungi/bacteria	Observations	References
Saprophytic fungi	(i) Predominance of ascomycetes in the early stages of fungal succession over Basidiomycetes. (ii) Changes in the diversity and biomass of fungi in decomposing wood and soil, as well as vertical partitioning in the soil and surprising fungal communities in places like deep-sea hydrothermal vents.	Voríšková and Baldrian (2013) Xu et al. (2018)
Mycorrhizal fungi	(i) Decreased diversity in ectomycorrhizal fungi inhabiting in decaying wood due to interrupted carbon fluxes following tree removal. (ii) Exhibition of unique mycorrhizal communities across habitats in orchids, whereas ericaceous plants shared their mycorrhizal symbionts with other plants.	Hartmann et al. (2012) Jacquemyn et al. (2016)
Plant pathogenic fungi	(i) For tracking of invasive species and fungal infections, spore trapping and metabarcoding is used. (ii) Drier soils and less human influence resulted in a decrease in fungal pathogen abundance.	Banchi et al. (2018) Hiiesalu et al. (2017); Yan et al. (2018)
Foliar endophytes	(i) Importance of local habitat conditions and spatial. (ii) Detection of hundreds of OTUs from single leaf samples, outperforming culture techniques.	Eusemann et al. (2016) Datlof et al. (2017)
Bacteria	Higher microbial diversity in maize rhizosphere due to nutrient-rich conditions favoring Proteobacteria, including Burkholderia and Pseudomonas.	Yang et al. (2017)
-do-	Improved soybean root microbiome analysis, requiring only 20 mg of sample provided high-quality data, effectively tracking microbial shifts under stress.	Kumaishi et al. (2022)
-do-	High rhizosphere microbial diversity in bulk soil than roots, with multiple Pseudomonas phylotypes identified through long-read sequencing.	Chiniquy et al. (2021)
-do-	Identification of genes for DNA repair and heavy metal tolerance in Proteobacteria, influenced by Lead (Pb), Cadmium (Cd) and Zinc (Zn).	Beattie et al. (2018)
-do-	Identification of Proteobacteria, Bacteroidetes, and Firmicutes as metal-resistant bacteria.	Chen et al. (2018)
-do-	Creation of beneficial resistant bacteria to manage trace metal contamination in agricultural lands.	Yan et al. (2020).
-do-	Identification of 575 core OTUs, with Proteobacteria being predominant in maize seeds by diversity and composition analysis of endophytic bacteria.	Liu et al. (2020)
-do-	Identification of 23 bacterial and 6 fungal phyla in soil, and 11 bacterial and 4 fungal phyla in flax plants and highlighted functional changes, including a shift from parasitic to saprophytic fungi during retting process.	Djemiel et al. (2017)

include everything from test selection to validation and routine implementation, are intended to help researchers and diagnosticians use HTS technology for pathogen identification in an efficient manner. These guidelines will improve the accuracy of pest diagnostics and encourage the incorporation of HTS in plant health management by highlighting crucial components such as risk analysis, sample panel design, appropriate controls, and performance evaluation.

7 Perspective

The integration of multi-omics approaches offers transformative potential to decode the intricate interactions among microbiota, host plants, pathogens, and environmental factors in plant disease dynamics. However, challenges persist, including the need for hypothesis-driven studies, advanced computational tools, and standardized methodologies for integrating diverse datasets. Developing functional ecology databases, enhancing metabolomics reference tools for compound detection, and reducing costs are critical steps toward broader application. Early successes in understanding plant defense, stress detection, and disease management highlight its potential in coming days. Future efforts must focus on collaborative networks and tools to harness multi-omics for sustainable plant disease management in a changing environment.

Conflict of Interest The authors declare no conflict of interest.

References

Afzal M, Khan QM, Sessitsch A (2014) Endophytic bacteria: prospects and applications for the phytoremediation of organic pollutants. Chemosphere 117:232–242

Ait Kaki A, Kacem Chaouche N, Dehimat L, Milet A, Youcef-Ali M, Ongena M, Thonart P (2013) Biocontrol and plant growth promotion characterization of Bacillus species isolated from Calendula officinalis rhizosphere. Indian J Microbiol 53:447–452

Akhter R, Sofi SA (2022) Precision agriculture using IoT data analytics and machine learning. J King Saud Univ-Comp Inform Sci 34(8):5602–5618

Aleklett K, Hart M, Shade A (2014) The microbial ecology of flowers: an emerging frontier in phyllosphere research. Botany 92(4):253–266

Aylward J, Steenkamp ET, Dreyer LL, Roets F, Wingfield BD, Wingfield MJ (2017) A plant pathology perspective of fungal genome sequencing. IMA Fungus 8(1):1–15

Bahram M, Netherway T, Hildebrand F, Pritsch K, Drenkhan R, Loit K, Tedersoo L (2020) Plant nutrient-acquisition strategies drive topsoil microbiome structure and function. New Phytol 227(4):1189–1199

Baldwin D, Crane V, Rice D (1999) A comparison of gel-based, nylon filter and microarray techniques to detect differential RNA expression in plants. Curr Opin Plant Biol 2:96

Banchi E, Ametrano CG, Stanković D, Verardo P, Moretti O, Gabrielli F, Lazzarin S, Borney MF, Tassan F, Tretiach M, Pallavicini A, Muggia L (2018) DNA metabarcoding uncovers fungal diversity of mixed airborne samples in Italy. PLOS One 13(3):e0194489. https://doi.org/10.1371/journal.pone.0194489

Barret M, Briand M, Bonneau S, Préveaux A, Valière S, Bouchez O, Jacques MA (2015) Emergence shapes the structure of the seed microbiota. Appl Environ Microbiol 81(4):1257–1266
Bashan Y, Salazar BG, Moreno M, Lopez BR, Linderman RG (2012) Restoration of eroded soil in the Sonoran Desert with native leguminous trees using plant growth-promoting microorganisms and limited amounts of compost and water. J Environ Manag 102:26–36
Bastida F, Hernández T, García C (2014) Metaproteomics of soils from semiarid environment: functional and phylogenetic information obtained with different protein extraction methods. J Proteome. https://doi.org/10.1016/j.jprot.2014.02.006
Beattie GA, Lindow SE (1995) The secret life of foliar bacterial pathogens on leaves. Annu Rev Phytopathol 33(1):145–172
Beattie RE, Henke W, Campa MF, Hazen TC, McAliley LR, Campbell JH (2018) Variation in microbial community structure correlates with heavy-metal contamination in soils decades after mining ceased. Soil Biol Biochem 12657–12663. https://doi.org/10.1016/j.soilbio.2018.08.011
Berendsen RL, Pieterse CM, Bakker PA (2012) The rhizosphere microbiome and plant health. Trends Plant Sci 17(8):478–486
Bhattacharyya PN, Jha DK (2012) Plant growth-promoting rhizobacteria (PGPR): emergence in agriculture. World J Microbiol Biotechnol 28:1327–1350
Birt HW, Pattison AB, Skarshewski A, Daniells J, Raghavendra A, Dennis PG (2023) The core fungal microbiome of banana (Musa spp). Front Microbiol 14:1127779
Blaya J, Marhuenda FC, Pascual JA, Ros M (2016) Microbiota characterization of compost using omics approaches opens new perspectives for phytophthora root rot control. PLoS One 11:0158048. https://doi.org/10.1371/journal
Brader G, Compant S, Vescio K, Mitter B, Trognitz F, Ma LJ, Sessitsch A (2017) Ecology and genomic insights into plant-pathogenic and plant-nonpathogenic endophytes. Annu Rev Phytopathol 55(1):61–83. https://doi.org/10.1146/annurev-phyto-080516-035641
Cao TND, Mukhtar H, Le LT, Tran DPH, Ngo MTT, Nguyen TB, Bui XT (2023) Roles of microalgae-based biofertilizer in sustainability of green agriculture and food-water-energy security nexus. Sci Total Environ 870:161927
Chai YN, Schachtman DP (2022) Root exudates impact plant performance under abiotic stress. Trends Plant Sci 27(1):80–91
Chanway CP (1997) Inoculation of tree roots with plant growth promoting soil bacteria: an emerging technology for reforestation. For Sci 43(1):99–112
Chaparro JM, Badri DV, Bakker MG, Sugiyama A, Manter DK, Vivanco JM (2013) Root exudation of phytochemicals in Arabidopsis follows specific patterns that are developmentally programmed and correlate with soil microbial functions. PLoS One 8:55731. https://doi.org/10.1371/journal.pone.0055731
Chen Y, Jiang Y, Huang H, Mou L, Ru J, Zhao J, Xiao S (2018) Long-term and high-concentration heavy-metal contamination strongly influences the microbiome and functional genes in Yellow River sediments. Sci Tot Environ 637-6381400-1412. https://doi.org/10.1016/j.scitotenv.2018.05.109
Chen S, Waghmode TR, Sun R, Kuramae EE, Hu C, Liu B (2019) Root-associated microbiomes of wheat under the combined effect of plant development and nitrogen fertilization. Microbiome 7(1):1–13
Chiniquy D, Barnes EM, Zhou J, Hartman K, Li X, Sheflin A, Pella A, Marsh E, Prenni J, Deutschbauer AM, Schachtman DP, Tringe SG (2021) Microbial Community Field Surveys Reveal Abundant Pseudomonas Population in Sorghum Rhizosphere Composed of Many Closely Related Phylotypes. Front in Microbiol 12. https://doi.org/10.3389/fmicb.2021.598180
Chow J, Tang H, Mazmanian SK (2011) Pathobionts of the gastrointestinal microbiota and inflammatory disease. Curr Opin Immunol 23:473–480. https://doi.org/10.1016/j.coi.2011.07.010
Compant S, Clément C, Sessitsch A (2010) Plant growth-promoting bacteria in the rhizo-and endosphere of plants: their role, colonization, mechanisms involved and prospects for utilization. Soil Biol Biochem 42(5):669–678

Cordier T, Robin C, Capdevielle X, Desprez-Loustau ML, Vacher C (2012) Spatial variability of phyllosphere fungal assemblages: genetic distance predominates over geographic distance in a European beech stand (Fagus sylvatica). Fungal Ecol 5(5):509–520

Correa-Galeote D, Bedmar EJ, Arone GJ (2018) Maize endophytic bacterial diversity as affected by soil cultivation history. Front Microbiol 9:484

Dam NM, Bouwmeester HJ (2016) Metabolomics in the rhizosphere: tapping into belowground chemical communication. Trends Plant Sci 21:256–265. https://doi.org/10.1016/j.tplants.2016.01.008

Datlof EM, Amend AS, Earl K, Hayward J, Morden CW, Wade R, Zahn G, Hynson NA (2017) Uncovering unseen fungal diversity from plant DNA banks. Peer J 5e3730. https://doi.org/10.7717/peerj.3730

Davison J, Moora M, Öpik M, Adholeya A, Ainsaar L, Bâ A, Burla S, Diedhiou AG, Hiiesalu I, Jairus T, Johnson NC, Kane A, Koorem K, Kochar M, Ndiaye C, Pärtel M, Reier Ü, Saks Ü, Singh R, Vasar M, Zobel M (2015) Global assessment of arbuscular mycorrhizal fungus diversity reveals very low endemism Cosmopolitan plant root symbionts. Science 349(6251): 970–973. https://doi.org/10.1126/science.aab1161

Desai S, Grover M, Amalraj ELD, Kumar GP, Ahmed SMH (2012) Exploiting plant growth promoting rhizomicroorganisms for enhanced crop productivity. Microorganisms in sustainable agriculture and biotechnology 227–241

Djemiel C, Grec S, Hawkins S (2017) Characterization of bacterial and fungal community dynamics by high-throughput sequencing (HTS) metabarcoding during Flax dew-retting. Front Microbiol 8. https://doi.org/10.3389/fmicb.2017.02052

Djemiel C, Dequiedt S, Bailly A, Tripied J, Lelièvre M, Horrigue W, Jolivet C, Bispo A, Saby N, Valé M, Maron PA, Ranjard L, Terrat S (2023) Biogeographical patterns of the soil fungal: bacterial ratio across France. Msphere 8(5):e00365–23. https://doi.org/10.1128/msphere.00365-23

Donn S, Kirkegaard JA, Perera G, Richardson AE, Watt M (2015) Evolution of bacterial communities in the wheat crop rhizosphere. Environ Microbiol 17(3):610–621

Dutta H, Dutta A (2016) The microbial aspect of climate change. Energy Ecol Environ 1:209–232

Edwards J, Johnson C, Santos-Medellín C, Lurie E, Podishetty NK, Bhatnagar S, Sundaresan V (2015) Structure, variation, and assembly of the root-associated microbiomes of rice. Proc Natl Acad Sci 112(8):911–920

Eusemann P, Schnittler M, Nilsson RH, Jumpponen A, Dahl MB, Würth DG, Buras A, Wilmking M, Unterseher M (2016) Habitat conditions and phenological tree traits overrule the influence of tree genotype in the needle mycobiome–P icea glauca system at an arctic treeline ecotone. Summary New Phytol 211(4):1221–1231. https://doi.org/10.1111/nph.13988

Fadiji AE, Babalola OO (2020) Metagenomics methods for the study of plant-associated microbial communities: a review. J Microbiol Methods 170:105860

Fitzpatrick CR, Copeland J, Wang PW, Guttman DS, Kotanen PM, Johnson MT (2018) Assembly and ecological function of the root microbiome across angiosperm plant species. Proc Natl Acad Sci 115(6):1157–1165

Friesen TL, Stukenbrock EH, Liu Z, Meinhardt S, Ling H, Faris JD, Rasmussen JB, Solomon PS, McDonald BA, Oliver RP (2006) Emergence of a new disease as a result of interspecific virulence gene transfer. Nat Genet 38(8):953–956

Gao M, Xiong C, Gao C, Tsui CK, Wang MM, Zhou X, Cai L (2021) Disease-induced changes in plant microbiome assembly and functional adaptation. Microbiome 9:1–18

Garrido-Oter R, Nakano RT, Dombrowski N, Ma KW, McHardy AC, Schulze-Lefert P (2018) Modular traits of the Rhizobiales root microbiota and their evolutionary relationship with symbiotic rhizobia. Cell Host Microbe 24(1):155–167. https://doi.org/10.1016/j.chom.2018.06.006

Ghorani S, Massumi H, Farhangi SH, Mansouri M, Heydarnejad J, Hosseinipour A (2024) Metatranscriptome analysis of symptomatic bitter apple plants revealed mixed viral infections with a putative novel polerovirus. BMC Genomics 25(1):181

Glassner H, Zchori-Fein E, Yaron S, Sessitsch A, Sauer U, Compant S (2018) Bacterial niches inside seeds of Cucumis melo L. Plant Soil 422:101–113

Gould AL, Zhang V, Lamberti L, Jones EW, Obadia B, Korasidis N, Gavryushkin A, Carlson JM, Beerenwinkel N, Ludington WB (2018) Microbiome interactions shape host fitness. Proc Natl Acad Sci 115(51):11951–11960

Grzyb A, Wolna-Maruwka A, Niewiadomska A (2021) The significance of microbial transformation of nitrogen compounds in the light of integrated crop management. Agronomy 11(7):1415

Gupta A, Singh UB, Sahu PK, Paul S, Kumar A, Malviya D, Saxena AK (2022) Linking soil microbial diversity to modern agriculture practices: a review. Int J Environ Res Public Health 19(5):3141

Hartmann M, Howes CG, VanInsberghe D, Yu H, Bachar D, Christen R, Nilsson RH, Hallam SJ, Mohn WW (2012) Significant and persistent impact of timber harvesting on soil microbial communities in Northern coniferous forests. Abstract The ISME Journal 6(12):2199–2218. https://doi.org/10.1038/ismej.2012.84

Hibbett DS, Ohman A, Kirk PM (2009) Fungal ecology catches fire. New Phytol 184:279–282

Hayden HL, Savin KW, Wadeson J, VVSR G, Mele PM (2018) Comparative metatranscriptomics of wheat rhizosphere microbiomes in disease suppressive and non-suppressive soils for Rhizoctonia solani AG8. Front Microbiol 9(859). https://doi.org/10.3389/fmicb.2018.00859

Heijden MG, Bardgett RD, Straalen NM (2008) The unseen majority: soil microbes as drivers of plant diversity and productivity in terrestrial ecosystems. Ecol Lett 11(3):296–310

Herbst FA, Lünsmann V, Kjeldal H, Jehmlich N, Tholey A, Bergen M, Nielsen JL, Hettich RL, Seifert J, Nielsen PH (2016) Enhancing metaproteomics—the value of models and defined environmental microbial systems. Proteomics 16:783

Hiiesalu I, Bahram M, Tedersoo L (2017) Plant species richness and productivity determine the diversity of soil fungal guilds in temperate coniferous forest and bog habitats. Abstract Mole Ecol 26(18):4846–4858. https://doi.org/10.1111/mec.14246

Hol WG, Boer W, Hollander M, Kuramae EE, Meisner A, Putten WH (2015) Context dependency and saturating effects of loss of rare soil microbes on plant productivity. Front Plant Sci 6:485

Hu L, Robert CA, Cadot S, Zhang XI, Ye M, Li B, Erb M (2018) Root exudate metabolites drive plant-soil feedbacks on growth and defense by shaping the rhizosphere microbiota. Nat Commun 9(1):2738. https://doi.org/10.1038/s41467-018-05122-7

Hugenholtz P, Tyson GW (2008) Microbiology: metagenomics. Nature 455(7212):481

Jacquemyn H, Waud M, Merckx VSFT, Brys R, Tyteca D, Hedrén M, Lievens B (2016) Habitat-driven variation in mycorrhizal communities in the terrestrial orchid genus Dactylorhiza. Abstract Sci Rep 6(1). https://doi.org/10.1038/srep37182

Jayaraman D, Forshey KL, Grimsrud PA, Ané JM (2012) Leveraging proteomics to understand plant–microbe interactions. Front Plant Sci 3:44. https://doi.org/10.3389/fpls.2012.00044

Jones P, Garcia BJ, Furches A, Tuskan GA, Jacobson D (2019) Plant host-associated mechanisms for microbial selection. Front Plant Sci 10:862

Jousset A, Bienhold C, Chatzinotas A, Gallien L, Gobet A, Kurm V, Hol WG (2017) Where less may be more: how the rare biosphere pulls ecosystems strings. ISME J 11(4):853–862

Kecskeméti E, Berkelmann-Löhnertz B, Reineke A (2016) Are epiphytic microbial communities in the carposphere of ripening grape clusters (Vitis vinifera L.) different between conventional, organic, and biodynamic grapes? PLoS One 11(8):0160852

Kim HJ, Choi SR, Cho IS, Jeong RD (2024) Viral Metatranscriptomic analysis to reveal the diversity of viruses infecting Satsuma Mandarin (Citrus unshiu) in Korea. Plant Pathol J 40(2):115

Knief C (2014) Analysis of plant microbe interactions in the era of next generation sequencing technologies. Front. Plant Sci 5(216). https://doi.org/10.3389/fpls.2014.00216

Köberl M, Müller H, Ramadan EM, Berg G (2011) Desert farming benefits from microbial potential in arid soils and promotes diversity and plant health. PLoS One 6(9):24452

Kumaishi K, Usui E, Suzuki K, Kobori S, Sato T, Toda Y, Takanashi H, Shinozaki S, Noda M, Takakura A, Matsumoto K, Yamasaki Y, Tsujimoto H, Iwata H, Ichihashi Y (2022) High throughput method of 16S rRNA gene sequencing library preparation for plant root microbial community profiling. Abstract Sci Rep 12(1). https://doi.org/10.1038/s41598-022-23943-x

Kumar A, Maurya BR, Raghuwanshi R (2014) Isolation and characterization of PGPR and their effect on growth, yield and nutrient content in wheat (Triticum aestivum L). Biocatal Agric Biotechnol 3(4):121–128

Kuypers MM, Marchant HK, Kartal B (2018) The microbial nitrogen-cycling network. Nat Rev Microbiol 16(5):263–276

Lareen A, Burton F, Schäfer P (2016) Plant microbial fuel: the energy-microbiome nexus. Trends Plant Sci 21(7):517–526

Lee GE, Lee HJ, Jeong RD (2024) Comprehensive metatranscriptomic analysis of plant viruses in imported frozen cherries and blueberries. Plant Pathol J 40(4):377

Leng L, Chang J, Geng K, Lu Y, Ma K (2015) Uncultivated methylocystis species in paddy soil include facultative methanotrophs that utilize acetate. Microb Ecol 70:88–96

Liu Y, Yan H, Zhang X, Zhang R, Li M, Xu T, Yang F, Zheng H, Zhao J (2020) Investigating the endophytic bacterial diversity and community structures in seeds of genetically related maize (Zea mays L.) genotypes 3. Biotech 10(1). https://doi.org/10.1007/s13205-019-2034-8

Liu D, Howell K (2021) Community succession of the grapevine fungal microbiome in the annual growth cycle. Environ Microbiol 23(4):1842–1857

Liu Y, Zuo S, Xu L, Zou Y, Song W (2012) Study on diversity of endophytic bacterial communities in seeds of hybrid maize and their parental lines. Arch Microbiol 194:1001–1012

Liu YX, Qin Y, Bai Y (2019) Reductionist synthetic community approaches in root microbiome research. Curr Opin Microbiol 49:97–102

Lugtenberg B, Kamilova F (2009) Plant-growth-promoting rhizobacteria: mechanisms and applications. Ann Rev Microbiol 63(1):541–556

Maestre FT, Delgado-Baquerizo M, Jeffries TC, Eldridge DJ, Ochoa V, Gozalo B, Quero JL, García-Gómez M, Gallardo A, Ulrich W, Bowker MA, Arredondo T, Barraza-Zepeda C, Bran D, Florentino A, Gaitán J, Gutiérrez JR, Huber-Sannwald E, Jankju M, Mau RL, Miriti M, Naseri K, Ospina A, Stavi I, Wang D, Woods NN, Yuan X, Zaady E, Singh BK (2015) Increasing aridity reduces soil microbial diversity and abundance in global drylands. Proc Natl Acad Sci 112(51):15684–15689. https://doi.org/10.1073/pnas.1516684112

Mallon CA, Poly F, Le Roux X, Marring I, Elsas JD, Salles JF (2015) Resource pulses can alleviate the biodiversity–invasion relationship in soil microbial communities. Ecology 96(4):915–926

Malusa E, Vassilev N (2014) A contribution to set a legal framework for biofertilisers. Appl Microbiol Biotechnol 98:6599–6607

Mapelli F, Marasco R, Balloi A, Rolli E, Cappitelli F, Daffonchio D, Borin S (2012) Mineral–microbe interactions: biotechnological potential of bioweathering. J Biotechnol 157(4):473–481

Marzano SYL, Nelson BD, Ajayi-Oyetunde O, Bradley CA, Hughes TJ, Hartman GL, Eastburn DM, Domier LL (2016) Identification of diverse mycoviruses through metatranscriptomics characterization of the viromes of five major fungal plant pathogens. J Virol 90(15):6846–6863

Masuda S, Gan P, Kiguchi Y, Anda M, Sasaki K, Shibata A, Iwasaki W, Suda W, Shirasu K (2024) Uncovering microbiomes of the rice phyllosphere using long-read metagenomic sequencing. Commun Biol 7:1–13. https://doi.org/10.1038/s42003-024-05998-w

Mendes R, Garbeva P, Raaijmakers JM (2013) The rhizosphere microbiome: significance of plant beneficial, plant pathogenic, and human pathogenic microorganisms. FEMS Microbiol Rev 37(5):634–663

Musweswe NL, Otun SO, Kalu CM, Memory T, Ntushelo K (2024) Tracking microbes from irrigation water to crops: the potential of metagenomics and meta-transcriptomics. Acta Microbiol Bulg 40(02). https://doi.org/10.59393/amb24400203

Nadarajah K, Abdul Rahman NSN (2021) Plant–microbe interaction: aboveground to belowground, from the good to the bad. Int J Mol Sci 22(19):10388

Newsham KK, Hopkins KW, Carvalhais LC, Fretwell PT, Rushton SP, O'Donnell AG, Dennis PG (2016) Relationship between soil fungal diversity and temperature in the maritime. Antarctic Nat Clim Chan 6(2):182–186. https://doi.org/10.1038/nclimate2806

Pfeiffer S, Mitter B, Oswald A, Schloter-Hai B, Schloter M, Declerck S, Sessitsch A (2017) Rhizosphere microbiomes of potato cultivated in the High Andes show stable and dynamic core microbiomes with different responses to plant development. FEMS Microbiol Ecol 93(2):242

Pieterse CM, Zamioudis C, Berendsen RL, Weller DM, Wees SC, Bakker PA (2014) Induced systemic resistance by beneficial microbes. Annu Rev Phytopathol 52:347–375

Pramanik K, Das A, Banerjee J, Das A, Chatterjee S, Sharma R, Gupta S (2020) Metagenomic insights into rhizospheric microbiome profiling in lentil cultivars unveils differential microbial nitrogen and phosphorus metabolism under rice-fallow ecology. Int J Mol Sci 21(23):8895

Qiu Z, Verma JP, Liu H, Wang J, Batista BD, Kaur S, Singh BK (2022) Response of the plant core microbiome to Fusarium oxysporum infection and identification of the pathobiome. Environ Microbiol 24(10):4652–4669

Raaijmakers JM, Mazzola M (2016) Soil immune responses. Science 352(6292):1392–1393

Raaijmakers JM, Bruijn I, Nybroe O (2010) Natural functions of Lipopeptides from Bacillus and Pseudomonas: more than surfactants and antibiotics. FEMS Microbiol Rev 34(6):1037–1062

Raddadi N, Cherif A, Daffonchio D, Fava F (2013) Halo-alkalitolerant and thermostable cellulases with improved tolerance to ionic liquids and organic solvents from Paenibacillus tarimensis isolated from the Chott El Fejej, Sahara desert, Tunisia. Bioresour Technol 150:121–128

Rogers AB (2012) Gastric Helicobacter spp. in animal models: pathogenesis and modulation by extragastric coinfections. Methods Mol Biol 921:175–188. https://doi.org/10.1007/978-1-62703-005-2_21

Saarenpää S, Shalev O, Ashkenazy H, Carlos V, Lundberg DS, Weigel D, Giacomello S (2024) Spatial metatranscriptomics resolves host–bacteria–fungi interactomes. Nat Biotechnol 42(9):1384–1393

Saritha M, Naorem AK, Shrvan Kumar KKM, Panwar NR (2023) Exploring the role of microorganisms in enhancing pearl millet growth and productivity. Ann Arid Zone 62(1):19–27

Sessitsch A, Kuffner M, Kidd P, Vangronsveld J, Wenzel WW, Fallmann K, Puschenreiter M (2013) The role of plant-associated bacteria in the mobilization and phytoextraction of trace elements in contaminated soils. Soil Biol Biochem 60(100):182–194

Shokralla S, Spall JL, Gibson JF, Hajibabaei M (2012) Next-generation sequencing technologies for environmental DNA research. Mol Ecol 21(8):1794–1805

Siddiqui ZA (2006) PGPR: prospective biocontrol agents of plant pathogens. In: PGPR: biocontrol and biofertilization, pp 111–142

Singh RP, Kothari R (2017) The omics era and host microbiomes. Understanding host-microbiome interactions—an omics approach. Springer

Sogin ML, Morrison HG, Huber JA, Welch DM, Huse SM, Neal PR, Herndl GJ (2006) Microbial diversity in the deep sea and the underexplored "rare biosphere". Proc Nat Acad Sci 103(32):12115–12120

Souza RSC, Okura VK, Armanhi JSL, Jorrín B, Lozano N, Silva MJ, Arruda P (2016) Unlocking the bacterial and fungal communities assemblages of sugarcane microbiome. Sci Rep 6(1):28774

Steven B, Huntley RB, Zeng Q (2018) The influence of flower anatomy and apple cultivar on the apple flower phytobiome. Phytobiomes 2(3):171–179

Sun B, Gu L, Bao L, Zhang S, Wei Y, Bai Z, Zhuang X (2020) Application of biofertilizer containing Bacillus subtilis reduced the nitrogen loss in agricultural soil. Soil Biol Biochem 148:107911

Tedersoo L, Bahram M, Põlme S, Kõljalg U, Yorou NS, Wijesundera R, Ruiz LV, Vasco-Palacios AM, Thu PQ, Suija A, Smith ME, Sharp C, Saluveer E, Saitta A, Rosas M, Riit T, Ratkowsky D, Pritsch K, Põldmaa K, Piepenbring M, Phosri C, Peterson M, Parts K, Pärtel K, Otsing E, Nouhra E, Njouonkou AL, Nilsson RK, Morgado LN, Mayor J, May TW, Majuakim L, Lodge DJ, Lee SS, Larsson K, Kohout P, Hosaka K, Hiiesalu I, Henkel TW, Harend H, Guo L, Greslebin A, Grelet G, Geml J, Gates G, Dunstan W, Dunk C, Drenkhan R, Dearnaley J, De Kesel A, Dang T, Chen X, Buegger F, Brearley FQ, Bonito G, Anslan S, Abell S, Abarenkov K (2014) Global diversity and geography of soil fungi Assessing fungal diversity worldwide. Science 346(6213). https://doi.org/10.1126/science.1256688

Tedersoo L, Sánchez-Ramírez S, Kõljalg U, Bahram M, Döring M, Schigel D, May T, Ryberg M, Abarenkov K (2018) High-level classification of the Fungi and a tool for evolutionary ecological analyses. Fungal Diversity 90(1):135–159 https://doi.org/10.1007/s13225-018-0401-0

Vasco-Palacios AM, Bahram M, Boekhout T, Tedersoo L (2020) Carbon content and pH as important drivers of fungal community structure in three Amazon forests. Plant Soil 450:111–131

Venkatachalam S, Ranjan K, Prasanna R, Ramakrishnan B, Thapa S, Kanchan A (2016) Diversity and functional traits of culturable microbiome members, including cyanobacteria in the rice phyllosphere. Plant Biol 18(4):627–637

Vorholt JA (2012) Microbial life in the phyllosphere. Nat Rev Microbiol 10(12):828–840

Voříšková J, Baldrian P (2013) Fungal community on decomposing leaf litter undergoes rapid successional changes. Abstract The ISME Journal 7(3):477–486. https://doi.org/10.1038/ismej.2012.116

Wallace JG, Kremling KA, Kovar LL, Buckler ES (2018) Quantitative genetics of the maize leaf microbiome. Phytobiom J 2(4):208–224

Xia Q, Rufty T, Shi W (2020) Soil microbial diversity and composition: links to soil texture and associated properties. Soil Biol Biochem 149:107953

Xu W, Gong L, Pang KL, Luo ZH (2018) Fungal diversity in deep-sea sediments of a hydrothermal vent system in the Southwest Indian Ridge. Deep Sea Research Part I: Oceanographic Research Papers 13116–26. https://doi.org/10.1016/j.dsr.2017.11.001

Xu J, Wang N (2019) Where are we going with genomics in plant pathogenic bacteria? Genomics 111(4):729–736

Yan DF, Mills JG, Gellie NJC, Bissett A, Lowe AJ, Breed MF (2018) High-throughput eDNA monitoring of fungi to track functional recovery in ecological restoration. Biol Conserv 217113–217120. https://doi.org/10.1016/j.biocon.2017.10.035

Yan C, Wang F, Geng H, Liu H, Pu S, Tian Z, Chen H, Zhou B, Yuan R, Yao J (2020) Integrating high-throughput sequencing and metagenome analysis to reveal the characteristic and resistance mechanism of microbial community in metal contaminated sediments. Sci Tot Environ 707136116. https://doi.org/10.1016/j.scitotenv.2019.136116

Yang Y, Wang N, Guo X, Zhang Y, Ye B (2017) Comparative analysis of bacterial community structure in the rhizosphere of maize by high-throughput pyrosequencing. PLOS ONE 12(5): e0178425. https://doi.org/10.1371/journal.pone.0178425

Yudina NY, Kozlova TN, Abramova TN, Arlyapov VA, Asulyan LD, Alferov VA, Reshetilov AN (2023) An immobilized microorganism-based biopreparation as a major component for new generation high-efficiency organic fertilizers. Appl Biochem Microbiol 59(2):223–230

Zarraonaindia I, Owens SM, Weisenhorn P, West K, Hampton-Marcell J, Lax S, Gilbert JA (2015) The soil microbiome influences grapevine-associated microbiota. MBio 6(2):10–1128

Zhalnina K, Louie KB, Hao Z, Mansoori N, Rocha UN, Shi S, Cho H, Karaoz U, Loqué D, Bowen BP, Firestone MK (2018) Dynamic root exudate chemistry and microbial substrate preferences drive patterns in rhizosphere microbial community assembly. Nat Microbiol 3:470–480. https://doi.org/10.1038/s41564-018-0129-3

Zhang W, Wang J, Xu L, Wang A, Huang L, Du H, Oelmüller R (2018) Drought stress responses in maize are diminished by Piriformospora indica. Plant Signal Behav 13(1):1414121. https://doi.org/10.1080/15592324.2017.1414121

Zuluaga MYA, Milani KML, Miras-Moreno B, Lucini L, Valentinuzzi F, Mimmo T, Pii Y, Cesco S, Rodrigues EP, de Oliveira ALM (2021) Inoculation with plant growth-promoting bacteria alters the rhizosphere functioning of tomato plants. Applied Soil Ecol 158103784. https://doi.org/10.1016/j.apsoil.2020.103784

Deciphering Metagenomics Approach for the Detection of Plant Pathogens: Challenges and Future Prospects

D. T. Nagrale, R. Sreenath Reddy, S. P. Gawande, N. S. Hiremani, Sunil Kumar, and M. D. Meshram

1 Introduction

Plant diseases caused by pathogenic microorganisms pose a significant threat to global agriculture, impacting crop productivity, food security, and economic stability (Singh et al. 2023). Traditional diagnostic techniques, while effective in certain contexts, are often limited by their reliance on culturable pathogens, subjective symptom observation, and the inability to detect mixed or low-abundance infections. These challenges underscore the need for advanced diagnostic tools capable of providing precise, comprehensive, and unbiased identification of plant pathogens. Metagenomics, an innovative and transformative approach, has emerged as a powerful solution to these challenges. By leveraging high-throughput sequencing technologies, metagenomics enables the simultaneous identification and characterization of entire microbial communities within a sample, bypassing the need for prior cultivation or knowledge of specific pathogens (Franzosa et al. 2015) and identification of novel biomolecules from soil ecosystems (Daniel 2005). This untargeted nature of metagenomics makes it uniquely suited to uncovering novel or poorly characterized pathogens and understanding their complex interactions within host plants and environmental contexts. The application of metagenomics in plant pathology represents a paradigm shift in diagnostic methodologies. It allows researchers to move beyond the constraints of targeted detection, offering insights into pathogen diversity, virulence factors, and co-infections. Moreover, advances in

D. T. Nagrale and R. Sreenath Reddy contributed equally with all other contributors.

D. T. Nagrale (✉) · S. P. Gawande · N. S. Hiremani · M. D. Meshram
ICAR- Central Institute for Cotton Research, Nagpur, India

R. Sreenath Reddy
ICAR-IARI Academic Hub, The Graduate School, ICAR-CICR, Nagpur, India

S. Kumar
ICAR- Indian Agricultural Statistics Research Institute, New Delhi, India

Renu et al. (eds.), *Meta-omics in Crop Improvement*,
https://doi.org/10.1007/978-981-95-2294-1_7

bioinformatics and sequencing technologies, including long-read sequencing platforms, have significantly enhanced the resolution and accuracy of metagenomic analyses, paving the way for real-time diagnostics and field-level applications (Ravisankar and Mathew 2022).

2 Conventional Approaches to Pathogen Detection

Currently, a variety of diagnostic techniques are employed to fulfil the requirements of screening, monitoring, and identifying pathogens. While certain host-pathogen interactions can be diagnosed effectively through symptom observation by trained experts, this approach is limited by the nonspecific nature of symptoms in many cases, necessitating the use of more precise identification methods. One such traditional method involves the use of mist chambers that create high-humidity conditions to promote the growth and sporulation of pathogens, such as fungi and oomycetes, which can then be isolated on selective culture media. However, this technique is effective only for facultative parasites but poses significant challenges when applied to root, soil, and water samples, as fast-growing saprophytic organisms often outcompete primary pathogens and obscure their presence (Abdelfattah et al. 2019). The baiting technique is commonly applied for detecting oomycetes like *Pythium* and *Phytophthora* species, which produce motile zoospores (Erwin and Ribeiro 1996). After isolating these pathogens into pure cultures, their identification is carried out based on features such as colony morphology, colour, cell structure, and growth characteristics. Despite their utility, traditional methods share significant limitations, including being labor-intensive, time-consuming, and heavily reliant on the analyst's expertise.

Due to these limitations, microbiological and microscopy-based techniques often yield inconclusive results. Consequently, molecular identification is frequently integrated with traditional methods, especially when initial identification efforts fail. This approach involves amplifying and sequencing specific genes that serve as diagnostic markers for identifying particular taxa, commonly referred to as DNA barcoding regions. PCR-based techniques, which are applicable to all microorganisms, including those that are unculturable, offer advantages such as speed, sensitivity, and specificity (Spadaro et al. 2020). Real-time quantitative PCR (qPCR) further enhances sensitivity, specificity, and reliability, enabling both sequence identification and quantitative analysis (Sanzani et al. 2014). Another innovative method is loop-mediated isothermal amplification (LAMP), introduced by Notomi et al. (2000). Although the design and validation of LAMP primers are labor-intensive, the technique is rapid, highly specific, and effective even with minimal pathogen DNA in samples. LAMP is particularly valuable when combined with on-site DNA extraction techniques, allowing for the direct molecular detection of plant pathogens in field conditions (Chomczynski and Rymaszewski 2006).

Recent advancements in nanotechnology have also introduced novel methods for detecting nucleic acid sequences (Lau et al. 2017). For instance, Yao et al. (2009) employed fluorescent silica nanoparticles conjugated with antibodies to detect

Xanthomonas axonopodis pv. *vesicatoria*, a pathogen responsible for bacterial spot disease in solanaceae crops. Additionally, certain diseases can be diagnosed by identifying specific microRNAs (miRNAs) associated with plant responses to pathogens, utilizing these small non-coding transcripts as reliable infection biomarkers (Chen et al. 2012). Despite their high sensitivity, molecular diagnostic methods are typically limited to detecting only one or a few target pathogens within a single reaction (Schena et al. 2004). Hence, they are often categorized as "targeted methods." This group also includes microarray-based approaches, which can simultaneously detect a wide spectrum of microorganisms. However, these methods have had restricted use because of their intricacy and the difficulties involved in analysing the outcomes (Anderson et al. 2004). Like other molecular techniques, microarray-based methods rely on target-specific primers or probes, restricting their use to identify well-characterized microorganisms and rendering them unsuitable for discovering novel species.

In contrast, metagenomic-based approaches offer the capability to identify virtually all microorganisms within a single analysis, making them untargeted by nature. This characteristic makes metagenomics highly effective for detecting quarantine pathogens in plant tissues, seeds, and food products. By enabling comprehensive identification, these approaches are invaluable for addressing challenges associated with unknown or poorly characterized microorganisms.

3 Overview of Metagenomics

Metagenomics, utilizing shotgun sequencing, enables comprehensive analysis of the entire DNA content within a sample (Quince et al. 2017). This process begins with extracting total DNA from the sample, fragmenting it into smaller segments, and sequencing these fragments simultaneously through high-throughput sequencing platforms (Sharpton 2014). The resulting sequencing reads are assembled into contiguous sequences (contigs) based on overlapping regions, ultimately reconstructing a metagenome that represents the microbial community of the sample. This metagenome is then analyzed to identify microorganisms and explore functional genes present in the sample (Semenov 2021). While highly useful, assembly-based metagenomics poses challenges such as significant computational demands and complex data analysis. Assembling sequences can be particularly difficult due to the genetic similarities between closely related species and insufficient sequencing coverage for low-abundance species, which can hinder the formation of large contigs (Quince et al. 2017). To mitigate these issues, assembly-free approaches have been developed, which directly compare individual sequencing reads to reference genome databases, bypassing the need for contig assembly (Lind and Pollard 2021). These methods are faster and improve the detection of low-abundance species, although their accuracy depends on the availability of extensive reference databases and they risk false positives due to misassignment of conserved sequences (Ayling et al. 2020).

The emergence of third-generation sequencing technologies, particularly long-read sequencing, has addressed many limitations associated with second-generation platforms.

In a metagenomic study with sequencing of cDNA from the tomato plants infected with virus PepMV, achieved identification of 97% with viral genome sequence (Adams et al. 2009). Metagenomic libraries with phylogenetic information of 16S-rDNA gene are useful in the identification of bacteria (Lara-Victoriano et al. 2011) and reclassified the bacteria with important taxonomic categories (Cottrell et al. 2005). The metagenomic studies are very crucial in the detection of plant pathogens which produce no visible disease symptoms on host plant like citrus Huanglongbing (HLB) disease (Lara-Victoriano et al. 2011). Shotgun and amplicon-based metagenomics methods are effective in the detection of soil borne plant pathogens namely *Rhizoctonia* spp., *Fusarium* spp., *Verticillium* spp., *Sclerotinia* spp., *Pythium* spp., and *Phytophthora* spp., infecting several crops (Català et al. 2015; Lewis and Papavizas 1991; Baysal-Gurel and Kabir 2019; Panth et al. 2020). The application of pyrosequencing techniques in plant pathology was studied in a metatranscriptomic analysis which reported that phytopathogenic virus inhabit plant tissues together with fungi and bacteria (Molina et al. 2012). Similarly, metagenomic tools helps to discover the novel functional gene and deciphering possible metabolic pathways associated with pathogenesis that suppresses the host immunity against plant pathogens (Aravind et al. 2009; Porras-Alfaro and Bayman 2011). The short reads produced by second-generation sequencers often fail to span large repetitive regions, leading to fragmented genome assemblies and complicating de novo assembly. In contrast, long-read sequencing can bridge these repetitive regions, facilitating genome assembly and enhancing the resolution needed to distinguish closely related species (Amarasinghe et al. 2020). Metagenomics provides distinct advantages for pathogen detection, including: enabling identification at the strain level with extensive information, facilitating unbiased, simultaneous detection of fungal, oomycete, and prokaryotic pathogens without PCR-related biases, eliminating the need for prior genetic knowledge of the pathogen and recovering genomes from non-culturable microorganisms (Aragona et al. 2022). These capabilities are especially beneficial for detecting plant viruses, which lack universal marker genes for amplicon-based sequencing. The unbiased nature of metagenomics also allows for the identification of previously unknown pathogens (Adams and Fox 2016). However, the routine application of shotgun metagenomics for plant pathogen detection faces several barriers. These include the high costs associated with the deep sequencing required, interference from contaminating sequences such as plant DNA, reduced sensitivity for detecting low-abundance species, and the incomplete reference genome databases for plant pathogens (Piombo et al. 2021).

Despite these challenges, metagenomics has shown considerable potential in pathogen detection. For instance, an Illumina-based metagenomics approach successfully characterized the microbiome of Arabidopsis leaves, pathogenic species such as *Protomyces* and *Peronospora* could be identified using a database comprising of 242 marker genes (Lind and Pollard 2021). Long-read sequencing has similarly been applied effectively to detect fungal pathogens in agricultural and forest ecosystems (Loit et al. 2019) and to identify novel plant viruses in crops such as wheat and maize (Lappe et al. 2022). Another study highlighted the utility of metagenomics in detecting *Calonectria pseudonaviculata*, the causal agent of boxwood blight, with taxonomic resolution reaching the strain level (Yang et al. 2022). Metagenomics has also been employed for pathogen detection directly in the field. For example, Boykin et al. (2019) utilized the portable Nanopore MinION sequencing platform to identify viral pathogens in cassava plants in Sub-Saharan Africa. This approach allowed for the detection of multiple viruses within three hours of sample collection, demonstrating its feasibility for rapid field diagnostics. Similarly, metagenomics has been used to investigate outbreaks of *Xylella fastidiosa*, achieving sensitivity levels comparable to qPCR while providing subspecies-level identification (Johnson et al. 2022). These findings emphasize the value of metagenomics as a powerful approach for the precise and early identification of plant pathogens, especially those that are challenging to culture or have not been previously characterized.

4 Application of Metagenomics in Plant Pathogen Detection

4.1 Workflow for Metagenomics-Based Plant Pathogen Detection

4.1.1 Sample Preparation and Laboratory Techniques

Effective application of metagenomics technology starts with the proper selection of plant samples. For the detection of latent pathogens, it is crucial to obtain samples from living plants that exhibit advanced disease symptoms. The plant chosen for sampling should clearly display the most noticeable signs of infection. The sampling strategy may vary depending on the study's goals, particularly when examining infections in plants grown under natural or unregulated environmental conditions. In such cases, the metagenomic approach should be tailored to align with the symptoms typical of the microorganism being studied. Observational techniques are essential to differentiate the abiotic stresses responsible for plant infections (Burdon and Zhan 2020). Samples collected for metagenomic analysis should be stored at low temperatures (around 4 °C) to maintain their integrity and transported to the laboratory under stable conditions. Prolonged storage under standard room conditions increases the risk of DNA contamination from other organisms, which can compromise the accuracy of metagenomic analyses and lead to erroneous interpretations of the data (Mzava et al. 2022). The next step in metagenomic analysis is the extraction of nucleic acids. Both commercial kits and standard manual methods can be used for this purpose, but commercial kits are often preferred as they minimize the risk of environmental contamination. Nucleic acid extraction should be conducted in a sterile environment to avoid contamination. The extracted nucleic acids, containing DNA from multiple species, are referred to as mixed DNA, or environmental DNA (eDNA) when obtained from environmental samples (Ruppert et al.

2019). To confirm latent infections, traditional culturing methods may serve as an optional supplementary step. By isolating the main pathogen in vitro, researchers can verify its involvement in the infection, thus providing a control for metagenomic sequencing. Alternatively, DNA extraction can be performed directly from infected leaf tissue using appropriate extraction kits.

4.1.2 Metagenomic Library Construction

Building metagenomic libraries involves generating a nucleic acid mixture that is suitable for sequencing while maintaining the diversity of DNA sequences present in the sample. This is particularly important in microbiota research. When pathogens are the target, selectively protecting or enriching specific sequences adds complexity to the library preparation process. In some metagenomic studies, the entire pool of extracted nucleic acids is sequenced, while in others, barcoding techniques are employed to focus on specific microbial populations. Since sequencing technologies have limitations and can only process a fraction of DNA or RNA (Chopra et al. 2020), it is crucial that the library accurately reflects the original sample's composition. In plant-related studies, most of the extracted nucleic acids are derived from the plant itself. However, specialized purification kits can effectively separate microbial or pathogen DNA from host DNA. To construct metagenomic libraries with minimal amounts of microbial nucleic acids, enrichment techniques targeting both DNA and RNA can be utilized. PCR amplification is often used to enrich pathogen-specific fragments, which increases the representation of pathogen-derived nucleic acids in the sample. After preparing the nucleic acids, sequencing adapters and barcodes are added. Barcoding involves attaching unique short sequences to DNA fragments to allow the pooling of multiple samples, each of which can be traced back using bioinformatics. High-sensitivity library preparation kits, such as Nextera XT (Illumina, San Diego), can create libraries from as little as one nanogram of DNA, ensuring precision even with minimal material (Gokdemir et al. 2022). While constructing metagenomic libraries is essential for identifying specific gene regions, a universal standard for library preparation is still lacking. Researchers typically rely on advanced kits from manufacturers like Illumina and NEB to achieve reliable results (Jiang et al. 2022).

4.1.3 High-Throughput Sequencing

High-throughput sequencing platforms play a critical role in analysing metagenomic samples, with Illumina sequencing, Nanopore sequencing, and Roche/454 pyrosequencing being the most commonly utilized techniques. Among these, Illumina sequencing is known for its higher sensitivity and ability to generate unique data, offering a read depth of 1 to 5 million reads with alignment lengths ranging from 75 to 100 base pairs. For detecting bacterial infections, specific 16S rRNA barcodes are employed, whereas barcodes targeting the ITS-23S rRNA region are commonly used for identifying fungal and viral pathogens respectively (Cusco et al. 2019). In certain studies, a combination of both barcodes is necessary. When the plant species involved in the study is unidentified, a method known as metabarcoding can be applied, which incorporates plant-specific barcodes into the analysis. Universally preferred plant barcodes include rbcL, trnL-trnF, rpc36-8, trnT2-rps4, and two mitochondrial genes, nad7 and atpA (Ward et al. 2009). For bacterial identification, 16S rRNA barcodes are predominantly used, while ITS and 18S rRNA genes are preferred for fungi and archaea.

4.1.4 Computational Framework for Metagenomic Analysis

After completing metagenomic next-generation sequencing, a series of bioinformatics analyses is essential for interpreting the data. The vast number of short reads generated during sequencing must first undergo a filtering process. This step aims to remove low-quality sequences and eliminate host genome data. To filter out short sequence reads, including those originating from the plant genome, a reference genome is employed to identify and isolate matching reads (Miller and Chiu 2022). Following the filtering step, the remaining sequences are aligned against microbial reference sequence databases. Among these, the NCBI database is widely utilized due to its comprehensive genomic data covering numerous organisms. The longer sequence reads are assembled de novo into clusters known as contigs, a term derived from "contiguous." In genomic sequencing, a contig represents a group of overlapping DNA sequences, providing a continuous representation of a specific genomic region and enabling links to physical maps. The objective of this process is to categorize as many reads as possible into appropriate taxonomic groups, such as species, genus, or phylum. Reads that do not align with known sequences are assembled de novo using specialized algorithms designed for metagenomic applications. Software tools like Velvet and Meta-IDBA are commonly used for de novo assembly (Peng et al. 2011). The various software tools used in metagenomics analysis and the process have been illustrated in Table 1 and Fig. 1.

4.2 Advantages over Traditional Detection Methods

Metagenomics represents a breakthrough in pathogen detection by overcoming the limitations of traditional diagnostic methods, which require prior knowledge of specific pathogens. Unlike conventional techniques, metagenomics enables the detection of a wide range of pathogens, including fungi, bacteria, and viruses. This approach is particularly valuable for identifying unculturable pathogens and offers strain-level resolution, improving our understanding of pathogen diversity and enhancing epidemiological studies. Its unbiased nature allows the discovery of both known and previously unidentified pathogens, including emerging or re-emerging

Table 1 Bioinformatics tools for post-metagenomics analysis

Analysis	Tools	References
Chimera detection	DECIPHER, Persus	Kim et al. (2013)
Phylogenetic classifications	BLAST, Phymm, CARMA	Brady et al. (2009)
Fungal detection	UNITE	Abarenkov et al. (2010)
Denoising	Denoiser, Pyronoise, acacia, DADA	Rosen et al. (2012)

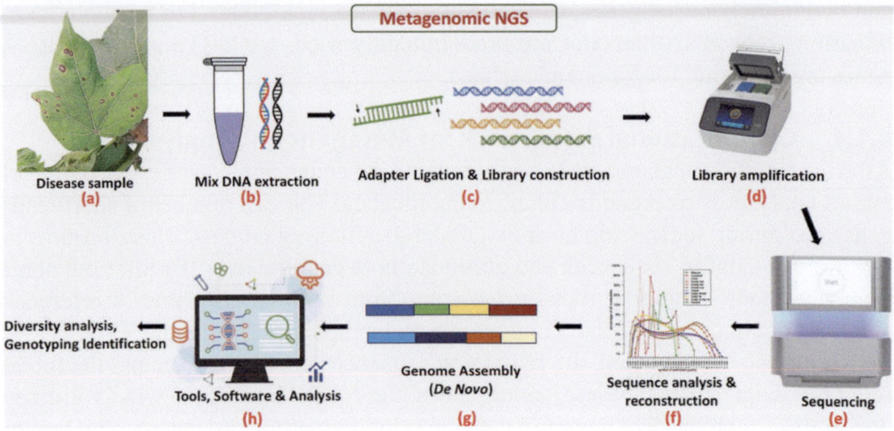

Fig. 1 Workflow for metagenomics-based pathogen detection. (**a**) Sample collection; (**b**) Mix DNA extraction; (**c**) library construction; (**d**) PCR amplification of specific gene regions in microbial pathogens using targeted gene barcodes (ITS, 18srRNA); (**e**) sequencing with Nanopore, Illumina etc.; (**f**) Sequence analysis and reconstruction; (**g**) bioinformatics analysis of metagenomic next-generation sequencing (mNGS) involves both de novo approaches and reference-based assembly methods and (**h**) bioinformatic analysis for assembling metabarcoding data, along with clustering and prediction tasks

ones. Additionally, metagenomics' high sensitivity makes it suitable for detecting low-abundance pathogens within complex microbial communities, where traditional methods may be ineffective. Beyond identification, it provides insights into pathogen gene functions and metabolic pathways, furthering our understanding of pathogen biology and host interactions. The development of portable sequencing devices like the Nanopore MinION has enabled real-time diagnostics in the field, facilitating the analysis of various samples, including plant tissues, soil, and water. Combined with automated sequencing and advanced data analysis, metagenomics minimizes human bias and uncovers new or underexplored pathogens, significantly enhancing plant disease management strategies.

5 Sensitivity, Specificity, and Real-Time Applicability

Metagenomics has transformed plant pathogen detection by providing exceptional sensitivity, specificity, and real-time applicability. Utilizing high-throughput sequencing technologies, it can detect even trace amounts of pathogenic DNA or RNA, including from organisms that cannot be cultured or are present in low abundance within complex microbial communities. This enhanced sensitivity plays a crucial role in the early diagnosis of diseases, as exemplified by its accurate identification of *Calonectria pseudonaviculata* in boxwood. The specificity of metagenomic approaches relies heavily on high-quality reference databases and bioinformatics tools. With the use of advanced algorithms and carefully curated genomic libraries, sequencing reads are accurately mapped to target pathogens, minimizing the occurrence of false positives and negatives. Markers such as the ITS region for fungi and the 16S rRNA gene for bacteria offer higher taxonomic resolution and allow for strain-level identification. However, challenges such as sequence similarity between closely related species may lead to occasional misclassifications. In such cases, combining metagenomics with confirmatory techniques like qPCR can improve reliability. Additionally, the introduction of portable sequencing platforms, such as Oxford Nanopore's MinION, has facilitated real-time pathogen detection in the field, with results available within hours. For example, metagenomics has been successfully used to detect cassava viral pathogens in Sub-Saharan Africa, enabling timely disease management actions. Despite the high costs and infrastructure requirements that limit its widespread use, ongoing advancements in sequencing technology and computational tools are expected to overcome these barriers, making metagenomics an increasingly integral tool in plant disease diagnostics and management.

6 Case Studies on the Diagnosis of Plant Pathogens Using Metagenomic Techniques

Metagenomics has become a valuable tool for identifying plant pathogens, enabling precise disease diagnosis by sequencing the complete genetic material present in infected plant tissues. Unlike conventional culturing methods, this approach bypasses the requirement for prior knowledge of the pathogen's genetic barcodes, as it does not rely on specific primers for sequencing. However, it is advisable to utilize databases containing fungal pathogen genomic information for effective result interpretation. For instance, Yang et al. (2022) employed metagenomic techniques to identify *Calonectria pseudonaviculata*, the fungal pathogen responsible for boxwood blight. Their analysis revealed that over 9% of the sequencing reads from severely infected plant tissues were associated with *C. pseudonaviculata*, based on data generated using different DNA extraction protocols and bioinformatics tools (Barroso-Bergada et al. 2022). This demonstrates the potential of metagenomics in studying plant fungal pathogens, marking it as a valuable tool for future research. Fungal infections are a major biotic stressor, significantly affecting crop yield and quality in agricultural systems. Unfortunately, molecular-level detection methods for plant-fungus interactions are advancing at a slower pace compared to methods for bacterial and viral pathogens. As mentioned, genomic databases for plant-pathogenic fungi remain less comprehensive. One of the foundational studies in fungal pathogen detection focused on the microbiomes of wheat plants infected by *Zymoseptoria tritici*. Behrens and Fischer (2022) conducted a metagenomic analysis of 450 wheat leaf samples, revealing distinct differences in the microbiota of healthy versus infected tissues. Interestingly, the microbiomes of infected leaves across different wheat cultivars exhibited remarkable similarity. These findings provide insights that can aid in managing *Zymoseptoria. tritici* infections and enhancing wheat health. Despite the increasing application of sequencing technologies in

Table 2 Case studies on the diagnosis of plant pathogens using metagenomic techniques

Crops	Study	Metagenomics techniques	References
Commercial crops			
Sugarcane	Detailed identification and characterization of viruses	si-RNA/VANA based metagenomic approach	Candresse et al. (2014)
Cereals			
Rice	General surveillance	Metagenomics-Ilumina	Franco Ortega et al. (2020)
Wheat	Detection of fungal microorganisms in the wheat phyllosphere.	Metabarcoding using ITS barcodes	Schiro et al. (2019)
Wheat	Identification of fungal microbiota in wheat residues	Metabarcoding	Kerdraon et al. (2019)
Maize, Wheat	To identify *Fusarium* species across different plant hosts	PaCBio SMRT Sequencing	Walder et al. (2017)
Maize	Analysis of fungal microbiota post-harvest	Metabarcoding	Cobo-Diaz et al. (2019)
Fruit crops			
Citrus	Pathogen identification	Shotgun metagenomics, complete pathogen genome sequencing.	
Grape	Identification of fungi in phyllosphere samples.	Metabarcoding	Behrens and Fischer (2022)
Grape	Exploration of the fungal microbiota associated with leaves and soil.	Metagenomics-Ilumina	Cureau et al. (2021)
Grape	Detection of fungal infections on the vine trunk	Metagenomics-Ilumina	Del Frari et al. (2019)
Grapevine	Detection of fungal pathogens associated to grapevine diseases	Metabarcoding	Morales-Cruz et al. (2018)
Banana	Impact of soil microbiota on *fusarium* wilt disease	Metabarcoding	Shen et al. (2018)
Strawberry	Analysis of microbial communities in strawberry cultivation soils with varying yield levels	Amplicon Based Metagenomics	Xu et al. (2015)

metagenomic research, many studies using this approach have primarily identified fungal pathogens as the causal agents. The outcomes of several such studies are summarized in Table 2.

7 Challenges and Limitations

Despite its transformative potential in plant pathogen detection, the metagenomics approach faces several challenges and limitations that must be addressed for broader applicability and efficiency.

7.1 Technical Challenges

1. Data handling and analysis: The massive datasets generated by high-throughput sequencing require advanced computational infrastructure and expertise in bioinformatics. Processing, analyzing, and interpreting these datasets is resource-intensive and time-consuming, often requiring specialized software and algorithms for assembly, alignment, and taxonomic classification.
2. Sequence contamination: Contamination from host DNA, soil microbes, or laboratory conditions can interfere with accurate pathogen identification. The lack of effective protocols to isolate pathogen DNA from plant or environmental samples often results in reduced sensitivity and specificity.
3. Low-abundance pathogens: Detecting low-abundance or latent pathogens remains challenging due to the limited sequencing depth and the overwhelming presence of host and non-target microbial DNA.
4. Reference database limitations: The success of metagenomics relies heavily on the availability of comprehensive and curated reference databases. However, existing databases lack sufficient genomic information for many plant pathogens, particularly fungi and viruses, leading to difficulties in taxonomic resolution and the potential for misclassification.

7.2 Economic and Logistical Barriers

1. High costs: The initial setup costs for sequencing equipment and computational infrastructure, as well as the recurring expenses for reagents and consumables, are prohibitively high for many laboratories in developing regions.
2. Skilled workforce: A shortage of trained personnel in metagenomics techniques and bioinformatics analysis hampers the routine adoption of this technology.
3. Field-level applications: Implementing metagenomics for in-field pathogen detection remains logistically challenging due to the need for portable sequencing tools, stable storage of samples, and real-time data analysis capabilities.

7.3 Need for Standardization

1. Lack of protocol uniformity: Variability in sample collection, DNA extraction, library preparation, and sequencing protocols leads to inconsistent results across studies, limiting the reproducibility and comparability of findings.
2. Validation and benchmarking: Standardized methods for validating metagenomics-based pathogen detection against traditional diagnostic techniques are lacking, raising concerns about reliability and accuracy.

7.4 Regulatory and Ethical Concerns

1. Data privacy and security: The generation of genetic data from environmental and crop samples raises concerns about data security and intellectual property, especially in collaborative or cross-border research.
2. Regulatory compliance: Adhering to biosafety and bioethics regulations when working with genetically diverse and potentially harmful pathogens adds to the complexity of metagenomics applications.

8 Future Prospects

As sequencing technologies evolve, the integration of high-throughput sequencing with artificial intelligence (AI) and machine learning (ML) will enhance data analysis, facilitating the identification of novel pathogens and providing insights into pathogen evolution. Portable diagnostic tools, like the Oxford Nanopore MinION, will enable real-time, on-site detection, particularly in resource-limited areas. Significant advances are expected from improved genomic reference databases for plant pathogens, which will improve the accuracy of metagenomic studies. This progress will support predictive agriculture by enabling early detection of disease outbreaks, based on plant and soil microbiomes, leading to timely interventions. Additionally, metagenomics will enhance our understanding of plant-microbe interactions, contributing to the development of microbial biocontrol agents and bioinoculants for sustainable farming. For broader accessibility, reducing sequencing costs and standardizing sample collection, data processing, and interpretation protocols are essential for consistency and reproducibility. Addressing ethical, biosafety, and data privacy concerns will be crucial in the responsible application of metagenomics. With climate change influencing pathogen dynamics, metagenomics will play a key role in monitoring shifts and formulating strategies to mitigate their impact on crops. Collaboration across disciplines like plant pathology, bioinformatics, molecular biology, and agricultural engineering will drive innovation. Metagenomic next-generation sequencing (mNGS) techniques hold promise for species identification, genotyping, metabolic pathway analysis, and microbiota profiling.

9 Conclusion

Metagenomics has emerged as a transformative tool in plant pathogens detection. Unlike traditional detection techniques that often rely on prior knowledge or culturing of microorganisms, metagenomics offers a comprehensive, unbiased approach to identify both known and previously unrecognized pathogens. This technique is not only valuable for pathogen detection but also enriches our understanding of microbial diversity, pathogen-host interactions, and the functional roles of microbes. While metagenomics holds immense promise, several challenges remain, including high costs, the need for specialized expertise, and gaps in available microbial databases. Nevertheless, with continuous advancements in sequencing technology, bioinformatics, and portable diagnostic platforms, metagenomics is becoming more accessible and cost-effective. By overcoming these barriers, this technique is poised to revolutionize plant disease diagnosis and management, leading to more accurate early detection and facilitating the development of integrated disease management approaches. In the future, combining metagenomics with precision agriculture, real-time diagnostics, and machine learning could pave the way for proactive, more efficient control of plant diseases, contributing to global food security and environmental sustainability.

Conflict of Interest The authors declare no conflict of interest.

References

Abarenkov K, Nilsson RH, Larsson KH et al (2010) The UNITE database for molecular identification of fungi recent updates and future perspectives. New Phytol 186:281–285

Abdelfattah A, Sanzani SM, Wisniewski M et al (2019) Revealing cues for fungal interplay in the plant–air interface in vineyards. Front Plant Sci 10:922

Adams I, Fox A (2016) Diagnosis of plant viruses using next-generation sequencing and metagenomic analysis. In: Wang A, Zhou X (eds) Current research topics in plant virology. Springer International Publishing

Adams IP, Glover RH, Monger WA, Mumford R, Jackeviciene E, Navalinskiene M, Samuitiene M, Boonham N (2009) Next-generation sequencing and metagenomic analysis: a universal diagnostic tool in plant virology. Mol Plant Pathol 10(4):537–545

Amarasinghe SL, Su S, Dong X et al (2020) Opportunities and challenges in long-read sequencing data analysis. Genome Biol 21(1):30

Anderson PK, Cunningham AA, Patel NG et al (2004) Emerging infectious diseases of plants: pathogen pollution, climate change and agrotechnology drivers. Trends Ecol Evol 19:535–544

Aragona M, Haegi A, Valente MT et al (2022) New-generation sequencing technology in diagnosis of fungal plant pathogens: a dream comes true? J Fungi 8(7):737

Aravind R, Kumar A, Eapen SJ, Ramana KV (2009) Endophytic bacterial flora in root and stem tissues of black pepper (*Piper nigrum* L.) genotype: isolation, identification and evaluation against Phytophthora capsici. Lett Appl Microbiol 48:58–64. https://doi.org/10.1111/lam.2009.48.issue-1

Ayling M, Clark MD, Leggett RM (2020) New approaches for metagenome assembly with short reads. Brief Bioinform 21(2):584–594

Barroso-Bergada D, Massot M, Vignolles N et al (2022) Metagenomic next-generation sequencing (mNGS) data reveals the phyllosphere microbiome of wheat plants infected by the fungal pathogen *Zymoseptoria tritici*. Phytobiomes J 7(2):281–287

Baysal-Gurel F, Kabir MN (2019) Evaluation of fungicides and biocontrol products for the control of phytophthora root rot of hydrangeas. Arch Phytopathol Plant Protect 52:481–496

Behrens FH, Fischer M (2022) Evaluation of different phyllosphere sample types for parallel metabarcoding of Fungi and Oomycetes in *Vitis vinifera*. Phytobiomes J 6:207–213

Boykin LM, Sseruwagi P, Alicai T et al (2019) Tree Lab: portable genomics for early detection of plant viruses and pests in Sub-Saharan Africa. Genes 10(9):632

Brady A, Salzberg SL, Phymm BL (2009) Metagenomic phylogenetic classification with interpolated Markov models. Nat Methods 6:673–676

Burdon JJ, Zhan J (2020) Climate change and disease in plant communities. PLoS Biol 18:e3000949

Candresse T, Filloux D, Muhire B, Julian C, Galzi S, Fort G et al (2014) Appearances can be deceptive: revealing a hidden viral infection with deep sequencing in a plant quarantine context. PLoS One 9(7):e102945. https://doi.org/10.1371/journal.pone.0102945

Català S, Pérez-Sierra A, Abad-Campos P (2015) The use of genus-specific amplicon pyrosequencing to assess Phytophthora species diversity using eDNA from soil and water in northern Spain. PLoS One 10:e0119311

Chen L, Ren Y, Zhang Y et al (2012) Genome-wide profiling of novel and conserved Populus microRNAs involved in pathogen stress response by deep sequencing. Planta 235:873–883

Chomczynski P, Rymaszewski M (2006) Alkaline polyethylene glycol-based method for direct PCR from bacteria, eukaryotic tissue samples, and whole blood. BioTechniques 40:454–458

Chopra RS, Chopra C, Sharma NR (eds) (2020) Metagenomics: techniques, applications, challenges and opportunities. Springer, Singapore

Cobo-Diaz JF, Baroncelli R, Le Floch G et al (2019) Combined metabarcoding and co-occurrence network analysis to profile the bacterial, fungal and fusarium communities and their interactions in maize stalks. Front Microbiol 10:261

Cottrell MT, Waidner LA, Yu L, Kirchman DL (2005) Bacterial diversity of metagenomic and PCR libraries from the Delaware River. Environ Microbiol 7(12):1883–1895

Cureau N, Threlfall R, Savin M et al (2021) Year, location, and variety impact on grape-, soil-and leaf-associated fungal microbiota of Arkansas-grown table grapes. Microb Ecol 82:73–86

Cusco A, Catozzi C, Vines J et al (2019) Microbiota profiling with long amplicons using nanopore sequencing: full-length 16S rRNA gene and the 16S-ITS-23S of the rrn operon. F1000Res 7:1755

Daniel R (2005) The metagenomics of soil. Nat Rev Microbiol 3(6):470–478

Del Frari G, Gobbi A, Aggerbeck MR et al (2019) Characterization of the wood mycobiome of *Vitis vinifera* in a vineyard affected by esca. Spatial distribution of fungal communities and their putative relation with leaf symptoms. Front Plant Sci 10:910

Erwin DC, Ribeiro OK (1996) Phytophthora diseases worldwide. American Phytopathological Society (APS Press), St. Paul

Franco Ortega S, Ferrocino I, Adams I et al (2020) Monitoring and surveillance of aerial mycobiota of rice paddy through DNA metabarcoding and qPCR. J Fungi 6:372

Franzosa EA, Hsu T, Sirota-Madi A et al (2015) Sequencing and beyond: integrating molecular 'omics' for microbial community profiling. Nat Rev Microbiol 13(6):360–372

Gokdemir FS, Iseri OD, Sharma A et al (2022) Metagenomics next generation sequencing (mNGS): an exciting tool for early and accurate diagnostic of fungal pathogens in plants. J Fungi 8:1195

Jiang H, Xing Z, Liu X et al (2022) Comparison and development of a metagenomic next-generation sequencing protocol for combined detection of DNA and RNA pathogens in cerebrospinal fluid. BMC Infect Dis 22:1–10

Johnson MA, Liu H, Bush E et al (2022) Investigating plant disease outbreaks with long-read metagenomics: sensitive detection and highly resolved phylogenetic reconstruction applied to *xylella fastidiosa*. Microbial Genomics 8(5):822

Kerdraon L, Barret M, Laval V et al (2019) Differential dynamics of microbial community networks help identify microorganisms interacting with residue-borne pathogens: the case of *Zymoseptoria tritici* in wheat. Microbiome 7:1–17

Kim M, Lee KH, Yoon SW et al (2013) Analytical tools and databases for metagenomics in the next-generation sequencing era. Genom Inform 11:102

Lappe RR, Elmore MG, Lozier ZR et al (2022) Metagenomic identification of novel viruses of maize and teosinte in north America. BMC Genomics 23(1):767

Lara-Victoriano F, Castillo-Reyes F, Flores-Gallegos C, Aguilar CN, Rodríguez-Herrera R (2011) Metagenomics in plant pathology. In: Herrera RR, Aguilar CN, Simpson-Williamson JK, Sanchez GG (eds) Phytopathology in the omics era. Research Signpost Publishing, 000-000 ISBN: 978-81-308-0438-5

Lau HY, Wu H, Wee EJH et al (2017) Specific and sensitive isothermal electrochemical biosensor for plant pathogen DNA detection with colloidal gold nanoparticles as probes. Sci Rep 7:1–7

Lewis JA, Papavizas GC (1991) Biocontrol of cotton damping-off caused by *Rhizoctonia solani* in the field with formulations of *Trichoderma* spp. and *Gliocladium virens*. Crop Prot 10:396–402

Lind AL, Pollard KS (2021) Accurate and sensitive detection of microbial eukaryotes from whole metagenome shotgun sequencing. Microbiome 9(1):58

Loit K, Adamson K, Bahram M et al (2019) Relative performance of MinION (Oxford nanopore technologies) versus sequel (Pacific biosciences) third-generation sequencing instruments in identification of agricultural and forest fungal pathogens. Appl Environ Microbiol 85(21):e01368–e01319

Miller S, Chiu C (2022) The role of metagenomics and next-generation sequencing in infectious disease diagnosis. Clin Chem 68:115–124

Molina LG, Cordenonsi da Fonseca G, Morais GL, de Oliveira LFV, Carvalho JB, Kulcheski FR, Margis R (2012) Metatranscriptomic analysis of small RNAs present in soybean deep sequencing libraries. Genet Mol Biol 35(1):292–303

Morales-Cruz A, Figueroa-Balderas R, Garcia JF et al (2018) Profiling grapevine trunk pathogens in planta: a case for community-targeted DNA metabarcoding. BMC Microbiol 18:1–14

Mzava O, Cheng AP, Chang A et al (2022) A metagenomic DNA sequencing assay that is robust against environmental DNA contamination. Nat Commun 13:1–10

Notomi T, Okayama H, Masubuchi H et al (2000) Loop-mediated isothermal amplification of DNA. Nucleic Acids Res 28:e63

Panth M, Hassler SC, Baysal-Gurel F (2020) Methods for management of soilborne diseases in crop production. Agriculture 10:16

Peng Y, Leung HC, Yiu SM, Chin FY (2011) Meta-IDBA: a de Novo assembler for metagenomic data. Bioinformatics 27(13):i94–i101

Piombo E, Abdelfattah A, Droby S et al (2021) Metagenomics approaches for the detection and surveillance of emerging and recurrent plant pathogens. Microorganisms 9(1):188

Porras-Alfaro A, Bayman P (2011) Hidden fungi, emergent properties: endophytes and microbiomes. Annu Rev Phytopathol 49:291–315. https://doi.org/10.1146/annurev-phyto-080508-081831

Quince C, Walker AW, Simpson JT et al (2017) Shotgun metagenomics, from sampling to analysis. Nat Biotechnol 35(9):833–844

Ravisankar V, Mathew D (2022) Multiomics bioinformatics approaches in horticultural crops. Academic Press, Omics in horticultural crops, pp 27–54

Rosen MJ, Callahan BJ, Fisher DS et al (2012) Denoising PCR-amplified metagenome data. BMC Bioinform 13:1–16

Ruppert KM, Kline RJ, Rahman MS (2019) Past, present, and future perspectives of environmental DNA (eDNA) metabarcoding: a systematic review in methods, monitoring, and applications of global eDNA. Glob Ecol Conserv 17:e00547

Sanzani SM, Li MG, Faedda R et al (2014) Use of quantitative PCR detection methods to study biocontrol agents and phytopathogenic fungi and oomycetes in environmental samples. J Phytopathol 162:1–13

Schena L, Nigro F, Ippolito A (2004) Real-time PCR detection and quantification of soilborne fungal pathogens: the case of *Rosellinia necatrix*, *Phytophthora nicotianae*, *P. citrophthora*, and *Verticillium dahliae*. Phytopathol Mediterr 43:273–280

Schiro G, Colangeli P, Müller ME (2019) A metabarcoding analysis of the mycobiome of wheat ears across a topographically heterogeneous field. Front Microbiol 10:2095

Semenov MV (2021) Metabarcoding and metagenomics in soil ecology research: achievements, challenges, and prospects. Biol Bull Rev 11:40–53

Sharpton TJ (2014) An introduction to the analysis of shotgun metagenomic data. Front Plant Sci 5:209

Shen Z, Penton CR, Lv N et al (2018) Banana *Fusarium* wilt disease incidence is influenced by shifts of soil microbial communities under different monoculture spans. Microb Ecol 75:739–750

Singh BK, Delgado-Baquerizo M, Egidi E et al (2023) Climate change impacts on plant pathogens, food security and paths forward. Nat Rev Microbiol 21(10):640–656

Spadaro D, Agustí N, Ortega SF et al (2020) Diagnostics and identification of diseases, insects and mites. In: Integrated pest and disease management in greenhouse crops. Springer International Publishing, Cham

Walder F, Schlaeppi K, Wittwer R et al (2017) Community profiling of *Fusarium* in combination with other plant-associated fungi in different crop species using SMRT sequencing. Front Plant Sci 8:2019

Ward J, Gilmore SR, Robertson J et al (2009) A grass molecular identification system for forensic botany: a critical evaluation of the strengths and limitations. J Forensic Sci 54:1254–1260

Xu X, Passey T, Wei F et al (2015) Amplicon-based metagenomics identified candidate organisms in soils that caused yield decline in strawberry. Hortic Res 2:15022

Yang S, Johnson MA, Hansen MA et al (2022) Metagenomic sequencing for detection and identification of the boxwood blight pathogen *Calonectria pseudonaviculata*. Sci Rep 12:1–14

Yao KS, Li SJ, Tzeng KC et al (2009) Fluorescence silica nanoprobe as a biomarker for rapid detection of plant pathogens. Adv Mater Res 79:513–516

Integrating Proteo-metabolomics Systems Biology Approach for Molecular Insights Driving Sustainable Crop Improvement

Dhananjaya Pratap Singh, Sudarshan Maurya, Ratna Prabha, Lovkush Satnami, and Nagendra Rai

1 Introduction

Modern crop improvement strategies are increasingly embracing multi-omics approaches, with proteo-metabolomics emerging as a crucial tool in understanding and modifying plant traits for improved performance (Yan et al. 2022). This integration of proteomics and metabolomics provides unprecedented insights into plant biological systems, offering a comprehensive view of cellular processes and their regulation that was previously unattainable through single-omics approaches. Proteo-metabolomics represents the convergence of two powerful analytical approaches: proteomics, which studies the complete protein complement of an organism, and metabolomics, which examines the entire range of metabolites present in biological systems (Baharum and Azizan 2018). In plant science, this integrated approach has revolutionized our understanding of complex biological processes, from development to stress responses, providing deeper insights into the molecular mechanisms underlying important crop traits. The field of plant proteomics has undergone significant evolution since its inception, progressing from simple protein profiling to sophisticated quantitative analyses that can simultaneously identify and quantify thousands of proteins, analyze post-translational modifications, study protein-protein interactions, and investigate subcellular protein localization (Yan et al. 2022). These capabilities have proven impact in crop improvement, allowing researchers to identify key proteins involved in desired traits such as stress tolerance, yield, and nutritional quality (Kausar et al. 2022). Modern

D. P. Singh (✉) · S. Maurya · L. Satnami · N. Rai
ICAR-Indian Institute of Vegetable Research, Shahanshahpur, Varanasi, India
e-mail: dhananjaya.singh@icar.org.in

R. Prabha
Agricultural Knowledge Management Unit (AKMU), ICAR-Indian Agricultural Research Institute, AKMU Division, New Delhi, Delhi, India

Renu et al. (eds.), *Meta-omics in Crop Improvement*,
https://doi.org/10.1007/978-981-95-2294-1_8

proteomic approaches have enabled the development of more precise breeding strategies by identifying specific protein markers associated with beneficial traits.

Complementing proteomics, plant metabolomics has emerged as an essential component in understanding crop biology and improvement. This approach provides critical insights into primary metabolism regulation, secondary metabolite production, stress response mechanisms, and nutritional quality determinants (Raza et al. 2024). The integration of metabolomic data with proteomic information has revealed complex regulatory networks and metabolic pathways crucial for crop improvement, enabling researchers to better understand the biochemical basis of important agricultural traits. The synergistic application of proteomics and metabolomics has enabled comprehensive pathway analysis, detailed characterization of plant responses to various stresses, and identification of key factors controlling nutritional and organoleptic properties (Katam et al. 2022; Martínez-Esteso et al. 2024). This integrated approach has also significantly enhanced breeding applications through the development of novel markers and selection criteria (Yan et al. 2022). Recent technological advances in mass spectrometry, chromatography, and bioinformatics have substantially improved the capabilities of proteo-metabolomic analyses, allowing for more sophisticated and detailed studies of plant systems (Rajczewski et al. 2022; Singh et al. 2022).

In stress response studies, proteo-metabolomic approaches have been particularly valuable in elucidating the complex mechanisms plants employ to cope with various environmental challenges. Research has shown that stress responses involve coordinated changes in both protein expression and metabolite accumulation, providing multiple targets for crop improvement (Bulut et al. 2025). Similarly, in quality enhancement studies, these approaches have identified key molecular factors controlling nutritional content and flavor components, leading to more targeted breeding strategies for improved crop varieties (Martinez-Esteso et al. 2024). The continued evolution of proteo-metabolomics clubbed with the chemoinformatics technologies promises even greater contributions to crop improvement through enhanced analytical capabilities, improved data integration methods, and advanced computational tools (Mashabela et al. 2022). High-throughput platforms now allow for the simultaneous analysis of thousands of proteins and metabolites, providing unprecedented insights into plant biological systems and their responses to various environmental conditions and developmental cues (Allwood et al. 2008; Roychowdhury et al. 2023). These developments are expected to accelerate the pace of crop improvement and contribute significantly to global food security.

2 Historical Development and Current Status

The journey of proteo-metabolomics in crop science spans several decades, marked by significant technological advances and evolving research paradigms. The foundations were laid in the 1970s with the development of two-dimensional gel electrophoresis, which enabled the first comprehensive analyses of plant proteins (O'Farrell

2008). This breakthrough technique, although time-consuming and labor-intensive, provided the first glimpses into the complexity of plant proteomes and remained the primary tool for protein separation and visualization for many years. The development of soft ionization methods, particularly electrospray ionization (ESI) and matrix-assisted laser desorption/ionization (MALDI), revolutionized protein analysis by enabling the identification of proteins from complex biological samples (Gelpí 2009; Di Girolamo et al. 2013). These advances, coupled with the completion of various plant genome sequences, particularly *Arabidopsis thaliana* in 2000, provided the essential framework for modern plant proteomics (Bevan and Walsh 2005).

Parallel to proteomics developments, plant metabolomics emerged as a distinct field in the late 1990s and early 2000s. The initial focus was on primary metabolites, utilizing gas chromatography-mass spectrometry (GC-MS) techniques (Fiehn 2002; Dettmer et al. 2007). The field expanded rapidly with the introduction of liquid chromatography-mass spectrometry (LC-MS) systems, which enabled the analysis of a broader range of metabolites, including secondary compounds crucial for crop quality and stress responses (Akyol et al. 2023). Integration of proteomics and metabolomics in crop science began taking shape in the early 2000s, driven by the recognition that single-omics approaches provided incomplete pictures of biological systems (Resurreccion and Fong 2022). Early studies demonstrated the power of combining proteomic and metabolomic data to understand complex traits in crops (Wienkoop et al. 2008). This integration was further facilitated by advances in bioinformatics, databases, and analysis tools, that enabled meaningful interpretation of large-scale, multi-omics datasets (Vasupalli et al. 2024).

Current status reflects remarkable technological and methodological advances. Modern proteomic approaches utilize high-resolution mass spectrometry (HR-MS), enabling the identification and quantification of thousands of proteins in a single experiment (Ramalingam et al. 2015). Targeted and un-targeted metabolomics approaches have become increasingly sophisticated, allowing comprehensive profiling of both primary and secondary metabolites (Singh et al. 2022). Integration of these datasets has been enhanced by machine learning and artificial intelligence approaches, providing deeper insights into crop biological systems (Cembrowska-Lech et al. 2023). Recent developments include the emergence of spatial proteomics and metabolomics, enabling the study of protein and metabolite distributions within plant tissues and organs (Liu et al. 2021b). Single-cell approaches are beginning to reveal cell-type specific proteo-metabolic profiles, providing unprecedented resolution in understanding crop development and stress responses (Hu et al. 2023). Additionally, the application of CRISPR-Cas9 technology in conjunction with proteo-metabolomics has opened new avenues for precision crop improvement (Liu et al. 2021a). The field has also witnessed significant advances in sample preparation and analysis techniques. Improved protein extraction methods, particularly for recalcitrant plant tissues, have enhanced coverage of membrane proteins and other challenging protein classes (Yan et al. 2022). Similarly, metabolite extraction and detection methods have become more comprehensive, enabling better coverage of the plant metabolome (Salem et al. 2020).

3 Proteo-metabolomics: Significance in Crop Improvement

Proteo-metabolomics has emerged as an indispensable tool in modern crop improvement strategies, offering unprecedented insights into plant biological systems and enabling more precise and effective breeding approaches. The significance of this integrated approach lies in its ability to bridge the gap between genotype and phenotype, providing detailed molecular information that directly influences crop performance and quality. This understanding has become particularly crucial as agriculture faces mounting challenges from climate change, population growth, and the need for sustainable production systems. In the realm of stress tolerance improvement, proteo-metabolomic approaches have revolutionized our understanding of plant responses to both biotic and abiotic stresses. These techniques have enabled the identification of key protein markers and metabolic signatures associated with drought tolerance, salt resistance, and disease resistance (Yan et al. 2022). For instance, recent studies using integrated proteo-metabolomic analyses have revealed novel stress-responsive pathways in wheat, leading to the development of more resilient varieties better adapted to changing climatic conditions (Naik et al. 2023).

The application of proteo-metabolomics in yield enhancement has provided crucial insights into the molecular mechanisms governing crop productivity. By analyzing protein-metabolite networks involved in photosynthesis, nutrient uptake, and resource allocation, researchers have identified critical targets for yield improvement (Kumar et al. 2017). This understanding of protein-protein and metabolite-metabolite networks has led to the development of crops with enhanced nutrient use efficiency and improved source-sink relationships, directly contributing to increased yields under the regime of emerging threats due to climate change (Smith et al. 2018). Nutritional quality improvement represents another significant area where proteo-metabolomic approaches have made substantial contributions. These techniques have enabled the identification and manipulation of key pathways involved in the synthesis and accumulation of essential nutrients, vitamins, and beneficial compounds in crops (Marchev et al. 2021). The ability to track changes in protein expression and metabolite profiles during crop development has facilitated the breeding of varieties with enhanced nutritional value, addressing global nutrition security challenges (Zenda et al. 2021; Khan et al. 2024).

In modern breeding programs, proteo-metabolomic data has become instrumental in developing more efficient selection strategies. The integration of these molecular markers with traditional breeding approaches has accelerated the development of improved varieties with desired traits (Heazlewood et al. 2022). This has been particularly valuable in cases where traditional phenotypic selection is challenging or time-consuming, such as in developing stress-tolerant varieties or improving complex quality traits. The significance of proteo-metabolomics extends to crop protection strategies as well. These approaches have enhanced our understanding of plant-pathogen interactions and defense mechanisms, leading to the development of more effective and sustainable crop protection methods (Singh et al. 2023a, b). By identifying key proteins and metabolites involved in disease resistance, researchers have been able to develop crops with enhanced natural defense capabilities, reducing the reliance on chemical pesticides. Post-harvest quality improvement has also benefited significantly from proteo-metabolomic insights. Understanding the molecular changes occurring during crop storage and processing has led to improved preservation methods and enhanced product quality (Marchev et al. 2021). This has particular relevance for reducing post-harvest losses and maintaining nutritional value during storage, contributing to food security objectives. Modern crop improvement strategies increasingly rely on integrating proteo-metabolomic data with other omics approaches and traditional breeding methods. This integration has enabled more comprehensive understanding of crop traits and more precise breeding strategies (Pott et al. 2021). The ability to combine multiple layers of molecular information has proven particularly valuable in addressing complex traits that involve multiple genes and pathways.

4 Fundamentals of Proteo-metabolomics

The fundamental aspects of proteo-metabolomics encompass a complex interplay of proteins and metabolites that govern plant growth, development, and responses to environmental conditions. Understanding these fundamentals is crucial for their effective application in crop improvement strategies. At its core, proteo-metabolomics integrates two distinct but complementary approaches: the study of the complete protein complement (proteome) and the analysis of small molecule metabolites (metabolome) within biological systems (Fig. 1) (Huang et al. 2018). Proteomics in crop science focuses on the systematic analysis of protein expression, post-translational modifications, protein-protein interactions, and protein localization patterns. This approach provides crucial insights into the functional molecules that execute and regulate cellular processes. Modern proteomic analyses employ various techniques, with mass spectrometry serving as the cornerstone technology. The workflow typically involves protein extraction, separation, identification, and quantification steps, each requiring specific methodological considerations for plant tissues (Kumar et al. 2017).

Protein extraction from plant tissues presents unique challenges due to the presence of rigid cell walls, high concentrations of proteases, and abundant secondary metabolites. Successful extraction requires careful optimization of protocols to ensure comprehensive protein recovery while minimizing interference from non-protein components. Various extraction methods have been developed, including sequential extraction procedures that enable access to different protein fractions, from soluble cytosolic proteins to membrane-bound proteins (Yan et al. 2022). Separation of complex protein mixtures traditionally relies on gel-based approaches such as two-dimensional gel electrophoresis (2-DE) or more modern gel-free techniques like liquid chromatography (LC). These separation methods are crucial for reducing sample complexity and enabling more accurate protein identification (Wijethunga and Emenike 2024). Protein identification and characterization employ

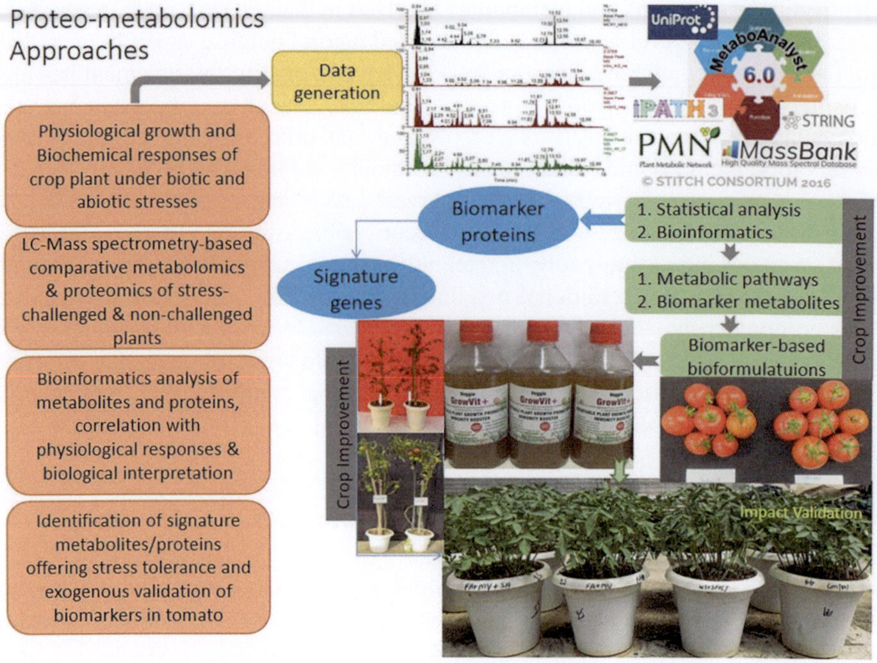

Fig. 1 Integrated proteo-metabolomics workflow for crop improvement and stress response analysis. This scheme illustrates comprehensive proteo-metabolomics pipeline used to study plant stress responses and identify biomarkers for crop improvement while investigating physiological growth and biochemical responses of crops under biotic and abiotic stresses. Liquid chromatography mass spectrometry technique is employed for comparative metabolomics and proteomics analysis of stressed and control plants, followed by bioinformatics analysis to correlate metabolites and proteins with physiological responses. The central components show data generation leading into the identification of signature genes and biomarker proteins and metabolites. Analytical approaches include statistical analysis, bioinformatics, metabolic pathway assessment, and biomarker metabolite identification. The right side shows biomarker-based bioformulations and impact validation with experimental plants demonstrating the practical application of identified markers. The figure incorporates database resources (PATH, STRING, PMN, MassBank) that support the analytical framework, culminating in validated biomarkers for tomato and other crops under stress conditions

mass spectrometry-based approaches, particularly tandem mass spectrometry (MS/MS). These techniques enable both the identification of proteins through peptide sequencing and the quantification of protein abundance. Recent advances in MS technology have dramatically improved the depth of proteome coverage and the accuracy of protein quantification. The metabolomics component of this integrated approach focuses on the comprehensive analysis of small molecule metabolites, including primary metabolites (such as sugars, amino acids, and organic acids) and secondary metabolites (including flavonoids, alkaloids, and other specialized compounds). Metabolomic analyses employ various analytical platforms, with gas chromatography-mass spectrometry (GC-MS) and liquid chromatography-mass spectrometry (LC-MS) being the most widely used (Zhan et al. 2022). Sample preparation for metabolomic analysis requires careful consideration of metabolite stability and extraction efficiency. Different extraction protocols may be necessary for various classes of metabolites, and the choice of extraction method can significantly impact the coverage and reliability of metabolomic analyses. The development of standardized protocols has been crucial for ensuring reproducibility and comparability across studies.

Data analysis and integration represent critical aspects of proteo-metabolomic studies. The large-scale data generated by these approaches require sophisticated bioinformatics tools and statistical methods for meaningful interpretation. Modern data analysis pipelines incorporate machine learning and artificial intelligence approaches to identify patterns and relationships within complex datasets (Hu et al. 2015; Mann et al. 2021). Integration of proteomic and metabolomic data provides a more comprehensive understanding of biological systems than either approach alone. This integration reveals functional relationships between proteins and metabolites, helping to elucidate regulatory networks and metabolic pathways. Network analysis approaches have become increasingly important for understanding these complex interactions and their implications for crop improvement (Zhan et al. 2022).

5 Proteomics: Basic Concepts

Proteomics represents the large-scale study of proteins, encompassing their expression, structure, functions, modifications, and interactions within biological systems. In crop science, proteomics has become an essential tool for understanding the molecular mechanisms underlying plant growth, development, and stress responses. The fundamental concepts are built upon the central dogma of molecular biology, where genes are transcribed to RNA and subsequently translated into proteins, which serve as the functional molecules in cells (Yan et al. 2022). Plant proteome is highly dynamic and complex, responding continuously to environmental conditions and developmental signals. Unlike the relatively static genome, the proteome exhibits temporal and spatial variations, with protein expression patterns changing throughout plant development and in response to various stimuli. This dynamic nature makes proteomics particularly valuable for understanding plant responses to environmental changes and stress conditions (Mergner and Kuster 2022). Protein synthesis and modification in plants involve several key processes that influence proteomic analyses. Following translation, proteins undergo various post-translational modifications (PTMs) that can significantly alter their function, localization, and stability. Common PTMs in plants include phosphorylation, glycosylation, ubiquitination, and SUMOylation, each playing crucial roles in protein regulation and cellular signaling (Muleya et al. 2022; Yan et al. 2022). Understanding these modifications is essential for comprehending protein function and regulation in crop improvement contexts.

The complexity of plant proteomes is further increased by alternative splicing of mRNA transcripts, which can generate multiple protein isoforms from a single gene. This phenomenon contributes to proteome diversity and functional complexity, enabling plants to fine-tune their responses to various environmental conditions. Recent studies have shown that alternative splicing plays crucial roles in crop stress responses and development (Reddy et al. 2013). Protein-protein interactions form another fundamental aspect of proteomics. These interactions create functional networks that regulate various cellular processes, from signal transduction to metabolic pathways. Modern proteomic approaches enable the identification and characterization of these interaction networks, providing insights into protein function and regulation in crop systems (Wu et al. 2025). Subcellular localization of proteins is critical for their function and regulation. Plants have complex cellular organizations with multiple compartments, each containing specific protein complements. Understanding protein localization patterns helps elucidate protein function and regulatory mechanisms. Advanced proteomic techniques now enable the systematic analysis of protein distribution across cellular compartments (Hein et al. 2024).

Protein turnover and degradation represent important regulatory mechanisms in plant cells. The balance between protein synthesis and degradation influences cellular protein composition and function. Proteomic approaches can measure protein abundance and turnover rates, providing insights into protein stability and regulation under various conditions (Liu et al. 2019). Quantitative aspects of proteomics are essential for understanding protein expression patterns and their changes under different conditions. Various approaches exist for protein quantification, including label-free and isotope-labeled methods. These techniques enable the measurement of relative and absolute protein abundances, crucial for comparing protein expression across different conditions or treatments (Rozanova et al. 2021). The structural aspects of proteins, including their three-dimensional conformations and domains, are fundamental to their function. Structural proteomics approaches provide insights into protein architecture and how structural changes influence protein function. This understanding is particularly relevant for engineering improved crop traits through protein modification (Beveridge and Calabrese 2021).

6 Proteome Analysis Techniques

Proteome analysis in crop science employs a diverse array of techniques, each offering specific advantages for different analytical objectives. The foundation of these analyses begins with protein extraction, a critical step that significantly influences the quality and comprehensiveness of subsequent analyses. Plant tissues present unique challenges for protein extraction due to their rigid cell walls, high levels of proteases, and abundant secondary metabolites that can interfere with protein analysis (Padula et al. 2017). Sample preparation typically involves tissue homogenization followed by protein extraction using appropriate buffers. The choice of extraction method depends on the target proteins and tissue type. Sequential extraction procedures have been developed to access different protein fractions, from easily soluble proteins to membrane-bound proteins. Recent advances include the development of pressure cycling technology and ultrasound-assisted extraction methods, which have improved protein recovery from recalcitrant plant tissues (Duong and Lee 2023).

Protein separation techniques form the next crucial step in proteome analysis. 2-DE remains a valuable tool, particularly for comparing protein expression patterns across different conditions. In 2-DE, proteins are separated based on their isoelectric points in the first dimension and molecular weights in the second dimension. Despite its limitations, such as poor resolution of hydrophobic and extreme pI proteins, 2-DE provides a visual representation of the proteome and is particularly useful for detecting post-translational modifications (Magdeldin et al. 2014). Gel-free approaches, particularly liquid chromatography (LC), have revolutionized proteome analysis (Yan et al. 2022). Various LC techniques, including reversed-phase, ion exchange, and size exclusion chromatography, can be used individually or in combination to achieve better separation of complex protein mixtures. Multi-dimensional protein identification technology (MudPIT) combines different separation modes to enhance proteome coverage significantly (Lohrig and Wolters 2009). Mass spectrometry (MS) serves as the cornerstone of modern proteome analysis. Two main ionization techniques, electrospray ionization (ESI) and matrix-assisted laser desorption/ionization (MALDI), are commonly used. ESI is particularly suitable for coupling with LC systems (LC-MS/MS), while MALDI is often used for targeted protein analysis and imaging applications. Recent advances in MS instrumentation, such as orbitrap and ion mobility MS, have dramatically improved sensitivity and resolution (Birhanu 2023). Quantitative proteomics has become increasingly important in crop research. Label-free quantification methods rely on spectral counting or intensity-based measurements and are relatively straightforward to implement. Isotope-labeled approaches, including chemical labeling (e.g., iTRAQ, TMT) and metabolic labeling (SILAC), provide more accurate quantification but may be more challenging to apply in plant systems (Jorrin-Novo 2014; Liu et al. 2019). The choice of quantification method depends on the experimental goals, sample complexity, and required precision (Aryal et al. 2014).

Protein identification typically employs database searching algorithms that match experimental MS/MS spectra with theoretical spectra generated from protein sequence databases. The availability of high-quality genome sequences for many crop species has greatly improved protein identification accuracy (Chapman et al. 2013). De novo sequencing approaches are also valuable, particularly for non-model crops lacking comprehensive sequence databases (Johnson et al. 2020). Post-translational modification (PTM) analysis requires specialized techniques due to the often substoichiometric nature of modified proteins. Enrichment strategies, such as immunoaffinity purification or chemical tagging, are commonly employed to concentrate modified proteins or peptides before analysis (Bobalova et al. 2023). Advanced MS techniques, including electron transfer dissociation (ETD), provide improved characterization of PTM sites (Kim and Pandey 2012). These techniques have revealed complex interaction networks in crop plants, providing insights into protein function and regulation.

7 Protein Identification and Characterization Methods

Protein identification and characterization in crop proteomics has evolved significantly with technological advancements, particularly in mass spectrometry (MS) and bioinformatics. These methods are crucial for understanding protein function, regulation, and modification in crop systems. Modern approaches combine sophisticated analytical techniques with powerful computational tools to provide comprehensive protein characterization (Szabo and Janaky 2015). Mass spectrometry-based protein identification represents the cornerstone of modern proteomics. The process typically begins with protein digestion, usually using trypsin, which generates peptides suitable for MS analysis. Bottom-up proteomics, where proteins are identified through peptide analysis, is the most common approach. The resulting peptides are analyzed using tandem mass spectrometry (MS/MS), which provides both peptide mass and sequence information through fragmentation patterns (Wu et al. 2025).

Database searching is fundamental to protein identification. MS/MS spectra are compared against theoretical spectra generated from protein sequence databases using various search algorithms. Popular search engines like Mascot, SEQUEST, and MS-GF+ employ different algorithms to match experimental spectra with database entries. The availability of high-quality genome sequences for many crop species has significantly improved identification accuracy. However, proper database curation and the inclusion of common contaminants remain crucial for reliable results (Yan et al. 2022). De novo sequencing approaches have become increasingly important, particularly for non-model crops lacking comprehensive sequence databases. These methods determine peptide sequences directly from MS/MS spectra without relying on sequence databases. Advanced algorithms and software tools have improved the accuracy of de novo sequencing, making it a valuable complement to database searching approaches in proteomics (Ma and Johnson 2012). Characterization of proteins extends beyond simple identification to include detailed analysis of protein structure and modifications. High-resolution MS techniques enable the determination of precise protein masses and the identification of sequence variants. Ion mobility MS has added another dimension to protein characterization by providing information about protein structure and conformation (Robb et al. 2023).

Post-translational modification (PTM) analysis represents a critical aspect of protein characterization. Modern MS instruments can identify various PTMs, including phosphorylation, glycosylation, ubiquitination, and acetylation. PTM analysis often requires enrichment strategies to concentrate modified proteins or peptides before analysis. Electron transfer dissociation (ETD) and electron capture dissociation (ECD) have proven particularly valuable for PTM site identification, as they preserve labile modifications during fragmentation (Yan et al. 2022; Wu et al. 2025). Targeted protein characterization employs selective reaction monitoring (SRM) or parallel reaction monitoring (PRM) approaches. These methods provide highly sensitive and specific detection of proteins of interest, making them valuable for validation studies and absolute quantification. The development of spectral libraries has facilitated the design of targeted assays for crop proteins (Deutsch et al.

2018). Structural characterization of proteins employs various complementary techniques. While X-ray crystallography and NMR spectroscopy provide high-resolution structural information, cryo-electron microscopy has emerged as a powerful tool for analyzing protein structure, particularly for large complexes. These structural insights are crucial for understanding protein function and engineering improved crop traits (Sikic et al. 2010; Rasheed et al. 2020).

Characterization of protein-protein interaction (PPI) uses techniques that employ various techniques, from traditional yeast two-hybrid assays to modern approaches like affinity purification-mass spectrometry (AP-MS) and proximity-dependent biotin identification (BioID) (Wu et al. 2025). These methods have revealed complex interaction networks in crop plants, providing insights into protein function and regulation. Cross-linking mass spectrometry (XL-MS) has emerged as a valuable tool for studying protein complex architecture and dynamics (Lee and O'Reilly 2023). Top-down proteomics, where intact proteins are analyzed without prior digestion, is gaining importance in crop research. This approach provides comprehensive characterization of protein forms, including sequence variants and combinations of PTMs. Although technically challenging, top-down analysis offers unique insights into protein proteoforms and their biological significance (Po and Eyers 2023).

8 Quantitative Proteomics Approaches

Quantitative aspects of proteo-metabolomics have become increasingly important in crop science. Various quantification strategies, including label-free and isotope-labeled approaches, enable the measurement of relative and absolute changes in protein and metabolite levels. These quantitative measurements are essential for understanding the magnitude and significance of molecular changes associated with different crop traits or stress responses (Fang et al. 2019). Quantitative proteomics represents a crucial approach in crop science for understanding protein abundance changes under different conditions, developmental stages, and stress responses. The field has evolved from simple comparative analyses to sophisticated methods enabling precise protein quantification. These approaches can be broadly categorized into label-free and label-based methods, each with distinct advantages and applications in crop research.

Label-free quantification (LFQ) has gained significant popularity due to its simplicity and cost-effectiveness. This approach relies on either spectral counting or intensity-based measurements. Spectral counting methods quantify proteins based on the number of identified peptide spectra, assuming that more abundant proteins will generate more spectra. Intensity-based methods utilize peptide ion intensities or peak areas for quantification, offering greater precision but requiring sophisticated data analysis (Arike and Peil 2014). Advanced LFQ approaches incorporate retention time alignment and normalization strategies to improve quantification accuracy. The development of data-independent acquisition (DIA) methods has further enhanced label-free quantification by providing comprehensive peptide fragmentation data and improved reproducibility (Li et al. 2021). Sequential Window Acquisition of all Theoretical Mass Spectra (SWATH-MS) has emerged as a powerful DIA approach, enabling systematic quantification of thousands of proteins across multiple samples (Sidoli et al. 2015). Isotope-labeled approaches offer higher quantification precision through the use of stable isotopes. Metabolic labeling techniques, such as SILAC (Stable Isotope Labeling by Amino acids in Cell culture), while challenging to implement in whole plants, have been successfully adapted for plant cell cultures and tissue-specific studies (Lewandowska et al. 2013). The development of plant-specific SILAC variants has expanded its applicability in crop research.

Chemical labeling methods provide versatile alternatives for quantitative proteomics in crops. Isotope-coded affinity tags (ICAT) enable selective labeling of cysteine-containing peptides, while isobaric tags for relative and absolute quantitation (iTRAQ) and tandem mass tags (TMT) allow multiplexed analysis of multiple samples (Gouw et al. 2010; Butler et al. 2010). These approaches have proven particularly valuable for comparing protein expression across different treatments or developmental stages in crop studies (Chen et al. 2021). Targeted quantitative proteomics employs selected reaction monitoring (SRM) or parallel reaction monitoring (PRM) for precise quantification of specific proteins of interest (Ronsein et al. 2015). These approaches offer high sensitivity and selectivity, making them ideal for validation studies and absolute quantification. The development of crop-specific SRM assay libraries has facilitated the widespread application of these techniques. Absolute quantification strategies often utilize synthetic peptide standards. Absolute quantification strategies often utilize synthetic peptide standards. Absolute QUAntification (AQUA) peptides, which are isotope-labeled versions of target peptides, enable precise absolute quantification of proteins (Kettenbach et al. 2011). Using stable isotope-labeled peptides serving as internal standards, AQUA method facilitates specific and sensitive targeted quantification of protein and post-translational modifications in complex protein mixtures. QconCAT technology, utilizing artificial proteins containing multiple quantification peptides, provides a cost-effective alternative for multiplexed absolute quantification (Scott et al. 2016).

Data analysis represents a critical component of quantitative proteomics. Sophisticated software tools and statistical approaches are required to process the complex datasets generated by these experiments. Machine learning algorithms have become increasingly important for data analysis, improving protein quantification accuracy and reliability. The integration of multiple quantitative approaches often provides the most comprehensive understanding of protein dynamics in crop systems (Avery et al. 2022). Emerging approaches in quantitative proteomics include single-cell proteomics and spatial proteomics (Paul et al. 2021). While technically challenging in plant systems, these methods promise to reveal cell-type-specific protein expression patterns and spatial distribution of proteins within plant tissues. Novel sample preparation methods and increased instrument sensitivity are making these approaches increasingly feasible.

9 Metabolomics: Core Concepts and Scope

Metabolomics represents the comprehensive study of small molecule metabolites in biological systems, providing crucial insights into cellular processes and biochemical phenotypes (Alexandrov 2023). In crop science, metabolomics has emerged as a powerful approach for understanding plant metabolism, stress responses, and quality traits (Hamany Djande et al. 2020). The scope of metabolomics extends from primary metabolites essential for growth and development to specialized secondary metabolites involved in plant defense and environmental adaptation. The plant metabolome is remarkably complex, encompassing thousands of diverse compounds with widely varying chemical properties (Manickam et al. 2023). Primary metabolites include sugars, amino acids, organic acids, and lipids that are directly involved in growth, development, and reproduction. Secondary metabolites, such as flavonoids, alkaloids, terpenoids, and phenolics, serve specialized functions in plant defense, signaling, and environmental adaptation (Elshafie et al. 2023). This chemical diversity presents both opportunities and challenges for metabolomic analysis (Salam et al. 2023).

A fundamental concept in plant metabolomics is metabolic flux, which describes the dynamic flow of metabolites through biochemical pathways (Xiong et al. 2020). Understanding metabolic flux is crucial for elucidating how plants respond to environmental changes and stress conditions. Modern metabolomic approaches enable the tracking of metabolite dynamics and pathway activities, providing insights into metabolic regulation and adaptation mechanisms (Yan et al. 2022). The spatial organization of metabolism represents another key concept. Plant metabolism is compartmentalized across different organelles, tissues, and organs, with specific metabolic processes occurring in distinct locations (Bar-Peled and Kory 2022). Advanced metabolomic techniques, including imaging mass spectrometry and single-cell approaches, are revealing the spatial distribution of metabolites and their roles in plant function (Alexandrov 2020). Metabolic networks form complex, interconnected systems that regulate plant biochemistry. These networks exhibit both robustness and plasticity, allowing plants to maintain essential functions while adapting to environmental changes. Network analysis approaches have become increasingly important for understanding metabolic regulation and identifying key control points in plant metabolism (Gao and Zhao 2024).

The scope of crop metabolomics extends to various applications in agriculture and biotechnology. Stress response analysis using metabolomic profiling reveals how plants respond to various biotic and abiotic stresses, identifying key metabolites and pathways involved in stress tolerance (Carrera et al. 2021; Singh et al. 2023a, b, 2024). For quality trait assessment, metabolomics enables the evaluation of crop quality traits, including nutritional value, flavor compounds, and beneficial phytochemicals (Sharma et al. 2021; Romero et al. 2021). Metabolomics supports crop breeding programs for enabling improved performance of varieties and hybrids. Metabolic biomarkers help in crop breeding programs, facilitating selection of varieties with desired traits and monitoring metabolic changes during crop improvement (Sakurai 2022). Biosynthetic pathway elucidation remains the integrated part of the metabolomic analysis, that helps uncover novel biosynthetic pathways and regulatory mechanisms, particularly for specialized metabolites (Sangwan et al. 2024). The technical scope of metabolomics encompasses various analytical approaches. Mass spectrometry-based methods, particularly when coupled with chromatographic separation, provide high sensitivity and broad metabolite coverage (Dettmer et al. 2007; Sarkar et al. 2025). Nuclear magnetic resonance (NMR) spectroscopy offers complementary capabilities for metabolite identification and structure elucidation (Emwas et al. 2019). The integration of multiple analytical platforms often provides the most comprehensive metabolomic analysis (Xu et al. 2013; Boccard and Rudaz 2018). Data analysis and interpretation represent crucial aspects of metabolomics. Sophisticated bioinformatics tools and statistical approaches are required to process and interpret complex metabolomic datasets (Cambiaghi et al. 2017). Machine learning and artificial intelligence approaches are increasingly employed for metabolite identification and pathway analysis.

10 Metabolite Analysis Techniques

Metabolite analysis techniques in crop science encompass a diverse array of analytical methods that provide comprehensive insights into plant metabolic processes. These techniques primarily include chromatography-based approaches such as Gas Chromatography-Mass Spectrometry (GC-MS), Liquid Chromatography-Mass Spectrometry (LC-MS), and Nuclear Magnetic Resonance (NMR) spectroscopy, each offering unique advantages for metabolite analysis (Manickam et al. 2023). GC-MS excels in analyzing volatile and derivatized compounds, while LC-MS is particularly effective for polar and high-molecular-weight metabolites (Box 1). NMR spectroscopy provides detailed structural information and can analyze compounds in their native state without separation. Advanced techniques like Ion Mobility-Mass Spectrometry (IM-MS) and Capillary Electrophoresis-Mass Spectrometry (CE-MS) offer enhanced separation capabilities and improved metabolite identification in comprehensive metabolomics of crop plants (Son et al. 2024). These analytical platforms are complemented by sophisticated bioinformatics tools that facilitate data processing, metabolite annotation, and pathway analysis. The integration of multiple analytical techniques allows researchers to capture a broader range of metabolites, from primary metabolites like sugars and amino acids to secondary metabolites such as flavonoids and alkaloids. Modern metabolomics approaches also incorporate stable isotope labeling techniques to track metabolic fluxes and understand dynamic metabolic processes in plants (Gessler et al. 2024). Furthermore, advances in high-resolution mass spectrometry and enhanced separation methods continue to improve the depth and accuracy of metabolite profiling in crop science (Deschamps et al. 2023). The selection of appropriate techniques depends on the chemical properties of target metabolites, required sensitivity, and specific research objectives. Modern metabolite analysis typically employs multiple complementary approaches to achieve comprehensive metabolome coverage. The application of these techniques has revolutionized our understanding of plant metabolism, stress responses, and crop improvement strategies.

Box 1 Mass Spectrometry (MS) in Modern Metabolomics

1. *Electron Impact (EI) ionization* for GC-MS is a widely accepted technique that involves strong beam for high-energy bombards on the sample molecules, causing them to release electrons and generate positively charged ions, that could be analyzed and identified based on their fragmentation patterns. It is the most commonly used ionization method of samples because of its ability of extensive fragmentation to facilitate structural determination of unknown molecules.
2. *Electrospray Ionization (ESI)* is a widely applicable ionization technique in Liquid Chromatography-Mass Spectrometry (LC-MS) that facilitates analysis of particularly large biomolecules like proteins and peptides. Due to a high voltage applied at a capillary, liquid sample converts to charged droplets that evaporates to release ions that get detected by the mass spectrometer. The method is preferred for analyzing complex mixtures due to "soft" ionization process that minimizes fragmentation of the analytes.
3. *Atmospheric pressure chemical ionization* (APCI) is a soft ionization method used for nonpolar to moderately polar anylates. It is a popular technique in mass spectrometry for the analysis of small molecules like drugs and pesticides.
4. *Matrix-assisted laser desorption/ionization* (MALDI) is a mass spectrometry for analyzing many types of biomolecules and organic compounds including proteins, peptides, nucleic acids, and carbohydrates. It has wide scale applications in medicine, biology, agriculture, and pharmacopathology. (Sarkar et al. 2025)

Chromatographic separation techniques play a vital role in metabolomics analysis by reducing sample complexity before detection (García-Pérez et al. 2024). Gas chromatography (GC) excels in separating, identifying and quantifying small molecule volatile compounds and those that can be derivatized to become volatile (Fiehn 2016). GC is particularly useful for analyzing primary metabolites like sugars, amino acids, and organic acids. Modern GC systems employ various column types and temperature programs to optimize separation. Liquid chromatography (LC) offers greater flexibility in analyzing a broader range of metabolites, particularly those that are thermally labile or difficult to derivatize (Zhou et al. 2012). Different LC modes serve specific purposes like (i) reverse-phase LC for moderately polar to non-polar compounds, (ii) hydrophilic interaction liquid chromatography (HILIC) for highly polar metabolites, (iii) ion chromatography for charged molecules, and (iv) size exclusion chromatography for large metabolites. MS serves as the primary detection method in modern metabolomics. Different ionization techniques like (i) Electron Impact (EI) ionization for GC-MS, (ii) Electrospray

Ionization (ESI) for LC-MS, (iii) Atmospheric Pressure Chemical Ionization (APCI) for less polar compounds, and (iv) Matrix-Assisted Laser Desorption/Ionization (MALDI) for specific applications are employed depending on the nature of analytes (Sawada and Hirai 2013). High-resolution mass spectrometers, including time-of-flight (TOF), Orbitrap, and Fourier transform ion cyclotron resonance (FT-ICR) instruments, provide accurate mass measurements crucial for metabolite identification (Maia et al. 2021).

Nuclear Magnetic Resonance (NMR) spectroscopy offers complementary capabilities for metabolite analysis. While less sensitive than MS, NMR provides (i) non-destructive analysis, (ii) absolute quantification without standards, (iii) detailed structural information, and (iv) analysis of complex mixtures without separation. Advanced NMR techniques, including 2D methods and hyphenated approaches, enhance metabolite identification and quantification capabilities in plant metabolomics experiments (Mahrous and Farag 2015). Capillary electrophoresis (CE) provides excellent separation of charged metabolites and has proven particularly valuable for analyzing ionic compounds and small molecules. CE-MS combinations offer high separation efficiency and detection sensitivity for polar and charged metabolites (Zhang and Ramautar 2021). Furthermore, imaging techniques including (i) MALDI imaging mass spectrometry, (ii) desorption electrospray ionization (DESI) imaging, (iii) secondary ion mass spectrometry (SIMS), and (iv) Raman microscopy for metabolite analysis were also advanced significantly were also advanced significantly (Buchberger et al. 2018). These techniques provide spatial information about metabolite distribution in plant tissues (Porta Siegel et al. 2018).

11 Sample Preparation in Metabolomics

The selection of appropriate extraction protocols represents a critical cornerstone in metabolomics research, fundamentally determining the quality and comprehensiveness of metabolite analysis. The inherent chemical diversity of plant metabolites necessitates carefully optimized extraction strategies that can effectively maintain metabolite stability while ensuring optimal recovery across different compound classes. Since sample preparation in metabolomics represents a crucial first step in metabolite profiling, the extraction protocol should be carefully chosen to extract as much compounds of interest as possible and maintain metabolite stability while achieving good recovery. For diverse metabolic compound classes like (i) polar metabolites (sugars, amino acids, organic acids), (ii) non-polar compounds (lipids, some secondary metabolites), and (iii) semi-polar metabolites, the extraction methods may require need for specific solvents or their combinations and specified conditions (Salem et al. 2016). Rapid quenching of real-time metabolism during sampling is essential to prevent metabolite degradation or transformation due to sample storage (Lu et al. 2017).

For polar metabolites, including sugars, amino acids, and organic acids, extraction typically employs cold aqueous systems or aqueous-organic solvent mixtures. The process requires strict temperature control (0–4 °C) to prevent metabolite degradation, often incorporating buffer systems for pH stability and antioxidants to prevent oxidative modifications. The addition of enzyme inhibitors proves crucial in preventing unwanted metabolite alterations during the extraction process, particularly for phosphorylated compounds and other labile metabolites (Kirkwood et al. 2013). Non-polar compounds (lipids and certain secondary metabolites) demand extraction protocols utilizing organic solvents like chloroform, hexane, or ethyl acetate. These extractions require particular attention to solvent purity and sequential extraction steps to maximize recovery. The prevention of oxidation during processing and appropriate storage conditions become especially critical for maintaining lipid integrity and preventing artifact formation (Mohammad et al. 2021). Moderately or semi-polar metabolites present unique challenges, often requiring mixed solvent systems such as methanol/water or acetonitrile/water mixtures (Chaves et al. 2020). The optimization of these protocols involves careful adjustment of solvent ratios, pH modification, and multiple extraction cycles to ensure complete recovery. Temperature control during extraction remains crucial for maintaining compound stability and ensuring reproducible results.

The development and validation of extraction protocols require comprehensive consideration of multiple parameters, including sample preparation methods, extraction duration and temperature, solvent-to-sample ratios, and appropriate cleanup procedures (Wieczorek et al. 2024). The incorporation of internal standards, quality control measures, and thorough assessment of matrix effects ensures reliable quantification and reproducible results. Method validation using standard compounds, recovery assessment studies, and stability testing under extraction conditions provide essential data for protocol optimization. Modern metabolomics approaches increasingly emphasize the importance of systematic method development, incorporating sophisticated experimental designs to evaluate extraction efficiency across different metabolite classes (Oh et al. 2023). This comprehensive approach, as demonstrated by Qiu et al. (2023), enables researchers to develop robust extraction protocols that can effectively capture the diverse array of metabolites present in plant systems while maintaining their native state and relative abundances.

12 Targeted Versus Un-targeted Approaches

Metabolomics in plant research encompasses two main analytical approaches namely targeted and non-targeted (un-targeted or unbiased) metabolomics, each with its distinct characteristics and applications. These approaches are often referred to by various synonymous terms in the scientific literature, reflecting different aspects of the methodology. Targeted metabolomics, also known as targeted profiling, quantitative metabolomics, or directed metabolomics, is a hypothesis-driven approach focusing on specific known metabolites or metabolite classes and is mainly focused on the quantification of specifically known metabolites providing concise and accurate data (Kapoore and Vaidyanathan 2016). The approach focuses on the quantitative analysis of a predefined set of metabolites, typically involving compounds of known identity and established biological significance. This approach is hypothesis-driven and aims to provide precise quantification of specific metabolites of interest. In crop science, targeted approaches are particularly valuable for studying well-characterized metabolic pathways, monitoring specific biomarkers, or analyzing compounds related to specific traits such as stress resistance or nutritional quality (Roberts et al. 2012). This method requires prior knowledge of metabolite structures and relies heavily on authentic standards for absolute quantification. In mass spectrometry-based analysis, it's frequently called SRM (Selected Reaction Monitoring) or MRM (Multiple Reaction Monitoring)-based metabolomics (Kontostathi et al. 2019). This method offers higher sensitivity and selectivity for the compounds of interest, along with better reproducibility, though it provides lower metabolome coverage. Specific characteristic features make targeted metabolomics particularly suitable for validation studies and routine analysis of known compounds (Box 2). Plant researchers often employ targeted metabolomics for pathway-specific analysis, biomarker validation, hypothesis testing, and metabolic flux analysis (Katam et al. 2022).

Box 2 Key Characteristics of Targeted Metabolomics Approach

(i) High specificity and sensitivity for selected compounds
(ii) Absolute quantification using authentic standards
(iii) Optimized extraction and analysis protocols
(iv) Focused data analysis procedures
(v) Lower complexity in data interpretation

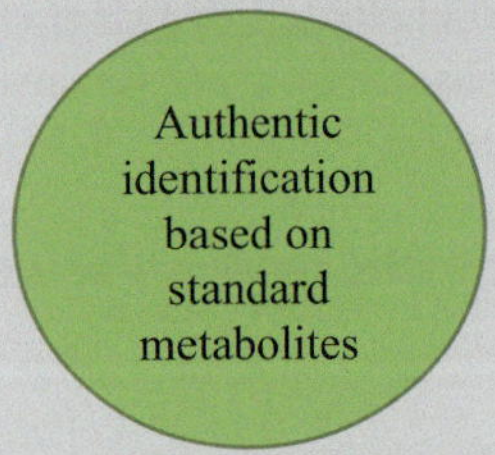

Untargeted (non-targeted) metabolomics, alternatively called un-targeted metabolomics, global metabolomics, or discovery metabolomics, takes a comprehensive, hypothesis-generating approach (Wang et al. 2010). This method aims to analyze all detectable metabolites in a biological sample simultaneously. It is sometimes referred to as metabolic fingerprinting, unbiased metabolomics, or full-scan metabolomics with multiple characteristics (Box 3). Unlike targeted approaches, non-targeted metabolomics doesn't require prior knowledge of metabolite identity and provides broader metabolome coverage, though it typically offers lower sensitivity for individual compounds (Yan et al. 2022). The approach generates

Box 3 Characteristics of Untargeted Metabolomics Approach

(i) Comprehensive metabolic profiling
(ii) Broad metabolome coverage
(iii) Discovery-driven research potential
(iv) Complex data analysis requirements
(v) Semi-quantitative results
(vi) Challenges in metabolite identification

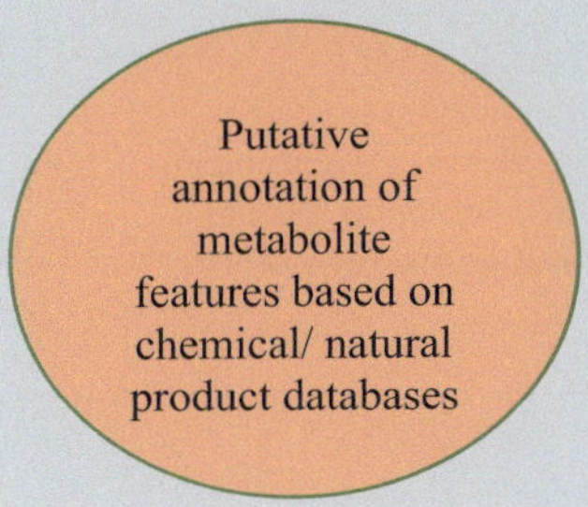

semi-quantitative or relative quantification data and requires more complex data analysis procedures. In plant research, this method is being used for metabolite discovery, chemical phenotyping, comparative metabolomics, and biomarker identification (García-Pérez et al. 2024). Untargeted metabolomics aims to analyze all detectable metabolites in a biological sample simultaneously, without prior knowledge of their identity. This comprehensive approach provides a global view of metabolic changes and can lead to the discovery of novel compounds and unexpected metabolic relationships. The analysis is particularly valuable in exploratory research, biomarker discovery, and understanding system-wide metabolic responses to environmental changes or genetic modifications (Bozza et al. 2021). Modern plant metabolomics often employs hybrid approaches that combine elements of both targeted and non-targeted analysis. These methods, known as semi-targeted, pseudo-targeted, widely targeted, or multi-targeted metabolomics, aim to balance comprehensive coverage with quantitative accuracy (Shen et al. 2023). They frequently utilize multiple analytical platforms for the consideration of method selection, technical analysis, and integration in plant metabolomics (Table 1).

In plant-specific applications, several specialized terms have emerged. Phytometabolomics specifically refers to plant metabolomics, while based on chemical classes, researchers use terms like lipidomics (for lipids), glycomics (for carbohydrates), or fluxomics (for metabolic fluxes) (Belhaj et al. 2021). Furthermore, proteo-metabolomics describes the combined analysis of proteins and metabolites.

Table 1 Considerations for method selection, technical analysis, and integration in plant metabolomics

Targeted metabolomics approach	Untargeted metabolomics approach
(i) Sample preparation.	
1. Sample optimization for compounds 2. Selective extraction procedures 3. Specialized analytical methods 4. Reference standards requirement 5. Target specific compounds required 6. Validated quantification protocols.	1. Broad-spectrum extraction methods 2. Multiple solvent selection 3. Generic separation techniques 4. High-resolution analytical platforms 5. Comprehensive data acquisition 6. Advanced bioinformatics tools
(ii) Technical considerations	
1. Method development and validation 2. Standard curve preparation 3. Matrix effect evaluation 4. Quality control procedures 5. Data processing protocols	1. Sample optimization 2. Analytical platform selection 3. Data preprocessing requirements 4. Metabolite annotation strategies 5. Statistical analysis approaches
(iii) Integration	
1. Initial broad screening via un-targeted analysis 2. Focused quantification of discovered biomarkers 3. Validation of hypotheses generated from global profiling 4. Detailed pathway analysis 5. Comprehensive metabolic phenotyping	
(iv) Applications	
1. Nutritional quality assessment 2. Stress marker monitoring 3. Pathway-specific analysis 4. Breeding program support 5. Quality control	1. Metabolic phenotyping 2. Biomarker discovery 3. Stress response profiling 4. Metabolic network analysis 5. Novel compound identification

The field has also developed terms based on analytical approaches, such as spatial metabolomics for location-specific analysis, single-cell metabolomics for cellular-level studies, and real-time metabolomics for time-resolved analysis. All these terms reflect the diverse methodological approaches and applications in plant metabolomics research, allowing researchers to choose the most appropriate technique based on their specific research questions and objectives. These approaches and their nomenclature shall continue to evolve as new analytical technologies and methodologies emerge, making plant metabolomics a dynamic and advancing field in plant science research.

13 Data Acquisition, Processing, and Analysis

Data acquisition, pre-processing, and processing represent crucial steps in plant metabolomics that fundamentally determine the quality and reliability of metabolomic analyses. The complexity of plant and microbial metabolomes, combined with the sophisticated analytical platforms used in their analysis, necessitates careful attention to data handling at every stage of the workflow. The approaches differ substantially between targeted and un-targeted metabolomics, each requiring specific considerations and methodologies. Understanding these differences and implementing appropriate data handling strategies is crucial for successful metabolomic studies in crop science.

The data acquisition phase begins with the generation of complex analytical signals from various platforms, primarily MS-based techniques coupled with chromatographic separation (Table 2). During LC-MS or GC-MS analysis, raw data

Table 2 Key considerations in data acquisition strategies in metabolomics

Targeted approaches	Untargeted approaches
Data acquisition	
1. Optimized acquisition parameters for known compounds 2. Selected reaction monitoring (SRM) or multiple reaction monitoring (MRM) 3. Predetermined retention time windows 4. Internal standard monitoring 5. Quality control sample integration	1. Full-scan MS acquisition 2. High-resolution mass measurements 3. Broad retention time monitoring 4. MS/MS data for structure elucidation 5. Data-dependent acquisition (DDA) 6. Data-independent acquisition (DIA)
Data Processing Workflows	
1. Peak detection and integration 2. Retention time alignment 3. Internal standard normalization 4. Calibration curve construction 5. Concentration calculation	1. Raw data conversion and formatting 2. Baseline correction and noise reduction 3. Peak detection and deconvolution 4. Alignment across samples 5. Feature extraction and annotation 6. Missing value imputation 7. Data normalization
Quality Control and Validation	
1. Method validation parameters 2. Linearity assessment 3. Precision and accuracy evaluation 4. Recovery studies 5. Stability testing	1. Technical replicate analysis 2. Quality control sample integration 3. Batch effect correction 4. Signal drift monitoring 5. Feature reliability assessment
Data Integration and Interpretation	
1. Pathway-specific analysis 2. Known metabolite relationships 3. Quantitative pathway modeling 4. Targeted biological interpretation 5. Hypothesis validation	1. Global pathway analysis 2. Network reconstruction 3. Multi-omics data integration 4. Systems biology approaches 5. Novel pathway discovery
Data Analysis Approaches	
1. Absolute quantification 2. Method validation parameters 3. Statistical significance testing 4. Quality control metrics 5. Pathway-specific interpretation	1. Multivariate statistical analysis Principal Component Analysis (PCA) Partial Least Squares Discriminant Analysis (PLS-DA) Hierarchical Clustering Analysis (HCA) 2. Feature selection methods Variable Importance in Projection (VIP) Random Forest Analysis Statistical significance testing 3. Pattern recognition algorithms 4. Network analysis approaches 5. Pathway mapping and enrichment analysis

is collected as three-dimensional measurements incorporating retention time, mass-to-charge ratios (*m/z*), and signal intensities (Sun and Xia 2024). This process requires careful optimization of instrumental parameters and continuous monitoring of system performance. Quality control samples, including pooled samples and standard mixtures, are regularly analyzed throughout analytical sequences to assess system stability and enable post-acquisition data normalization. The acquisition phase must account for potential variations in instrumental response, matrix effects, and signal drift, which can significantly impact data quality (Gorrochategui et al. 2016).

In targeted metabolomics, data acquisition follows predetermined protocols focused on specific metabolites of interest. This approach typically employs selected reaction monitoring (SRM) or multiple reaction monitoring (MRM) methods, with optimized acquisition parameters for known compounds. The use of predetermined retention time windows and careful internal standard monitoring ensures high precision and accuracy in quantifying known metabolites (Anh et al. 2024). Quality control samples are regularly integrated into the analytical sequence to monitor system performance and ensure data reliability throughout the acquisition process. Un-targeted metabolomics takes a broader approach to data acquisition, aiming to capture as much metabolite information as possible. This comprehensive strategy typically employs full-scan MS acquisition with high-resolution mass measurements, coupled with broad retention time monitoring. MS/MS data collection for structure elucidation is often incorporated, using either data-dependent or data-independent acquisition modes. Furthermore, the approach maximizes metabolome coverage while maintaining data quality through careful experimental design and quality control measures (Souza and Patti 2021; Ivanisevic and Want 2019).

Raw metabolomics data undergoes extensive preprocessing to transform complex analytical signals into reliable, interpretable data matrices. Peak detection and deconvolution represent critical initial steps, where true metabolite signals are distinguished from background noise and system artifacts (Pinto et al. 2022). Advanced algorithms process overlapping peaks, perform baseline corrections, and align signals across multiple samples. This stage proves particularly challenging in plant metabolomics due to the presence of numerous isobaric compounds and complex matrix effects. Data filtering and cleaning procedures then remove artifacts, contaminant signals, and unreliable measurements based on predefined quality criteria. The handling of missing values, often encountered in metabolomics datasets, requires sophisticated computational approaches including various imputation methods or statistical treatments (Sun and Xia 2024).

Data processing workflows in metabolomics vary significantly between the two approaches. In targeted metabolomics, processing follows a relatively streamlined pathway, including peak detection and integration, retention time alignment, and internal standard normalization. Calibration curves are constructed for absolute quantification, and the entire process is often automated through dedicated software platforms (Chen et al. 2022; Amer et al. 2023). This focused approach allows for highly precise quantification of specific metabolites of interest. Processing of untargeted metabolomics data presents greater complexity, requiring sophisticated computational tools and careful parameter optimization (Perez de Souza and Fernie 2024). The workflow typically begins with raw data conversion and formatting, followed by baseline correction and noise reduction. Peak detection and deconvolution processes identify individual metabolite signals, which are then aligned across samples to ensure comparability. Feature extraction, annotation, and normalization steps follow, with careful attention paid to missing value imputation and quality control assessment (Parker et al. 2023).

Data normalization in metabolomics represents a crucial preprocessing step that addresses both technical and biological variations in the dataset. Multiple normalization strategies may be employed, including internal standard normalization, total ion current normalization, or biomass-based normalization. Each approach offers distinct advantages and limitations, necessitating careful consideration of the experimental context and research objectives. Batch effect correction and signal drift compensation become particularly important in large-scale studies where samples are analyzed over extended periods (Perez de Souza et al. 2017). Further processing of pre-processed data focuses on metabolite identification and quantification, representing one of the most challenging aspects of metabolomics. Identification relies on multiple levels of evidence, including accurate mass measurements, fragmentation patterns, and retention time indices. Database matching against established metabolite libraries provides initial annotations, while more sophisticated approaches may employ molecular networking or in-silico fragmentation prediction. The assignment of confidence levels to identifications follows standardized protocols, ensuring transparency and reproducibility in metabolite annotation (Guo et al. 2022).

Quantification procedures vary depending on the analytical approach and research objectives, ranging from relative quantification based on peak areas to absolute quantification using calibration curves. Statistical validation of quantitative results incorporates error estimation and assessment of technical and biological variability. The final stages of data processing employ multivariate statistical analyses to extract biological meaning from the complex datasets. Multivariate analysis including Principal component analysis (PCA), partial least squares discriminant analysis (PLS-DA), and other machine learning approaches help identify patterns and relationships within the data (Ramadan et al. 2006). Pathway analysis and network mapping integrate the metabolomics data with existing biological knowledge, providing insights into metabolic regulation and responses to environmental conditions (Chen et al. 2022). The comprehensive data handling pipeline requires careful optimization and validation at each step, ensuring the generation of reliable and biologically meaningful results. The continuous development of new computational tools and statistical approaches continues to enhance our ability to extract valuable information from complex plant metabolomics datasets (Worley and Powers 2013).

14 Data Analysis in Metabolomics

Data analysis in metabolomics involves a systematic approach to transform complex analytical data into biologically meaningful results. The process encompasses several critical stages that require careful consideration and implementation. The initial stage involves processing raw instrumental data to obtain clean, high-quality datasets. This includes baseline correction, peak detection, deconvolution, and alignment of signals across multiple samples. Sophisticated algorithms handle noise reduction and feature extraction, ensuring reliable data for subsequent analysis. Before statistical analysis, data undergoes preprocessing steps including normalization, scaling, and missing value imputation (Santamaria and Pinto 2024). Certain most widely applied bioinformatics tools for metabolomics data analysis are listed in Table 3.

Data analysis strategies also differ markedly between targeted and untargeted approaches. Targeted metabolomics analysis focuses on absolute quantification, method validation, and specific pathway interpretation. Statistical testing is typically straightforward, addressing specific research questions about known metabolites (Wang et al. 2010). Quality control metrics are carefully monitored to ensure reliable quantification throughout the analytical process (Schrimpe-Rutledge et al. 2016). In untargeted metabolomics, data analysis employs a broader range of statistical and bioinformatic approaches. Multivariate statistical analyses, including Principal Component Analysis (PCA) and Partial Least Squares Discriminant Analysis (PLS-DA), are commonly used to explore patterns in the data (Bartel et al. 2013). Feature selection methods, such as Variable Importance in Projection (VIP) and Random Forest Analysis, help identify significant metabolic changes. Pattern recognition algorithms and network analysis approaches provide insights into metabolic relationships, while pathway mapping and enrichment analysis help interpret the biological significance of observed changes (Pudjihartono et al. 2022).

Quality control and validation procedures are essential in both approaches but differ in their implementation. Targeted analysis requires rigorous method validation, including linearity assessment, precision and accuracy evaluation, and stability testing. Untargeted approaches rely more heavily on technical replicates, quality control samples, and careful monitoring of batch effects and signal drift (Begou et al. 2025). Integration and interpretation of metabolomics data also vary between approaches. Targeted metabolomics typically focuses on pathway-specific analysis and known metabolite relationships, often employing quantitative pathway modeling to test specific hypotheses (Khanijou et al. 2022). Untargeted metabolomics takes a broader view, incorporating global pathway analysis, network reconstruction, and systems biology approaches to discover novel metabolic relationships and pathways (Amara et al. 2022).

Table 3 Overview of certain software tools most frequently applied in LC-MS based metabolomics data analysis. The tools collectively address various analytical needs from raw data processing to pathway analysis, though methodological differences exist among their approaches

Tool name	Functions	Web access/URL	Citation
MetaboAnalyst	Comprehensive metabolomics data analysis, statistical analysis, pathway analysis	www.metaboanalyst.ca	Xia et al. (2015) Nucleic Acids Research, 43:W251-W257
MassHunter	LC-MS data acquisition and analysis, qualitative and quantitative analysis, metabolite identification	www.agilent.com/massHunter	Agilent Technologies (Commercial)
SIEVE	Differential analysis of small molecules, biomarker discovery, statistical analysis	www.thermofisher.com/sieve	Thermo Scientific (Commercial)
Progenesis QI	Label-free quantification, metabolite identification, pathway mapping	www.waters.com/progenesis	Waters Corporation (Commercial)
XCMS	Feature detection, retention time correction, peak alignment, statistical analysis	https://xcmsonline.scripps.edu	Smith et al. (2006) Analytical Chemistry, 78:779–787
Metabox	Metabolomics data analysis, pathway analysis, integration with genomics data	https://github.com/kwanjeeraw/metabox	Wanichthanarak et al. (2017) Bioinformatics, 33:2617–2619
metaX	Data preprocessing, quality control, statistical analysis	https://github.com/wenbostar/metaX	Wen et al. (2017) Bioinformatics, 33:2588–2590
CAMERA	Annotation of LC/MS data, compound identification	https://bioconductor.org/packages/CAMERA	Kuhl et al. (2012) Analytical Chemistry, 84:283–289
MAIT	LC-MS data processing, statistical analysis, metabolite identification	http://b2slab.upc.edu/software-and-downloads/metabolite-automatic-identification-toolkit/	Fernández-Albert et al. (2014) Analytical Chemistry, 86:2320–2325
Workflow4Metabolomics	Integrated metabolomics data analysis platform	https://workflow4metabolomics.org	Giacomoni et al. (2015) Bioinformatics, 31:1493–1495
MZmine	LC-MS data processing, peak detection, alignment, visualization	http://mzmine.github.io	Pluskal et al. (2010) BMC Bioinformatics, 11:395
MetAlign	Preprocessing and comparison of LC-MS data	www.metalign.nl	Lommen (2009) Analytical Chemistry, 81:3079–3086

15 Data Integration

The integration of proteomic and metabolomic data represents a powerful approach for understanding complex biological systems in crop science, providing insights that cannot be obtained from either technique alone. The integration enables a more comprehensive understanding of cellular processes, regulatory networks, and phenotypic outcomes in crop plants. The challenge lies in effectively combining these distinct data types while accounting for their different temporal and spatial scales, as well as their inherent biological relationships (Sanches et al. 2024). Modern approaches to proteo-metabolomic data integration begin with careful experimental design to ensure compatibility between proteomic and metabolomic datasets. This includes considerations of sampling strategies, temporal coordination, and appropriate biological and technical replication. Synchronized sampling and analysis help capture the dynamic relationships between proteins and metabolites, particularly important when studying stress responses or developmental processes in crops (Wienkoop et al. 2008).

Statistical integration of proteo-metabolomic data employs various approaches, from simple correlation analyses to sophisticated multivariate methods. Initial steps often involve data normalization and scaling to make the different data types comparable. Correlation networks can reveal relationships between proteins and metabolites, while more advanced approaches like partial least squares regression and canonical correlation analysis help identify meaningful associations across the datasets (Jendoubi 2021). Machine learning algorithms increasingly play a role in discovering complex patterns and relationships within integrated datasets (Mehrotra and Mendes 2006). Pathway-based integration approaches leverage existing knowledge of biochemical pathways and regulatory networks. This strategy is particularly valuable in crop improvement studies, where understanding metabolic regulation is crucial. By mapping both protein and metabolite data onto known pathways, researchers can identify regulatory points and potential targets for crop enhancement. Modern pathway analysis tools incorporate both proteomics and metabolomics data, enabling visualization and analysis of complex biological processes (Jamil et al. 2020).

Network analysis has emerged as a powerful tool for integrating proteo-metabolomic data. These approaches construct biological networks incorporating both proteins and metabolites, revealing functional modules and regulatory hubs. Network topology analysis can identify key regulatory points and potential intervention targets for crop improvement. Time-series data integration is particularly valuable, allowing the examination of dynamic network responses to environmental changes or developmental cues (Karimpour-Fard et al. 2015). Biological interpretation of integrated proteo-metabolomic data requires consideration of multiple cellular processes and their interactions (Jendoubi 2021). This includes understanding enzyme-substrate relationships, regulatory cascades, and feedback mechanisms. Advanced visualization tools help researchers interpret complex relationships between proteins and metabolites, while systems biology approaches provide frameworks for understanding emergent properties of biological systems (Charitou et al. 2016). Computational challenges in data integration include handling different data scales, missing values, and technical variations between platforms. Modern bioinformatics tools increasingly incorporate specialized algorithms for proteo-metabolomic data integration, including methods for dealing with these technical challenges (Marinka et al. 2024). Cloud-based platforms and specialized software packages facilitate the processing and integration of large-scale proteo-metabolomic datasets (Koppad et al. 2021).

The application of integrated proteo-metabolomic analysis in crop improvement has revealed new insights into stress responses, developmental processes, and quality traits (Ajjamada and Kushalappa 2013; Naik et al. 2023). For example, studies of drought stress responses have identified coordinated changes in proteins and metabolites that contribute to stress tolerance. Similarly, investigations of crop quality traits have revealed complex relationships between proteins involved in biosynthetic pathways and their metabolic products (Juarez-Escobar et al. 2021). Trends in proteo-metabolomic data integration include the development of machine learning approaches for pattern recognition and prediction, the incorporation of spatial information through imaging techniques, and the integration of additional omics data types. These advances promise to further enhance our understanding of crop biology and accelerate crop improvement efforts (Iqbal et al. 2021).

16 Multi-omics Data Integration

A comprehensive approach to understand biological systems in the crop science research and integrate the information from different molecular levels including genomics, transcriptomics, proteomics, and metabolomics is known as multi-omics data integration (Iqbal et al. 2021; Naik et al. 2023). The effort reflects holistic view of cellular processes and regulatory networks within the tissues or cells for generating deeper molecular insights into the crop biology to facilitate more effective crop improvement methods. The complexity of the study lies in creating an effective combination of data types that could create contrast difference in their biological relationships, measurement scales, and temporal dynamics (Kausar et al. 2022). For a successful multi-omics integration methodology, careful experimental design, data collection, and continuous monitoring of the experiments are essential. The studies on studying dynamic processes and experimental continuum, e.g., plant stress responses or developmental transitions, need synchronization of sampling materials, collection timing, and data generation across different omics platforms (Subramanian et al. 2020). In such sensitive cases, considerable attention is needed to address specific tissue conditions, temporal resolution and appropriate biological replications to ensure meaningful integration of data types in a particular experimental setup. Modern experimental strategies including high-throughput sample collection, screening, and extraction methods have facilitated simultaneous incorporation of multi-omics measurements from a single biological tissue. It not only minimizes experimental error due to variability but enhances the integration potential at the same time (Graw et al. 2021).

Network-based integration approaches have proven particularly valuable in crop science, enabling the construction of comprehensive biological networks that span multiple molecular levels. These networks can reveal regulatory relationships between genes, proteins, and metabolites, identifying key regulatory hubs and potential intervention points for crop improvement. Modern network analysis tools incorporate methods for handling multi-omics data, including algorithms for identifying functional modules and regulatory circuits that span different molecular levels (Yan et al. 2022). Machine learning and artificial intelligence approaches are increasingly important in multi-omics integration. Deep learning models can handle the complexity and high dimensionality of multi-omics data, discovering patterns and relationships that might not be apparent through traditional statistical approaches (Flores et al. 2023). These methods are particularly valuable for predictive modeling in crop improvement, enabling the identification of key molecular features associated with desired traits. Pathway-based integration strategies leverage existing knowledge of biological pathways and regulatory networks. By mapping multi-omics data onto known pathways, researchers can identify coordinated changes across molecular levels and understand how different cellular processes are regulated (Ballard et al. 2024). Modern pathway analysis tools incorporate methods for visualizing and analyzing multi-omics data in the context of biological pathways, facilitating the interpretation of complex datasets (Reimand et al. 2019).

Temporal and spatial aspects of multi-omics data present unique integration challenges. Time-series experiments require methods that can account for different temporal dynamics across molecular levels, while spatial studies must consider tissue-specific variations and cellular heterogeneity. Advanced computational methods have been developed to handle these complexities, including approaches for temporal alignment and spatial registration of multi-omics data (Chu et al. 2024). Application of integrated multi-omics approaches in crop improvement has revealed new insights into complex traits and processes. For example, studies of drought tolerance have identified coordinated responses across transcriptional, protein, and metabolic levels, leading to the identification of key regulatory networks and potential targets for improvement (Kim et al. 2024). Similarly, investigations of crop quality traits have benefited from multi-omics integration, revealing complex molecular mechanisms underlying important agricultural characteristics (Yu et al. 2023). Data management and computational infrastructure represent significant challenges in multi-omics integration. Sophisticated data management systems and computational resources are essential for the storage and analysis of highly complex voluminous multi-omics data. Cloud-based platforms and specialized software packages have been developed to handle these requirements, facilitating the storage, processing, and analysis of large-scale multi-omics datasets (Koppad et al. 2021). Current trends in multi-omics integration include the development of single-cell multi-omics approaches, the incorporation of epigenomic data, and the integration of environmental and phenotypic data (Baysoy et al. 2023). Such advances promise to provide even more comprehensive understanding of crop biology and accelerate crop improvement efforts.

17 Bioinformatics Tools and Platforms

Bioinformatics tools and web-based platforms are important modern proteomics and metabolomics research, making it feasible the raw data processing, analysis, and interpretation of complex molecular data generated during the crop science studies. These tools evolved in parallel to the sophisticated equipments and tools that has enhances the advancement of analytical technologies. The present platforms are offering unparalleled sophisticated capabilities for engaging multi-faceted large-scale multi-dimensional datasets. The selection and implementation of appropriate bioinformatics tools significantly influence the quality and depth of insights obtained from proteo-metabolomic studies. For proteomics data analysis, several specialized platforms have become essential tools in the research workflow (Calderón-González et al. 2016; Subba et al. 2019). MaxQuant (https://maxquant.org), one of the most widely used platforms, offers comprehensive functionality for raw MS data processing, protein identification, and quantification. The platform incorporates advanced algorithms for peptide search and validation, providing robust statistical frameworks for protein quantification. Similarly, Proteome Discoverer software (https://www.thermofisher.com) provides an integrated environment for proteomics data analysis, with capabilities for both database searching and de novo sequencing approaches. Open-source platforms have gained significant traction in the proteomics community. OpenMS (https://openms.de) offers a flexible framework for developing and implementing proteomics data analysis workflows. Its modular architecture allows researchers to customize analysis pipelines according to specific research needs. The Trans-Proteomic Pipeline (TPP) (www.tppms.org) provides a comprehensive suite of tools for protein identification and quantification, with particular strength in statistical validation of results.

Metabolomics data analysis requires specialized tools capable of handling the complexity and diversity of metabolite data. XCMS (https://xcmsonline.scripps.edu) has emerged as a prominent platform for processing LC-MS based metabolomics data, offering advanced features for peak detection, alignment, and relative quantification. MZmine (https://mzio.io/mzmine) provides a user-friendly interface for metabolomics data processing, with capabilities for both targeted and untargeted analysis approaches. These platforms incorporate various algorithms for peak detection and alignment, enabling robust metabolite profiling across multiple samples. Database resources play a vital role in proteo-metabolomic analysis. UniProt (https://www.uniprot.org) serves as a primary resource for protein information, while specialized plant databases like The Arabidopsis Information Resource (TAIR) (https://www.arabidopsis.org) and Gramene (https://gramene.org) provide crop-specific protein annotations. For metabolomics, databases such as KEGG (https://www.genome.jp/kegg), MetaCyc (https://metacyc.org), and PlantCyc (https://plantcyc.org) offer comprehensive information about metabolic pathways and compound annotations. The integration of these resources with analysis platforms enhances the biological interpretation of experimental results (Gardinassi et al. 2017).

Tools for pathway analysis and visualization have become increasingly important in proteomics and metabolomics data analysis (Chen et al. 2022). Cytoscape (https://cytoscape.org), with its extensive plugin ecosystem, enables complex network analysis and visualization of proteo-metabolomic data. (https://mapman.gabipd.org) provides plant-specific pathway visualization capabilities, while PathVisio (https://pathvisio.org) offers flexible tools for pathway creation and data mapping. These platforms facilitate the biological interpretation of integrated proteo-metabolomic data within the context of known metabolic and regulatory pathways (Wanichthanarak et al. 2015). Statistical analysis platforms have evolved to handle the complexity of proteo-metabolomic data. R-based tools, including various Bioconductor packages (https://www.bioconductor.org), provide powerful capabilities for statistical analysis and visualization. Specialized packages like limma and DESeq2, originally developed for genomics, have been adapted for proteo-metabolomic data analysis. These tools offer robust methods for differential expression analysis, pattern recognition, and multivariate statistics.

Cloud-based platforms that address the flexible yet important demand of data analysis in proteo-metabolomic research are becoming increasingly important (Warth et al. 2017). Galaxy server (https://usegalaxy.org) provides a web-based environment for data analysis, with specific tools for proteomics and metabolomics workflows. The platform offers advantages in terms of computational resources and collaboration capabilities, though considerations of data security and privacy remain important. Machine learning frameworks have become essential tools in proteo-metabolomic data analysis (Neely and Palmblad 2022). Platforms like TensorFlow (https://www.tensorflow.org) and PyTorch (https://pytorch.org), coupled with specialized libraries for biological data analysis, enable the development and implementation of advanced analytical approaches. These tools are particularly valuable for pattern recognition, feature selection, and predictive modeling in crop improvement studies. Integration platforms that can handle multiple omics data types are proving them increasingly important. Tools like mixOmics (https://mixomics.org) and MOFA (Multi-Omics Factor Analysis) (Argelaguet et al. 2018) provide sophisticated capabilities for integrating proteomic and metabolomic data with other omics datasets. These platforms implement various statistical approaches for data integration and interpretation, enabling comprehensive analysis of biological systems.

18 Network Analysis

Network analysis has emerged as a powerful approach for understanding complex molecular interactions and regulatory relationships in biological systems. In the context of proteo-metabolomic studies, network analysis provides frameworks for visualizing and analyzing the complex relationships between proteins and metabolites, offering insights into cellular organization and regulation. This approach has become particularly valuable in crop science, where understanding system-level responses is crucial for crop improvement strategies. Protein-protein interaction networks form the foundation of proteomics network analysis (Jordán et al. 2012). These networks are constructed using experimental data from various sources, including yeast two-hybrid screens, co-immunoprecipitation studies, and computational predictions. In crop systems, these networks reveal functional modules and regulatory hubs that control important biological processes. Modern approaches incorporate both physical and functional interactions, providing a more comprehensive view of protein relationships. Network topology analysis helps identify key proteins that may serve as potential targets for crop improvement (Kiouri et al. 2025).

Metabolic networks are a complex web representation of biochemical reactions in cellular systems. These networks are being constructed using known metabolic pathways and experimental metabolomics data (Ravikrishnan et al. 2018). In crop research, metabolic network analysis helps understand resource allocation, stress responses, and yield-related processes. Advanced approaches incorporate flux analysis to understand the dynamic behavior of metabolic systems. The integration of enzyme information with metabolite data provides insights into metabolic regulation and potential points for manipulation in crop improvement (Rao and Liu 2025). Integrated networks can reveal how changes in protein abundance or activity influence metabolite levels and vice versa. These networks often incorporate different types of edges representing various relationships: enzymatic reactions, regulatory interactions, and correlative associations. Network construction methods have evolved significantly. Modern analysis approaches use weighted networks to represent the strength or confidence of different interactions, providing more nuanced views of molecular relationships (Perez De Souza et al. 2020). Correlation-based approaches remain important for initial network construction, but more sophisticated methods now incorporate causal inference and dynamic relationships. Gaussian graphical models help identify direct associations between molecules, while Bayesian networks can represent causal relationships (Ravikrishnan et al. 2018).

A key aspect of molecular network investigation is the modularity analysis, in which various algorithms identify functional modules or communities within networks, revealing groups of molecules that work together in biological processes (Alcalá-Corona et al. 2021). In crop studies, these modules often correspond to specific pathways or responses to environmental conditions. Advanced approaches can detect overlapping modules and hierarchical organization, providing insights into the multiple roles of proteins and metabolites in different cellular contexts (Khanijou et al. 2022). Network visualization tools have become increasingly sophisticated. Modern platforms offer interactive visualization capabilities, allowing researchers to explore complex networks at different levels of detail. These tools often incorporate biological annotation data, pathway information, and experimental results, facilitating the interpretation of network analysis results (Ehlers et al. 2025). STRING (https://string-db.org), a database and tool for the analysis and visualization of protein-protein interactions and functions, and STITCH (http://stitch.embl.de/cgi/), a web resource to explore and predict interactions of proteins and chemicals, are offering useful solutions. Dynamic visualization approaches can represent temporal changes in network structure, particularly valuable for understanding system responses to environmental changes.

Differential network analysis is now becoming a powerful tool for comparatively analyzing molecular networks in biological systems in which interactions among various components majorly change over the time under different conditions (Shojaie 2021). This method reveals how network structure and properties change in response to environmental conditions, developmental stages, or genetic modifications. In crop research, differential network analysis helps identify condition-specific interactions and regulatory relationships, providing insights into adaptive responses and potential intervention points (Zeng et al. 2022). Machine learning approaches increasingly complement traditional network analysis methods. Deep learning models can predict network properties and interactions, while graph neural networks analyze complex network structures (Khemani et al. 2024). These approaches are facilitate handling of large-scale complex networks and identifying patterns in molecular interactions in a given tissue. Integration with experimental data continues to improve the accuracy and biological relevance of network predictions. Current trends in the network analysis include the development of multi-layer network models that can represent different types of molecular interactions simultaneously. They also provide comprehensive views of cellular organization, regulation, and localization. Additionally, the integration of spatial information into network analysis is revealing how molecular interactions vary across different cellular compartments and tissues (Panditrao et al. 2022).

19 Statistical Approaches

Statistical approaches in proteo-metabolomic data integration have evolved significantly to address the challenges of combining complex, multi-dimensional datasets generated through modern analytical platforms. These approaches encompass a range of methods from basic statistical tests to advanced machine learning algorithms, each serving specific purposes in data analysis and interpretation. The selection of appropriate statistical methods is crucial for extracting meaningful biological insights from integrated proteo-metabolomic studies (Yan et al. 2022). Fundamental statistical considerations begin with data preprocessing and normalization. Given the different scales and distributions inherent in proteomic and metabolomic data, appropriate normalization strategies are essential for meaningful integration (Sun and Xia 2024). Common approaches include quantile normalization, scaling methods (e.g., z-score transformation, Pareto scaling), and variance stabilizing transformations. These preprocessing steps ensure comparability between different data types while preserving biological relevance (Kohl et al. 2012). Correlation-based methods represent a primary approach in integrative analysis. Pearson and Spearman correlations help identify relationships between proteins and metabolites, while partial correlations can reveal direct associations by controlling for confounding variables (Deek et al. 2024). More sophisticated correlation approaches, such as canonical correlation analysis (CCA), examine relationships between multiple variables simultaneously, identifying patterns of association across the integrated dataset. These methods are particularly valuable for understanding functional relationships between proteins and their associated metabolites (Jiang et al. 2023).

Multivariate statistical methods play a crucial role in proteo-metabolomic integration. Principal Component Analysis (PCA) provides initial dimensionality reduction and visualization of integrated datasets, while Partial Least Squares (PLS) methods enable the modeling of relationships between protein and metabolite data. Orthogonal Projections to Latent Structures (OPLS) further refine these analyses by separating predictive from orthogonal variation. These approaches are particularly valuable for identifying key features driving biological responses in crop systems (García-Pérez et al. 2024). Clustering approaches help identify groups of co-regulated proteins and metabolites. Hierarchical clustering, k-means clustering, and self-organizing maps can reveal patterns in integrated datasets. Advanced clustering methods, such as biclustering and triclustering, are particularly useful for identifying modules of coordinated molecular responses across different experimental conditions or time points (Chaudhry et al. 2023). These methods help elucidate functional relationships and regulatory networks in crop systems.

Network-based statistical approaches have become increasingly important in proteo-metabolomic integration. Gaussian graphical models and weighted gene correlation network analysis (WGCNA) help construct networks of molecular interactions. Network topology analysis can identify hub molecules and key regulatory points, while community detection algorithms reveal functional modules within the integrated network (Perez De Souza et al. 2020). These approaches are particularly valuable for understanding system-level organization and regulation (Shen et al. 2019). Machine learning methods have revolutionized proteo-metabolomic data integration. Supervised learning approaches, including Random Forests and Support Vector Machines, help identify features associated with specific phenotypes or conditions. Deep learning methods, particularly autoencoders and neural networks, can capture complex, non-linear relationships in integrated datasets. These approaches are particularly powerful for predictive modeling and feature selection in crop improvement studies (Taye 2023). Time-series analysis methods address the temporal aspects of proteo-metabolomic data. Dynamic time warping algorithms align temporal profiles, while state-space models capture system dynamics. Methods for differential equation modeling help understand the kinetics of molecular responses. These approaches are crucial for studying developmental processes and stress responses in crops. Statistical validation and quality assessment remain crucial aspects of data integration. Cross-validation approaches help assess model reliability, while permutation tests evaluate statistical significance. Methods for handling missing data and outliers are particularly important in proteo-metabolomic studies, where technical variations can impact data quality. Robust statistical approaches help ensure the reliability of integrated analyses.

20 Conclusion

The field of proteo-metabolomics has transformed our understanding of plant biology and revolutionized crop improvement strategies by providing unprecedented molecular insights. Through the integration of proteomics and metabolomics, researchers have developed a more comprehensive view of plant cellular processes and regulatory networks that control growth, development, and responses to environmental stresses. These approaches have revealed complex molecular mechanisms underlying important agronomic traits, enabling more targeted and efficient breeding strategies. While current technological capabilities allow for extensive proteome coverage and precise metabolite profiling, significant challenges remain. The analysis of certain protein classes, particularly membrane proteins, low-abundance regulatory proteins, and specific post-translational modifications continues to present technical difficulties. Similarly, the comprehensive annotation of metabolites, especially secondary metabolites unique to specific crop species, requires further development. These limitations highlight important areas for future research and technological innovation.

Key priorities for advancing the field include developing more sensitive detection methods, enhancing computational tools for data integration and interpretation, improving temporal and spatial resolution in analyses, and creating more accessible, cost-effective analytical platforms. Additionally, greater emphasis on field validation of laboratory findings will be essential for translating proteo-metabolomic discoveries into practical agricultural applications. The practical implications of proteo-metabolomic research have been significant, enhancing breeding program efficiency through improved selection of stress-tolerant varieties, better understanding of crop quality traits, and more precise phenotyping approaches. Industry adoption continues to increase, with notable implementation in commercial breeding programs, crop protection product development, and quality control processes.

Acknowledgement DPS is thankful to the Indian Council of Agricultural Research (ICAR) India for funding support under Network Project on Agricultural Bioinformatics (CABin, ICAR-IASRI).

Conflict of Interest The authors declare no conflict of interest in any form.

References

Ajjamada C, Kushalappa RG (2013) Metabolo-proteomics to discover plant biotic stress resistance genes. Trends Plant Sci 18:522–531. https://doi.org/10.1016/j.tplants.2013.05.002

Akyol S, Ashrafi N, Yilmaz A, Turkoglu O, Graham SF (2023) Metabolomics: an emerging "omics" platform for systems biology and its implications for Huntington Disease Research. Metabolites 13(12):1203. https://doi.org/10.3390/metabo13121203

Alcalá-Corona SA, Sandoval-Motta S, Espinal-Enríquez J, Hernández-Lemus E (2021) Modularity in biological networks. Front Genet 12:701331. https://doi.org/10.3389/fgene.2021.701331

Alexandrov T (2020) Spatial metabolomics and imaging mass spectrometry in the age of artificial intelligence. Annu Rev Biomed Data Sci 3:61–87. https://doi.org/10.1146/annurev-biodatasci-011420-031537

Alexandrov T (2023) Spatial metabolomics: from a niche field towards a driver of innovation. Nat Metab 5:1443–1445. https://doi.org/10.1038/s42255-023-00881-0

Allwood JW, Ellis DI, Goodacre R (2008) Metabolomic technologies and their application to the study of plants and plant-host interactions. Physiol Plant 132(2):117–135. https://doi.org/10.1111/j.1399-3054.2007.01001.x

Amara A, Frainay C, Jourdan F, Naake T, Neumann S, Novoa-del-Toro EM, Salek RM, Salzer L, Scharfenberg S, Witting M (2022) Networks and graphs discovery in metabolomics data analysis and interpretation. Front Mol Biosci 9:841373. https://doi.org/10.3389/fmolb.2022.841373

Amer B, Deshpande RR, Bird SS (2023) Simultaneous quantitation and discovery (SQUAD) analysis: combining the best of targeted and untargeted mass spectrometry-based metabolomics. Meta 13(5):648. https://doi.org/10.3390/metabo13050648

Anh NK, Thu NQ, Tien NTN, Long NP, Nguyen HT (2024) Advancements in mass spectrometry-based targeted metabolomics and lipidomics: implications for clinical research. Molecules 29(24):5934. https://doi.org/10.3390/molecules29245934

Argelaguet R, Velten B, Arnol D, Dietrich S, Zenz T, Marioni JC, Buettner F, Huber W, Stegle O (2018) Multi-Omics Factor Analysis-a framework for unsupervised integration of multi-omics data sets. Mol Syst Biol 14(6):e8124. https://doi.org/10.15252/msb.20178124

Arike L, Peil L (2014) Spectral counting label-free proteomics. Methods Mol Biol 1156:213–222. https://doi.org/10.1007/978-1-4939-0685-7_14

Aryal UK, Xiong Y, McBride Z, Kihara D, Xie J, Hall MC, Szymanski DB (2014) A proteomic strategy for global analysis of plant protein complexes. Plant Cell 26(10):3867–3882. https://doi.org/10.1105/tpc.114.127563

Avery C, Patterson J, Grear T, Frater T, Jacobs DJ (2022) Protein function analysis through machine learning. Biomolecules 12(9):1246. https://doi.org/10.3390/biom12091246

Baharum SN, Azizan KA (2018) Metabolomics in systems biology. Adv Exp Med Biol 1102:51–68. https://doi.org/10.1007/978-3-319-98758-3_4

Ballard J, Wang Z, Li W, Shu X, Zhao L, Zhu M, Qian X, Tang G, Dong L (2024) Deep learning-based approaches for multi-omics data integration and analysis. BioData Mining 17:38. https://doi.org/10.1186/s13040-024-00391-z

Bar-Peled L, Kory N (2022) Principles and functions of metabolic compartmentalization. Nat Metab 4(10):1232–1244. https://doi.org/10.1038/s42255-022-00645-2

Bartel J, Krumsiek J, Theis FJ (2013) Statistical methods for the analysis of high-throughput metabolomics data. Comput Struct Biotechnol J 4:e201301009. https://doi.org/10.5936/csbj.201301009

Baysoy A, Bai Z, Satija R, Andersen MS, Gierahn TM, Kadash-Edmondson KE, Vieth B, Bai M, Kishi JY, Weissman JS, Chen F (2023) The technological landscape and applications of single-cell multi-omics. Nat Rev Mol Cell Biol 24:695–713. https://doi.org/10.1038/s41580-023-00615-w

Begou O, Gika HG, Theodoridis G, Wilson ID (2025) Quality control and validation issues in LC-MS-based metabolomics. Methods Mol Biol 2891:53–66. https://doi.org/10.1007/978-1-0716-4334-1_3

Belhaj MR, Lawler NG, Hoffman NJ (2021) Metabolomics and lipidomics: expanding the molecular landscape of exercise biology. Meta 11(3):151. https://doi.org/10.3390/metabo11030151

Bevan M, Walsh S (2005) The Arabidopsis genome: a foundation for plant research. Genome Res 15:1632–1642. https://doi.org/10.1101/gr.3723405

Beveridge R, Calabrese AN (2021) Structural proteomics methods to interrogate the conformations and dynamics of intrinsically disordered proteins. Front Chem 9:603639. https://doi.org/10.3389/fchem.2021.603639

Birhanu AG (2023) Mass spectrometry-based proteomics as an emerging tool in clinical laboratories. Clin Proteom 20(1):32. https://doi.org/10.1186/s12014-023-09424-x

Bobalova J, Strouhalova D, Bobal P (2023) Common post-translational modifications (PTMs) of proteins: analysis by up-to-date analytical techniques with an emphasis on barley. J Agric Food Chem 71(41):14825–14837. https://doi.org/10.1021/acs.jafc.3c00886

Boccard J, Rudaz S (2018) Integration of metabolomic data from multiple analytical platforms: towards extensive coverage of the metabolome. Adv Bot Res 89:203–228. https://doi.org/10.1016/bs.coac.2018.06.003

Bozza D, Barboni V, Spadafora ND, Pirone F, Simonetti G, Morselli L, Guidi L (2021) Untargeted metabolomics approaches for the characterization of cereals and their derived products by means of liquid chromatography coupled to high resolution mass spectrometry. J Chromatogr Open 6:100168. https://doi.org/10.1016/j.jcoa.2024.100168

Buchberger AR, DeLaney K, Johnson J, Li L (2018) Mass spectrometry imaging: a review of emerging advancements and future insights. Anal Chem 90(1):240–265. https://doi.org/10.1021/acs.analchem.7b04733

Bulut M, Karakas E, Fernie AR (2025) Metabolic responses to multi-stress: An update. Plant Stress 15:100729. https://doi.org/10.1016/j.stress.2024.100729

Butler GS, Dean RA, Morrison CJ, Overall CM (2010) Identification of cellular MMP substrates using quantitative proteomics: isotope-coded affinity tags (ICAT) and isobaric tags for relative and absolute quantification (iTRAQ). Methods Mol Biol 622:451–470. https://doi.org/10.1007/978-1-60327-299-5_26

Calderón-González KG, Hernández-Monge J, Herrera-Aguirre ME, Luna-Arias JP (2016) Bioinformatics tools for proteomics data interpretation. Adv Exp Med Biol 919:281–341. https://doi.org/10.1007/978-3-319-41448-5_16

Cambiaghi A, Ferrario M, Masseroli M (2017) Analysis of metabolomic data: tools, current strategies and future challenges for omics data integration. Brief Bioinform 18(3):498–510. https://doi.org/10.1093/bib/bbw031

Carrera FP, Noceda C, Maridueña-Zavala MG, Cevallos-Cevallos JM (2021) Metabolomics, a powerful tool for understanding plant abiotic stress. Agronomy 11(5):824. https://doi.org/10.3390/agronomy11050824

Cembrowska-Lech D, Krzemińska A, Miller T, Nowakowska A, Adamski C, Radaczyńska M, Mikiciuk G, Mikiciuk M (2023) An integrated multi-omics and artificial intelligence framework for advance plant phenotyping in horticulture. Biology (Basel) 12(10):1298. https://doi.org/10.3390/biology12101298

Chapman B, Castellana N, Apffel A, Ghan R, Cramer GR, Bellgard M, Haynes PA, Van Sluyter SC (2013) Plant proteogenomics: from protein extraction to improved gene predictions. Methods Mol Biol 1002:267–294. https://doi.org/10.1007/978-1-62703-360-2_21

Charitou T, Bryan K, Lynn DJ (2016) Using biological networks to integrate, visualize and analyze genomics data. Genet Sel Evol 48:27. https://doi.org/10.1186/s12711-016-0205-1

Chaudhry M, Shafi I, Mahnoor M, Vargas DLR, Thompson EB, Ashraf I (2023) A systematic literature review on identifying patterns using unsupervised clustering algorithms: a data mining perspective. Symmetry 15(9):1679. https://doi.org/10.3390/sym15091679

Chaves JO, de Souza MC, da Silva LC, Lachos-Perez D (2020) Extraction of flavonoids from natural sources using modern techniques. Front Chem 8:507887. https://doi.org/10.3389/fchem.2020.507887

Chen X, Sun Y, Zhang T, Shu L, Roepstorff P, Yang F (2021) Quantitative proteomics using isobaric labeling: a practical guide. Genomics Proteomics Bioinformatics 19:689–706. https://doi.org/10.1016/j.gpb.2021.08.012

Chen Y, Li EM, Xu LY (2022) Guide to metabolomics analysis: a bioinformatics workflow. Meta 12(4):357. https://doi.org/10.3390/metabo12040357

Chu LX, Wang WJ, Gu XP, Wu P, Gao C, Zhang Q, Wu J, Jiang DW, Huang JQ, Ying XW, Shen JM, Jiang Y, Luo LH, Xu JP, Ying YB, Chen HM, Fang A, Feng ZY, An SH, Li XK, Wang ZG (2024) Spatiotemporal multi-omics: exploring molecular landscapes in aging and regenerative medicine. Mil Med Res 11(1):31. https://doi.org/10.1186/s40779-024-00537-4

Deek RA, Ma S, Lewis J, Li H (2024) Statistical and computational methods for integrating microbiome, host genomics, and metabolomics data. eLife 13:e88956. https://doi.org/10.7554/eLife.88956

Deschamps E, Calabrese V, Schmitz I, Hubert-Roux M, Castagnos D, Afonso C (2023) Advances in ultra-high-resolution mass spectrometry for pharmaceutical analysis. Molecules 28(5):2061. https://doi.org/10.3390/molecules28052061

Dettmer K, Aronov PA, Hammock BD (2007) Mass spectrometry-based metabolomics. Mass Spectrom Rev 26(1):51–78. https://doi.org/10.1002/mas.20108

Deutsch EW, Perez-Riverol Y, Chalkley RJ, Wilhelm M, Tate S, Sachsenberg T, Walzer M, Käll L,Delanghe B, Böcker S, Schymanski EL, Wilmes P, Dorfer V, Kuster B, Volders PJ, Jehmlich N,Vissers JPC, Wolan DW, Wang AY, Mendoza L, Shofstahl J, Dowsey AW, Griss J, Salek RM,Neumann S, Binz PA, Lam H, Vizcaíno JA, Bandeira N, Röst H (2018) Expanding the use of spectral librariesin proteomics. J Proteome Res 17(12):4051–4060. https://doi.org/10.1021/acs.jproteome.8b00485

Di Girolamo F, Lante I, Muraca M, Putignani L (2013) The role of mass spectrometry in the "Omics" era. Curr Org Chem 17(23):2891–2905. https://doi.org/10.2174/1385272817888131118162725

Duong V-A, Lee H (2023) Bottom-up proteomics: advancements in sample preparation. Int J Mol Sci 24(6):5350. https://doi.org/10.3390/ijms24065350

Ehlers H, Brich N, Krone M, Fleming A, Hahn U, Leitte H (2025) An introduction to and survey of biological network visualization. Comput Graph 126:104115. https://doi.org/10.1016/j.cag.2024.104115

Elshafie HS, Camele I, Mohamed AA (2023) A comprehensive review on the biological, agricultural and pharmaceutical properties of secondary metabolites based-plant origin. Int J Mol Sci 24(4):3266. https://doi.org/10.3390/ijms24043266

Emwas AH, Roy R, McKay RT, Tenori L et al (2019) NMR spectroscopy for metabolomics research. Metabolites 9:123. https://doi.org/10.3390/metabo9070123

Fang C, Fernie AR, Luo J (2019) Exploring the diversity of plant metabolism. Trends Plant Sci 24:83–98. https://doi.org/10.1016/j.tplants.2018.09.006

Fiehn O (2002) Metabolomics—the link between genotypes and phenotypes. Plant Mol Biol 48:155–171. https://doi.org/10.1023/A:1013713905833

Fiehn O (2016) Metabolomics by gas chromatography-mass spectrometry: combined targeted and untargeted profiling. Curr Protoc Mol Biol 114:30.4.1–30.4.32. https://doi.org/10.1002/0471142727.mb3004s114

Flores JE, Claborne DM, Weller ZD, Webb-Robertson B-JM, Waters KM, Bramer LM (2023) Missing data in multi-omics integration: recent advances through artificial intelligence. Front Artif Intell 6:1098308. https://doi.org/10.3389/frai.2023.1098308

Gao Y, Zhao C (2024) Development and applications of metabolic models in plant multi-omics research. Front Plant Sci 15:1361183. https://doi.org/10.3389/fpls.2024.1361183

García-Pérez P, Becchi PP, Zhang L, Rocchetti G, Lucini L (2024) Metabolomics and chemometrics: the next-generation analytical toolkit for the evaluation of food quality and authenticity. Trends Food Sci Technol 147:104481. https://doi.org/10.1016/j.tifs.2024.104481

Gardinassi LG, Xia J, Safo S, Yu G (2017) Bioinformatics tools for the interpretation of metabolomics data. Curr Pharmacol Rep 3:374–383. https://doi.org/10.1007/s40495-017-0107-0

Gelpí E (2009) From large analogical instruments to small digital black boxes: 40 years of progress in mass spectrometry and its role in proteomics. Part II 1985–2000. J Mass Spectrom 44(8):1137–1161. https://doi.org/10.1002/jms.1621

Gessler A, Wieloch T, Saurer M, Ma H, Schleucher J, Grams TEE, Ubierna N, Tcherkez G (2024) The marriage between stable isotope ecology and plant metabolomics—new perspectives for metabolic flux analysis and the interpretation of ecological archives. New Phytol 244:21–31. https://doi.org/10.1111/nph.19973

Gorrochategui E, Jaumot J, Lacorte S, Tauler R (2016) Data analysis strategies for targeted and untargeted LC-MS metabolomic studies: overview and workflow. TrAC Trends Anal Chem 82:425–442. https://doi.org/10.1016/j.trac.2016.07.004

Gouw JW, Krijgsveld J, Heck AJ (2010) Quantitative proteomics by metabolic labeling of model organisms. Mol Cell Proteomics 9(1):11–24. https://doi.org/10.1074/mcp.R900001-MCP200

Graw S, Chappell K, Washam CL, Gies A, Bird J, Robeson MS 2nd, Byrum SD (2021) Multi-omics data integration considerations and study design for biological systems and disease. Mol Omics 17(2):170–185. https://doi.org/10.1039/d0mo00041h

Guo J, Yu H, Xing S, Huan T (2022) Addressing big data challenges in mass spectrometry-based metabolomics. Chem Commun 58:9979–9990. https://doi.org/10.1039/D2CC03598G

Hamany Djande CY, Pretorius C, Tugizimana F, Piater LA, Dubery IA (2020) Metabolomics: a tool for cultivar phenotyping and investigation of grain crops. Agronomy 10(6):831. https://doi.org/10.3390/agronomy10060831

Heazlewood JL, Wallace IS, Xu S-L (2022) Editorial: advances in plant proteomics. Front Plant Sci 13:1072217. https://doi.org/10.3389/fpls.2022.1072217

Hein MY, Peng D, Todorova V, McCarthy F (2024) Global organelle profiling reveals subcellular localization and remodeling at proteome scale. Cell. https://doi.org/10.1016/j.cell.2024.11.028

Hu J, Rampitsch C, Bykova NV (2015) Advances in plant proteomics toward improvement of crop productivity and stress resistance. Front Plant Sci 6:209. https://doi.org/10.3389/fpls.2015.00209

Hu T, Allam M, Cai S, Henderson W, Yueh B, Garipcan A, Ievlev AV, Afkarian M, Beyaz S, Coskun AF (2023) Single-cell spatial metabolomics with cell-type specific protein profiling for tissue systems biology. Nat Commun 14(1):8260. https://doi.org/10.1038/s41467-023-43917-5

Huang H, van Dullemen LFA, Akhtar MZ et al (2018) Proteo-metabolomics reveals compensation between ischemic and non-injured contralateral kidneys after reperfusion. Sci Rep 8(1):8539. https://doi.org/10.1038/s41598-018-26804-8

Iqbal Z, Iqbal MS, Khan MIR, Ansari MI (2021) Toward integrated multi-omics intervention: Rice trait improvement and stress management. Front Plant Sci 12:741419. https://doi.org/10.3389/fpls.2021.741419

Ivanisevic J, Want EJ (2019) From samples to insights into metabolism: uncovering biologically relevant information in LC-HRMS metabolomics data. Meta 9(12):308. https://doi.org/10.3390/metabo9120308

Jamil IN, Remali J, Azizan KA, Nor Muhammad NA, Arita M, Goh H-H, Aizat WM (2020) Systematic multi-omics integration (MOI) approach in plant systems biology. Front Plant Sci 11:944. https://doi.org/10.3389/fpls.2020.00944

Jendoubi T (2021) Approaches to integrating metabolomics and multi-omics data: a primer. Meta 11(3):184. https://doi.org/10.3390/metabo11030184

Jiang MZ, Aguet F, Ardlie K, Chen J, NHLBI Trans-Omics for Precision Medicine (TOPMed) Consortium, TOPMed Analysis Working Group (2023) Canonical correlation analysis for multi-omics: application to cross-cohort analysis. PLoS Genet 19(5). https://doi.org/10.1371/journal.pgen.1010517

Johnson RS, Searle BC, Nunn BL, Gilmore JM, Phillips M, Amemiya CT, Heck M, MacCoss MJ (2020) Assessing protein sequence database suitability using De novo sequencing. Mol Cell Proteomics 19(1):198–208. https://doi.org/10.1074/mcp.TIR119.001752

Jordán F, Nguyen T-P, Liu W-c (2012) Studying protein–protein interaction networks: a systems view on diseases. Brief Funct Genomics 11:497–504. https://doi.org/10.1093/bfgp/els035

Juarez-Escobar J, Bojórquez-Velázquez E, Elizalde-Contreras JM, Guerrero-Analco JA et al (2021) Current proteomic and metabolomic knowledge of zygotic and somatic embryogenesis in plants. Int J Mol Sci 22(21):11807. https://doi.org/10.3390/ijms222111807

Jorrin-Novo JV (2014) Plant proteomics methods and protocols. Methods Mol Biol 1072:3–13. https://doi.org/10.1007/978-1-62703-631-3_1

Kapoore RV, Vaidyanathan S (2016) Towards quantitative mass spectrometry-based metabolomics in microbial and mammalian systems. Philos Trans A Math Phys Eng Sci 374(2079):20150363. https://doi.org/10.1098/rsta.2015.0363

Karimpour-Fard A, Epperson LE, Hunter LE (2015) A survey of computational tools for downstream analysis of proteomic and other omic datasets. Hum Genomics 9:28. https://doi.org/10.1186/s40246-015-0050-2

Katam R, Lin C, Grant K, Katam CS, Chen S (2022) Advances in plant metabolomics and its applications in stress and single-cell biology. Int J Mol Sci 23(13):6985. https://doi.org/10.3390/ijms23136985

Kausar R, Wang X, Komatsu S (2022) Crop proteomics under abiotic stress: from data to insights. Plants 11(21):2877. https://doi.org/10.3390/plants11212877

Kettenbach AN, Rush J, Gerber SA (2011) Absolute quantification of protein and post-translational modification abundance with stable isotope-labeled synthetic peptides. Nat Protoc 6(2):175–186. https://doi.org/10.1038/nprot.2010.196

Khan KY, Ali B, Ghani HU (2024) Metabolomics combined with proteomics reveals phytotoxic effects of norfloxacin under drought stress on Oryza sativa. Plant Physiol Biochem 216:109130. https://doi.org/10.1016/j.plaphy.2024.109130

Khanijou JK, Kulyk H, Bergès C (2022) Metabolomics and modelling approaches for systems metabolic engineering. Metabolic Eng Commun 15. https://doi.org/10.1016/j.mec.2022.e00209

Khemani B, Patil S, Kotecha K, Abraham A, Quiles JM, Jha CK, Prabhune AA, Kamat RK (2024) A review of graph neural networks: concepts, architectures, techniques, challenges, datasets, applications, and future directions. J Big Data 11:18. https://doi.org/10.1186/s40537-023-00876-4

Kim MS, Pandey A (2012) Electron transfer dissociation mass spectrometry in proteomics. Proteomics 12(4–5):530–542. https://doi.org/10.1002/pmic.201100517

Kim J-S, Kidokoro S, Yamaguchi-Shinozaki K, Shinozaki K (2024) Regulatory networks in plant responses to drought and cold stress. Plant Physiol 195:170–189. https://doi.org/10.1093/plphys/kiae105

Kiouri DP, Batsis GC, Chasapis CT (2025) Structure-based approaches for protein–protein interaction prediction using machine learning and deep learning. Biomolecules 15(1):141. https://doi.org/10.3390/biom15010141

Kirkwood JS, Maier C, Stevens JF (2013) Simultaneous, untargeted metabolic profiling of polar and nonpolar metabolites by LC-Q-TOF mass spectrometry. Curr Protoc Toxicol Chapter 4:39. https://doi.org/10.1002/0471140856.tx0439s56

Kohl SM, Klein MS, Hochrein J, Oefner PJ, Spang R, Gronwald W (2012) State-of-the art data normalization methods improve NMR-based metabolomic analysis. Metabolomics 8(Suppl 1):146–160. https://doi.org/10.1007/s11306-011-0350-z

Kontostathi G, Makridakis M, Bitsika V, Tsolakos N, Vlahou A, Zoidakis J (2019) Development and validation of multiple reaction monitoring (MRM) assays for clinical applications. Methods Mol Biol 1959:205–223. https://doi.org/10.1007/978-1-4939-9164-8_14

Koppad S, Basava A, Gkoutos G, Acharjee A (2021) Cloud computing enabled big multi-omics data analytics. Bioinform Biol nsights 15:117793222110359. https://doi.org/10.1177/11779322211035921

Kumar R, Bohra A, Pandey AK, Pandey MK, Kumar A (2017) Metabolomics for plant improvement: status and prospects. Front Plant Sci 8:1302. https://doi.org/10.3389/fpls.2017.01302

Lee K, O'Reilly FJ (2023) Cross-linking mass spectrometry for mapping protein complex topologies in situ. Essays Biochem 67(2):215–228. https://doi.org/10.1042/EBC20220168

Lewandowska D, ten Have S, Hodge K, Tillemans V, Lamond AI, Brown JW (2013) Plant SILAC: stable-isotope labelling with amino acids of arabidopsis seedlings for quantitative proteomics. PLoS One 8(8). https://doi.org/10.1371/journal.pone.0072207

Li J, Smith LS, Zhu HJ (2021) Data-independent acquisition (DIA): An emerging proteomics technology for analysis of drug-metabolizing enzymes and transporters. Drug Discov Today Technol 39:49–56. https://doi.org/10.1016/j.ddtec.2021.06.006

Liu Y, Lu S, Liu K, Wang S, Huang L, Guo L (2019) Proteomics: a powerful tool to study plant responses to biotic stress. Plant Methods 15:135. https://doi.org/10.1186/s13007-019-0515-8

Liu Q, Yang F, Zhang J, Liu H, Rahman S, Islam S, Ma W, She M (2021a) Application of CRISPR/Cas9 in crop quality improvement. Int J Mol Sci 22(8):4206. https://doi.org/10.3390/ijms22084206

Liu W, Zhao H, Miao C, Jin W (2021b) Integrated proteomics and metabolomics analysis of transgenic and gene-stacked maize line seeds. GM Crops Food 12(1):361–375. https://doi.org/10.1080/21645698.2021.1934351

Lohrig K, Wolters D (2009) Multidimensional protein identification technology. Methods Mol Biol 564:143–153. https://doi.org/10.1007/978-1-60761-157-8_8

Lu W, Su X, Klein MS, Lewis IA, Fiehn O, Rabinowitz JD (2017) Metabolite measurement: pitfalls to avoid and practices to follow. Annu Rev Biochem 86:277–304. https://doi.org/10.1146/annurev-biochem-061516-044952

Ma B, Johnson R (2012) De novo sequencing and homology searching. Mol Cell Proteomics 11(2):O111.014902. https://doi.org/10.1074/mcp.O111.014902

Magdeldin S, Enany S, Yoshida Y et al (2014) Basics and recent advances of two dimensional-polyacrylamide gel electrophoresis. Clin Proteomics 11(1):16. https://doi.org/10.1186/1559-0275-11-16

Mahrous EA, Farag MA (2015) Two dimensional NMR spectroscopic approaches for exploring plant metabolome: a review. J Adv Res 6(1):3–15. https://doi.org/10.1016/j.jare.2014.10.003

Maia M, Figueiredo A, Cordeiro C, Silva MS (2021) FT-ICR-MS-based metabolomics: a deep dive into plant metabolism. Mass Spectrometry Rev 42:1535–1556. https://doi.org/10.1002/mas.21738

Manickam S, Rajagopalan VR, Kambale R, Rajasekaran R, Kanagarajan S, Muthurajan R (2023) Plant metabolomics: current initiatives and future prospects. Curr Issues Mol Biol 45(11):8894–8906. https://doi.org/10.3390/cimb45110558

Mann M, Kumar C, Zeng W-F, Strauss MT (2021) Artificial intelligence for proteomics and biomarker discovery. Cell Syst 12:759–770. https://doi.org/10.1016/j.cels.2021.06.006

Marchev AS, Vasileva LV, Amirova KM, Savova MS, Balcheva-Sivenova ZP, Georgiev MI (2021) Metabolomics and health: from nutritional crops and plant-based pharmaceuticals to profiling of human biofluids. Cell Mol Life Sci 78(19–20):6487–6503. https://doi.org/10.1007/s00018-021-03918-3

Marinka Z, Michelle ML, Aydin W, Kimberly G, Deisy MG (2024) Current and future directions in network biology. Bioinform Adv 4:vbae0994. https://doi.org/10.1093/bioadv/vbae099

Martínez-Esteso MJ, Morante-Carriel J, Samper-Herrero A, Martínez-Márquez A, Sellés-Marchart S, Nájera H, Bru-Martínez R (2024) Proteomics: an essential tool to study plant-specialized metabolism. Biomolecules 14(12):1539. https://doi.org/10.3390/biom14121539

Mashabela MD, Masamba P, Kappo AP (2022) Metabolomics and chemoinformatics in agricultural biotechnology research: complementary probes in unravelling new metabolites for crop improvement. Biology 11(8):1156. https://doi.org/10.3390/biology11081156

Mehrotra B, Mendes P (2006) Bioinformatics approaches to integrate metabolomics and other systems biology data. In: Saito K, Dixon RA, Willmitzer L (eds) Plant metabolomics. biotechnology in agriculture and forestry, vol 57. Springer, Berlin, Heidelberg, pp 105–115. https://doi.org/10.1007/3-540-29782-0_8

Mergner J, Kuster B (2022) Plant proteome dynamics. Annu Rev Plant Biol 73:67–92. https://doi.org/10.1146/annurev-arplant-102620-031308

Mohammad Z, Ehsan D, Subhasha N, Mahdiyeh G, Dariush S, William R, Meisam T (2021) The effect of solvents polarity and extraction conditions on the microalgal lipids yield, fatty acids profile, and biodiesel properties. Bioresour Technol:126303. https://doi.org/10.1016/j.biortech.2021.126303

Muleya V, Lois LM, Chahtane H, Thomas L, Chiapello M, Marondedze C (2022) (De) dctivation (Ir) reversibly or degradation: dynamics of post-translational protein modifications in plants. Life (Basel) 12(2):324. https://doi.org/10.3390/life12020324

Naik B, Kumar V, Rizwanuddin S, Chauhan M, Choudhary M, Gupta AK, Kumar P, Kumar V, Saris PEJ, Rather MA, Bhuyan S, Neog PR, Mishra S, Rustagi S (2023) Genomics, proteomics, and metabolomics approaches to improve abiotic stress tolerance in tomato plant. Int J Mol Sci 24(3):3025. https://doi.org/10.3390/ijms24033025

Neely BA, Palmblad M (2022) Machine learning in proteomics and metabolomics. J Proteome Res 21:2553–2554. https://doi.org/10.1021/acs.jproteome.2c00566

O'Farrell PH (2008) The pre-omics era: the early days of two-dimensional gels. Proteomics 8(23–24):4842–4852. https://doi.org/10.1002/pmic.200800719

Oh S-W, Imran M, Kim E-H, Park S-Y, Lee S-G, Park H-M, Jung J-W, Ryu T-H (2023) Approach strategies and application of metabolomics to biotechnology in plants. Front Plant Sci 14:1192235. https://doi.org/10.3389/fpls.2023.1192235

Padula MP, Berry IJ, O'Rourke MB, Raymond BBA, Santos J, Djordjevic SP (2017) A comprehensive guide for performing sample preparation and top-down protein analysis. Proteomes 5(2):11. https://doi.org/10.3390/proteomes5020011

Panditrao G, Bhowmick R, Meena C, Sarkar RR (2022) Emerging landscape of molecular interaction networks: opportunities, challenges and prospects. J Biosci 47(2):24. https://doi.org/10.1007/s12038-022-00253-y

Parker EJ, Billane KC, Austen N, Cotton A, George RM, Hopkins D, Lake JA, Pitman JK, Prout JN, Walker HJ, Williams A, Cameron DD (2023) Untangling the complexities of processing and analysis for untargeted LC-MS data using open-source tools. Meta 13(4):463. https://doi.org/10.3390/metabo13040463

Paul I, White C, Turcinovic I, Emili A (2021) Imaging the future: the emerging era of single-cell spatial proteomics. FEBS J 288:6990–7001. https://doi.org/10.1111/febs.15685

Perez de Souza L, Fernie AR (2024) Computational methods for processing and interpreting mass spectrometry-based metabolomics. Essays Biochem 68(1):5–13. https://doi.org/10.1042/EBC20230019

Perez de Souza L, Naake T, Tohge T, Fernie AR (2017) From chromatogram to analyte to metabolite. How to pick horses for courses from the massive web resources for mass spectral plant metabolomics. GigaScience 6(7). https://doi.org/10.1093/gigascience/gix037

Perez De Souza L, Alseekh S, Brotman Y, Fernie AR (2020) Network-based strategies in metabolomics data analysis and interpretation: from molecular networking to biological interpretation. Expert Rev Proteomics 17(4):243–255. https://doi.org/10.1080/14789450.2020.1766975

Pinto RC, Karaman I, Lewis MR (2022) Finding correspondence between metabolomic features in untargeted liquid chromatography–mass spectrometry metabolomics datasets. Anal Chem 94:5493–5503. https://doi.org/10.1021/acs.analchem.1c03592

Po A, Eyers CE (2023) Top-down proteomics and the challenges of true proteoform characterization. J Proteome Res 22:3663–3675. https://doi.org/10.1021/acs.jproteome.3c00416

Porta Siegel T, Hamm G, Bunch J, Cappell J, Fletcher JS, Schwamborn K (2018) Mass spectrometry imaging and integration with other imaging modalities for greater molecular understanding of biological tissues. Mol Imaging Biol 20:888–901. https://doi.org/10.1007/s11307-018-1267-y

Pott DM, Durán-Soria S, Osorio S, Vallarino JG (2021) Combining metabolomic and transcriptomic approaches to assess and improve crop quality traits. CABI Agric Biosci 2:1. https://doi.org/10.1186/s43170-020-00021-8

Pudjihartono N, Fadason T, Kempa-Liehr AW, O'Sullivan JM (2022) A review of feature selection methods for machine learning-based disease risk prediction. Front Bioinform 2:927312. https://doi.org/10.3389/fbinf.2022.927312

Qiu S, Cai Y, Yao H, Yang Z, Liu S, Gong Y, Yu S, Xiao B, Li L, Jia J, Lin J, Du G, Ye J (2023) Small molecule metabolites: discovery of biomarkers and therapeutic targets. Sig Transduct Target Ther 8:132. https://doi.org/10.1038/s41392-023-01399-3

Rajczewski AT, Jagtap PD, Griffin TJ (2022) An overview of technologies for MS-based proteomics-centric multi-omics. Expert Rev Proteomics 19(3):165–181. https://doi.org/10.1080/14789450.2022.2070476

Ramadan Z, Jacobs D, Grigorov M, Kochhar S (2006) Metabolic profiling using principal component analysis, discriminant partial least squares, and genetic algorithms. Talanta 68:1683–1691. https://doi.org/10.1016/j.talanta.2005.08.042

Ramalingam A, Kudapa H, Pazhamala LT, Weckwerth W, Varshney RK (2015) Proteomics and metabolomics: two emerging areas for legume improvement. Front Plant Sci 6:1116. https://doi.org/10.3389/fpls.2015.01116

Rao X, Liu W (2025) A guide to metabolic network modeling for plant biology. Plants 14(3):484. https://doi.org/10.3390/plants14030484

Rasheed F, Markgren J, Hedenqvist M, Johansson E (2020) Modeling to understand plant protein structure-function relationships-implications for seed storage proteins. Molecules 25(4):873. https://doi.org/10.3390/molecules25040873

Ravikrishnan A, Nasre M, Raman K (2018) Enumerating all possible biosynthetic pathways in metabolic networks. Sci Rep 8:9932. https://doi.org/10.1038/s41598-018-28007-7

Raza A, Salehi H, Bashir S, Malhi AI, Ayub A, Jamil S, Jabbar B, Iqbal N, Khan MB, Ahmad G, Athar HUR, Aslam A, Castro PH, Azeem F, Aksoy E, Batool M, Wang X, Zulfiqar MKZ, Calderón-Domínguez G, Khan SH, Scafaro AP, Furbank RT, Prasad PVV (2024) Transcriptomics, proteomics, and metabolomics interventions prompt crop improvement against metal(loid) toxicity. Plant Cell Rep 43:80. https://doi.org/10.1007/s00299-024-03153-7

Reddy AS, Marquez Y, Kalyna M, Barta A (2013) Complexity of the alternative splicing landscape in plants. Plant Cell 25(10):3657–3683. https://doi.org/10.1105/tpc.113.117523

Reimand J, Isserlin R, Voisin V, Kucera M, Tannus-Lopes C, Rostamianfar A, Wadi L, Meyer M, Wong J, Xu C, Merico D, Bader GD (2019) Pathway enrichment analysis and visualization of omics data using GSEA, Cytoscape and EnrichmentMap. Nat Protoc 14(2):482–517. https://doi.org/10.1038/s41596-018-0103-9

Resurreccion EP, Fong KW (2022) The integration of metabolomics with other omics: insights into understanding prostate cancer. Meta 12(6):488. https://doi.org/10.3390/metabo12060488

Robb CG, Dao TP, Ujma J, Castañeda CA, Beveridge R (2023) Ion mobility mass spectrometry unveils global protein conformations in response to conditions that promote and reverse liquid-liquid phase separation. J Am Chem Soc 145(23):12541–12549. https://doi.org/10.1021/jacs.3c00756

Roberts LD, Souza AL, Gerszten RE, Clish CB (2012) Targeted metabolomics. Curr Protoc Mol Biol Chapter 30(2):1–24. https://doi.org/10.1002/0471142727.mb3002s98

Romero H, Pott DM, Vallarino JG, Osorio S (2021) Metabolomics-based evaluation of crop quality changes as a consequence of climate change. Meta 11(7):461. https://doi.org/10.3390/metabo11070461

Ronsein GE, Pamir N, von Haller PD, Kim DS, Oda MN, Jarvik GP, Vaisar T, Heinecke JW (2015) Parallel reaction monitoring (PRM) and selected reaction monitoring (SRM) exhibit comparable linearity, dynamic range and precision for targeted quantitative HDL proteomics. J Proteome 113:388–399. https://doi.org/10.1016/j.jprot.2014.10.017

Roychowdhury R, Das SP, Gupta A, Parihar P, Chandrasekhar K, Sarker U, Kumar A, Ramrao DP, Sudhakar C (2023) Multi-omics pipeline and omics-integration approach to decipher plant's abiotic stress tolerance responses. Genes (Basel) 14(6):1281. https://doi.org/10.3390/genes14061281

Rozanova S, Barkovits K, Nikolov M, Schmidt C, Urlaub H, Marcus K (2021) Quantitative mass spectrometry-based proteomics: An overview. In: Marcus K, Eisenacher M, Sitek B (eds) Quantitative methods in proteomics. Methods in molecular biology, vol 2228. Humana, New York, pp 1–33. https://doi.org/10.1007/978-1-0716-1024-4_8

Sakurai N (2022) Recent applications of metabolomics in plant breeding. Breed Sci 72(1):56–65. https://doi.org/10.1270/jsbbs.21065

Salam U, Ullah S, Tang ZH, Elateeq AA, Khan Y, Khan J, Khan A, Ali S (2023) Plant metabolomics: An overview of the role of primary and secondary metabolites against different environmental stress factors. Life (Basel) 13(3):706. https://doi.org/10.3390/life13030706

Salem MA, Jüppner J, Bajdzienko K, Giavalisco P (2016) Protocol: a fast, comprehensive and reproducible one-step extraction method for the rapid preparation of polar and semi-polar metabolites, lipids, proteins, starch and cell wall polymers from a single sample. Plant Methods 12:45. https://doi.org/10.1186/s13007-016-0146-2

Salem MA, Perez de Souza L, Serag A, Fernie AR, Farag MA, Ezzat SM, Alseekh S (2020) Metabolomics in the context of plant natural products research: from sample preparation to metabolite analysis. Meta 10(1):37. https://doi.org/10.3390/metabo10010037

Sanches PHG, de Melo NC, Porcari AM, de Carvalho LM (2024) Integrating molecular perspectives: strategies for comprehensive multi-omics integrative data analysis and machine learning

applications in transcriptomics, proteomics, and metabolomics. Biology 13(11):848. https://doi.org/10.3390/biology13110848
Sangwan NS, Jha S, Mitra A (2024) Unlocking nature's treasure trove: biosynthesis and elicitation of secondary metabolites from plants. Plant Growth Regul 104:1–4. https://doi.org/10.1007/s10725-024-01184-4
Santamaria G, Pinto FR (2024) Bioinformatic analysis of metabolomic data: from raw spectra to biological insight. Biochemist 4(2):90–114. https://doi.org/10.3390/biochem4020005
Sarkar J, Singh R, Chandel S (2025) Understanding LC/MS-based metabolomics: a detailed reference for natural product analysis. Proteomics Clin Appl 19. https://doi.org/10.1002/prca.202400048
Sawada Y, Hirai MY (2013) Integrated LC-MS/MS system for plant metabolomics. Comput Struct Biotechnol J 4. https://doi.org/10.5936/csbj.201301011
Schrimpe-Rutledge AC, Codreanu SG, Sherrod SD, McLean JA (2016) Untargeted metabolomics strategies-challenges and emerging directions. J Am Soc Mass Spectrom 27(12):1897–1905. https://doi.org/10.1007/s13361-016-1469-y
Scott KB, Turko IV, Phinney KW (2016) Chapter 11: QconCAT: internal standard for protein quantification. In: Kelman Z (ed) Methods in enzymology, vol 566. Academic Press, pp 289–303. https://doi.org/10.1016/bs.mie.2015.09.022
Sharma V, Gupta P, Priscilla K, Kumar S, Hangargi B, Veershetty A, Kumar P, Dash PK, Bandopadhyay R, Basu U, Jha CK, Kumar A, Bohra A, Sirari A, Singh AK, Singh NP, Pareek A, Varshney RK, Bhat KV, Mantri S (2021) Metabolomics intervention towards better understanding of plant traits. Cells 10:346. https://doi.org/10.3390/cells10020346
Shen X, Wang R, Xiong X, Yin Y, Cai Y, Ma Z, Liu N, Zhu ZJ (2019) Metabolic reaction network-based recursive metabolite annotation for untargeted metabolomics. Nat Commun 10(1):1516. https://doi.org/10.1038/s41467-019-09550-x
Shen S, Zhan C, Yang C, Fernie AR, Luo J (2023) Metabolomics-centered mining of plant metabolic diversity and function: past decade and future perspectives. Mol Plant 16:43–63. https://doi.org/10.1016/j.molp.2022.09.007
Shojaie A (2021) Differential network analysis: a statistical perspective. Wiley Interdiscip Rev Comput Stat 13(2). https://doi.org/10.1002/wics.1508
Sidoli S, Lin S, Xiong L, Bhanu NV, Karch KR, Johansen E, Hunter C, Mollah S, Garcia BA (2015) Sequential Window Acquisition of all Theoretical Mass Spectra (SWATH) analysis for characterization and quantification of histone post-translational modifications. Mol Cell Proteomics 14(9):2420–2428. https://doi.org/10.1074/mcp.O114.046102
Sikic K, Tomic S, Carugo O (2010) Systematic comparison of crystal and NMR protein structures deposited in the protein data bank. Open Biochem J 4:83–95. https://doi.org/10.2174/1874091X01004010083
Singh DP, Bisen MS, Shukla R, Prabha R, Maurya S, Reddy YS et al (2022) Metabolomics-driven mining of metabolite resources: applications and prospects for improving vegetable crops. Intl J Mol Sci 23(20):12062. https://doi.org/10.3390/ijms232012062
Singh DP, Bisen MS, Prabha R, Maurya S et al (2023a) Untargeted metabolomics of *Alternaria solani*-challenged wild tomato species *Solanum cheesmaniae r*evealed key metabolite biomarkers and insight into altered metabolic pathways. Metabolites 13(5):585. https://doi.org/10.3390/metabo13050585
Singh DP, Maurya S, Yerasu SR et al (2023b) Metabolomics of early blight (*Alternaria solani*) susceptible tomato (*Solanum lycopersicum*) unfolds key biomarker metabolites and involved metabolic pathways. Sci Rep 13:21023. https://doi.org/10.1038/s41598-023-48269-0
Singh DP, Rai N, Farag MA, Aftab T, Abdelbacki AM, Jiang CJ, Chi SQ, Fan XW, Naveed M, Verma KK, Singh P, Muhammad I, Datta R, Zulfiqar S, Misra P, Kumar D, Swamy KN, Ramakrishna A, Pant RP, Hakeem KR, Sharma M (2024) Metabolic diversity, biosynthetic pathways, and metabolite biomarkers analysed via untargeted metabolomics and the antioxidant potential reveal for high temperature tolerance in tomato hybrid. Plant Stress 11:100420. https://doi.org/10.1016/j.stress.2024.100420

Smith MR, Rao IM, Merchant A (2018) Source-sink relationships in crop plants and their influence on yield development and nutritional quality. Front Plant Sci 9:1889. https://doi.org/10.3389/fpls.2018.01889

Son A, Kim W, Park J, Park Y, Lee W, Lee S, Kim H (2024) Mass spectrometry advancements and applications for biomarker discovery, diagnostic innovations, and personalized medicine. Int J Mol Sci 25(18):9880. https://doi.org/10.3390/ijms25189880

Souza AL, Patti GJ (2021) A protocol for untargeted metabolomic analysis: from sample preparation to data processing. Methods Mol Biol 2276:357–382. https://doi.org/10.1007/978-1-0716-1266-8_27

Subba P, Narayana Kotimoole C, Prasad TSK (2019) Plant proteome databases and bioinformatic tools: an expert review and comparative insights. OMICS 23(4):190–206. https://doi.org/10.1089/omi.2019.0024

Subramanian I, Verma S, Kumar S, Jere A, Anamika K (2020) Multi-omics data integration, interpretation, and its application. Bioinform Biol Insights 14:1177932219899051. https://doi.org/10.1177/1177932219899051

Sun J, Xia Y (2024) Pretreating and normalizing metabolomics data for statistical analysis. Metabolomics 20:62. https://doi.org/10.1007/s11306-024-02115-0

Szabo Z, Janaky T (2015) Challenges and developments in protein identification using mass spectrometry. TrAC Trends Anal Chem 69:76–87. https://doi.org/10.1016/j.trac.2015.03.007

Taye MM (2023) Understanding of machine learning with deep learning: architectures, workflow, applications and future directions. Computers 12(5):91. https://doi.org/10.3390/computers12050091

Vasupalli N, Bhat JA, Jain P, Sri T, Yadav P, Mandapaka M, Adinarayana VVRC, Bhat RS, Joshi P, Kadam US (2024) Omics big data for crop improvement: opportunities and challenges. Crop J 12:1517–1532. https://doi.org/10.1016/j.cj.2024.10.007

Wang JH, Byun J, Pennathur S (2010) Analytical approaches to metabolomics and applications to systems biology. Semin Nephrol 30(5):500–511. https://doi.org/10.1016/j.semnephrol.2010.07.007

Wanichthanarak K, Fahrmann JF, Grapov D (2015) Genomic, proteomic, and metabolomic data integration strategies. Biomark Insights 10(Suppl 4):1–6. https://doi.org/10.4137/BMI.S29511

Warth B, Levin N, Rinehart D, Teijaro J, Benton HP, Siuzdak G (2017) Metabolizing data in the cloud. Trends Biotechnol 35(6):481–483. https://doi.org/10.1016/j.tibtech.2016.12.010

Wieczorek MN, Zhou W, Pawliszyn J (2024) Perspective on sample preparation fundamentals. Adv Sample Prep 10:100114. https://doi.org/10.1016/j.sampre.2024.100114

Wienkoop S, Morgenthal K, Wolschin F et al (2008) Integration of metabolomic and proteomic phenotypes: analysis of data covariance dissects starch and RFO metabolism from low and high temperature compensation response in Arabidopsis thaliana. Mol Cell Proteomics 7(9):1725–1736. https://doi.org/10.1074/mcp.M700273-MCP200

Wijethunga AM, Emenike C (2024) Separation methods for food protein purification and analysis. Explor Foods Foodomics 2:391–407. https://doi.org/10.37349/eff.2024.00043

Worley B, Powers R (2013) Multivariate analysis in metabolomics. Curr Metabolomics 1(1):92–107. https://doi.org/10.2174/2213235X11301010092

Wu S, Zhang S, Liu C-M, Fernie AR, Yan S (2025) Recent advances in mass spectrometry-based protein interactome studies. Mol Cell Proteomics 24:100887. https://doi.org/10.1016/j.mcpro.2024.100887

Xiong W, Jiang H, Maness P (2020) Dynamic flux analysis: An experimental approach of fluxomics. Methods Mol Biol 2096:179–196. https://doi.org/10.1007/978-1-0716-0195-2_14

Xu Y, Correa E, Goodacre R (2013) Integrating multiple analytical platforms and chemometrics for comprehensive metabolic profiling: application to meat spoilage detection. Anal Bioanal Chem 405(15):5063–5074. https://doi.org/10.1007/s00216-013-6884-3

Yan S, Bhawal R, Yin Z, Meng X, Wang Z, Li H, Wang F, Li Y, Wang B, Zhao Y, Wang P, Luan H, Zhang Y (2022) Recent advances in proteomics and metabolomics in plants. Mol Horticult 2:17. https://doi.org/10.1186/s43897-022-00038-9

Yu X, Liu Z, Sun X (2023) Single-cell and spatial multi-omics in the plant sciences: technical advances, applications, and perspectives. Plant Commun 4:100508. https://doi.org/10.1016/j.xplc.2022.100508

Zenda T, Liu S, Dong A, Li J, Wang Y, Liu X, Wang N, Duan H (2021) Omics-facilitated crop improvement for climate resilience and superior nutritive value. Front Plant Sci 12:774994. https://doi.org/10.3389/fpls.2021.774994

Zeng Z, Zhang S, Li W, Chen B, Li W (2022) Gene-coexpression network analysis identifies specific modules and hub genes related to cold stress in rice. BMC Genomics 23(1):251. https://doi.org/10.1186/s12864-022-08438-3

Zhan C, Shen S, Yang C, Liu Z, Fernie AR, Graham IA, Luo J (2022) Plant metabolic gene clusters in the multi-omics era. Trends Plant Sci 29. https://doi.org/10.1016/j.tplants.2022.03.002

Zhang W, Ramautar R (2021) CE-MS for metabolomics: developments and applications in the period 2018-2020. Electrophoresis 42(4):381–401. https://doi.org/10.1002/elps.202000203

Zhou B, Xiao JF, Tuli L, Ressom HW (2012) LC-MS-based metabolomics. Mol BioSyst 8:470–481. https://doi.org/10.1039/c1mb05350g

The Core Plant Microbiome: Roles, Functions, and Importance in Plant Health

Sindhunath Chakraborty and Murugan Kumar

1 Introduction

The under- (rhizoplane and rhizosphere) and aboveground (phyllosphere and endosphere) parts of plants present a large number of niches where microorganisms belonging to different communities thrive and proliferate (Trivedi et al. 2020; Bulgarelli et al. 2012; Lundberg et al. 2012). These plant-associated microorganisms not only play crucial roles in the growth, development, and physiology of their respective hosts but also protect them from environmental stresses (Li et al. 2019; Agler et al. 2016a, b; Mendes et al. 2013). While the healthy plants encompass unique microbial communities, stress exposure sometimes causes slight or drastic alterations in the microbial community composition. These changes are generally brought forth by the plant itself through a strategy called "cry for help." In this process, the plants experiencing environmental stresses secrete various biochemical compounds like polysaccharides, hormones, proteins, secondary metabolites, etc. into the rhizosphere, which in turn attract some specific groups of beneficial microorganisms from the nearby environments (Carrion et al. 2019; Bakker et al. 2018). Upon attachment with the plant root surfaces and in the rhizosphere, these microorganisms grow luxuriantly by consuming secreted plant metabolites and simultaneously perform their functions according to the nature of the stresses so as to provide necessary protection to the host plants (Badri and Vivanco 2009). These microbial communities, due to their direct involvement in plant stress management process, offer many exceptional agricultural applications, especially in the alleviation of environmental stress-induced damages and growth inhibition of plants. However, for exploring these microorganisms in agricultural applications, precise

S. Chakraborty · M. Kumar (✉)
ICAR-National Bureau of Agriculturally Important Microorganisms, Kushmaur, Mau Nath Bhanjan, Uttar Pradesh, India
e-mail: kumar.m@icar.org.in

Renu et al. (eds.), *Meta-omics in Crop Improvement*,
https://doi.org/10.1007/978-981-95-2294-1_9

identification of microbial communities persistently crucial for the host plants is essential (Carrion et al. 2019). The microbial members consistently found in different compartments of their respective host plants irrespective of the alterations in environment, plant genotypes, and time are referred to as the core microbiota (Stopnisek et al. 2021; Hamonts et al. 2018; Vandenkoornhuyse et al. 2015). According to the existing reports, the core microbiota may constitute dominant taxa carrying genes with central roles in maintaining the growth, development, and subsistence of plants (Lemanceau et al. 2017). Moreover, the core microbiota is also known to establish mutualistic interactions with their respective host plants (Vandenkoornhuyse et al. 2015), and this particular property can be put to use for enhancing the performance of plants (Toju et al. 2018). The core microbiota executes their function with high efficiency when they include taxonomically related keystone taxa as intrinsic members. This is so because the keystone taxa, either separately or in a group, mediate the crosstalk between host plants and their accompanied microbiome (Banerjee et al. 2018; Shetty et al. 2017; Agler et al. 2016a, b). So far, most of the core plant microbiome-related studies were performed on a specific plant species and mainly dealt with the taxonomic characterization of the microbial members. There are only few reports available which tried to explore the drivers responsible for the recruitment of core microbiota and their functional attributes triggering host efficacy (Stopnisek et al. 2021; Schlatter et al. 2020; Yeoh et al. 2017; Xu et al. 2018). Despite the availability of a considerable number of reports on structures and functions of core plant microbiomes, only a very few review articles/chapters have so far been published on this topic. In this context, the present chapter on "core plant microbiome" is based on the recently published reports. The chapter provides an in-depth exploration of the core plant microbiome delving into its definition, historical evolution, and conceptual framework. Further, the chapter proceeds by discussing the driving forces responsible for the recruitment of core microbiota to different plant parts and then elaborates current methods used by researchers for identifying and quantifying core plant microbiome. The discussion culminates by emphasizing the functional significance and ecological roles of the core plant microbiome, underscoring its pivotal contributions to plant health and productivity.

2 Core Plant Microbiome

The core plant microbiome represents the assemblages of microbial communities that evolve and associate with a specific plant species or plant population and remain unchanged with variations in genotype. For a given plant population, the microorganismal segment common for all the analysed plants denotes the core microbiota of that population, whereas the microbial fraction common for a particular plant species belonging to distinct populations reflects the core microbiome of that plant species. From the above statement, it is apparent that the broader the ecological scale, the less complicated will be the core microbiome composition. Therefore, in order to enumerate the core microbiome of any individual plant species or

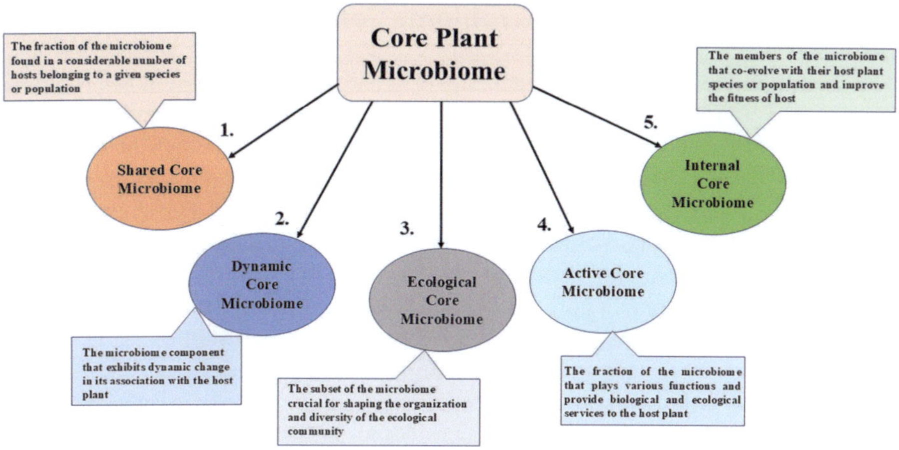

Fig. 1 Types and definitions of different core microbiomes

population, ecological boundaries must always be mentioned with the analytically procured data, otherwise the information may be misleading. Apart from this, other ecological complexities like changes occurring in core microbiome composition with time and space are needed to be mentioned as well to decipher core microbiota (Vandenkoornhuyse et al. 2015). A total of five different types of core microbiomes have been mentioned in the literature (Risely 2019) (Fig. 1).

Moreover, both taxonomic and functional approaches are invested to define core microbiome. While the taxonomic approach help deduce the phylogenetic relationship of core microbiota of a given plant species or population based on genomics fingerprints, the functional approach uses the sequence data obtained from metagenome analyses to interpret functional importance of the core microbiome. Hence, taxonomic study gives information about the shared microbiota of a given species or population, whereas functional approach unveils the shared functions. Combined application of these two approaches helps not only to correlate the diversity and functions of a core microbiome but also to perform theoretical evaluation of this correlation. There are many different hypotheses concerning the link between core microbiome diversity and functions (Naeem et al. 2002). The key species hypothesis (Paine 1969) describes that a particular function is governed by an individual species, whereas the functional redundancy hypothesis (Walker 1992) entails that a particular microbial diversity is responsible for a given function. Therefore, as per the key species hypothesis, disappearance of a key species results in the forfeiture of a function, and according to the functional redundancy hypothesis, loss of a function has no relation with the variation in microbiome diversity. The above-mentioned hypotheses elucidating the link between diversity and functions thus clearly indicate the flexibility of core microbiome composition. The combined taxonomic and functional approach-based core microbiome analysis also assumes that the dependency of plants on core microbiota increases as the ecological boundaries and complexities are increased.

3 Evolution of Core Microbiome Concept

The core microbiome concept was actually originated from the Human Microbiome Project. The aim of this project was to determine the core microbiome or factors common for microbiomes of large numbers of humans (Turnbaugh et al. 2007). Hence, the "core microbiome" appears to be a broad term and may reflect myriads of similar characteristics between distinct human microbiomes. These characteristics may include some particular genes, microbial taxa, metabolic routes, or some other machineries mediating cross talks between the host and its microbiome. The initial researches on human core microbiomes were generally based on taxonomic analyses, mainly in terms of 16S rRNA gene sequences (Willing et al. 2010; Turnbaugh et al. 2007). However, the obtained results were not satisfying as these studies showed large variations in their efficiencies to pinpoint a specific set of human associated microbes. For gut microbiome, in few cases, certain microbial taxa were found to be common for different individuals, whereas in other cases, no common taxa were identified (Turnbaugh et al. 2007). Contrarily, when the oral microbiomes of healthy humans were examined, a considerable number of consistent microbial taxa were identified (Eren et al. 2014; Zaura et al. 2009). The inconsistency of these outcomes ultimately led to different speculations such as the following: (1) the human core microbiomes may be similar for some selected populations and may not be universal, (2) the core microbiome can only be distinguished at a higher taxonomic level, and (3) the core microbiome may be entirely operational due to the presence of effective genes in place of individual microorganism. Although the aforesaid theories are not absolute, they deliver distinct perception about the compositional and functional characteristics of the core human microbiome. Moreover, the preliminary works on core human microbiomes laid the foundation for future researches on core microbiomes of other organisms. However, this search for core microbiome remained at its infancy until the report of Shade and Handelsman (Shade and Handelsman 2012). They, for the first time, mentioned that the core microbiome-related studies are supposed to be appropriate as they mostly rely on Venn diagram-based identification of operational taxonomic units (OTUs). To solve this issue, they suggested to evaluate relative abundances of microbial taxa and also advocated to analyze the consistent and changing core microbiomes measured at different timepoints. These suggestions helped researchers to not only identify the persistent microbial strains from the guts of separate human hosts (Faith et al. 2013) but also to disclosed the ability of few microbial taxa to persist even after a major environmental shift, for example, salt to freshwater migration of Atlantic salmon (Dehler et al. 2017). The identification of core microbiota may also be helpful in enumerating their functional roles using omics approaches, as shown in the case of honey bees (Engel et al. 2013; Sabree et al. 2012). Many other researchers also gave emphasis to the functional aspects of core microbiome rather than taxonomic approach (Lemanceau et al. 2017). It is worth mentioning that studies on the core microbiome of plant hosts have gained significant pace in the last years and majority of these studies have been performed using only taxonomic approach rather than functional. The core microbiomes of a large number of economically and medicinally valuable plant species have so far been explored by many groups, and their functional roles in maintaining the fitness of plants under stressful (biotic and abiotic) conditions have also been elaborated. The core microbiomes of some important cereal crops are described in Table 1. Overall, there were time-to-time amendments of new suggestions which upgraded the whole concept of core microbiome.

Table 1 Core microbiome compositions of some important cereal crops

Host plants	Compositions of core microbiome (phylum)	Examples of some representative microbes (orders)	References
Rice (leaf endosphere) *Oryza sativa*	*Proteobacteria*, *Bacteroidetes*, *Firmicutes*, *Actinobacteria*	*Burkholderiales*, *Enterobacterales*, *Pseudomonadales*, *Rhizobiales*, *Sphingomonadales* *Xanthomonadales*	Kumar et al. (2021)
Rice (phyllosphere) *Oryza sativa*	*Proteobacteria*, *Actinobacteria*, *Firmicutes*	*Burkholderiales*, *Actinomycetales*, *Bacillales*, *Clostridiales*, *Enterobacterales*, *Deinococcales*, *Hyphomicrobiales*,	Sahu et al. (2022)
Maize (Xylem Sap)*Zea Mays*	*Gammaproteobacteria* *Firmicutes*, *Alphaproteobacteria*, *Actinobacteria*	*Bacillales*, *Lactobacillales*, *Enterobacterales*, *Enterobacterales*, *Enterobacterales*, *Enterobacterales*, *Rhodospirillales*, *Rhizobiales*, *Rhizobiales*, *Sphingomonadales*, *Sphingomonadales*, *Sphingomonadales*, *Enterobacterales*, *Enterobacterales*, *Enterobacterales*, *Pseudomonadales*	Zhang et al. (2022)
Wheat (rhizosphere) *Triticum aestivum*	*Proteobacteria*, *Chloroflexi*, *Actinobacteria*, *Bacteroidetes*, *Acidobacteria*, *Gemmatimonadetes*, *Planctomycetes*, *Verrucomicrobia*, *Firmicutes*, and *cyanobacteria*	*Chloroflexales*, *Flavobacteriales*, *Gemmatimonadales*, *Myxobacteria*, *Acidimicrobiales*, *Chitinophagales*, *Herpetosiphonales* *Flavobacteriales*, *Xanthomonadales*, *Burkholderiales*	Kumar et al. (2023)
Pearl millet (rhizosphere) *Pennisetum glaucum*	*Ascomycota*, *Basidiomycota*, *Chytridiomycota*, and *Glomeromycota*	*Sordariales*, *Sebacinales*, *Pleosporales*, *Xylariales*, *Hypocreales*	Mofini et al. (2022)

4 Factors for the Recruitment of Core Plant Microbiota

The assembly of core microbiome in the rhizosphere, phyllosphere, or endosphere results from the interactions between plants and soil microbiota. There are several crucial factors which instigate these plant-microbe interactions and act as driving forces for the recruitment of core microbiota to different plant compartments.

Rhizosphere environment is considered as a vital factor responsible for the recruitment of core microbiota to plant roots. Soil microbial communities present around the roots are the major sources of core microbiota. A number of metagenomics studies on the rhizospheric soils of *Arabidopsis thaliana* showed that there is a strong correlation between the compositions of soil bacterial communities and rhizospheric bacterial communities (Bulgarelli et al. 2012; Lundberg et al. 2012; Schlaeppi et al. 2014). Other studies demonstrated that soil property is another crucial factor which has strong influence on the composition of soil as well as root microbial communities (Shakya et al. 2013; Schreiter et al. 2014). Therefore, the soil properties and microbial communities are two important drivers that govern the structure of rhizospheric microbial communities rather than the plant species or genotype (Bulgarelli et al. 2012; Lundberg et al. 2012; Shakya et al. 2013; Schlaeppi et al. 2014). While the structure of root-associated bacterial communities is dependent on the evolutionary background of the host plant especially when the same soil is used for cultivating different genotypes and species (Manter et al. 2010; Ofek et al. 2013; Bouffaud et al. 2014), the origin of soil is more crucial than species types in determining the composition of fungal assemblages (Bonito et al. 2014).

Rhizoplane or root surface represents another unique environment suitable for microbial growth. This is the habitat where rhizospheric soil microorganisms get into the contact of plant root for the first time. The region between rhizospheric soil and root surface is a highly dynamic environment as it is influenced by the root exudates containing carbon-rich molecules and antibiotics. Because of this, differentiation of microbial communities is frequently observed in this environment (Peiffer et al. 2013; Schreiter et al. 2014). The most common rhizoplane-associated microbes are fungi that form ectomycorrhiza which grow by firmly attaching to the root surface (Courty et al. 2010; Cairney 2011).

The endosphere or internal tissues of plants organs also present space for the growth of microbes. The biochemical composition of this habitat attracts unique and specific microbial taxa. Microorganisms capable of invading internal tissues constitute the endospheric microbiota. As compared to the external plant parts, microbial diversity of endosphere is very low (Bulgarelli et al. 2012; Schlaeppi et al. 2014). The most commonly identified endospheric microbes are fungi that form arbuscular mycorrhiza (Smith and Read 2008). Apart from this, other fungal species, bacteria, and archaea are also found to colonize endospheric regions of plants (Vandenkoornhuyse et al. 2002; Reinhold-Hurek and Hurek 2011; Sun et al. 2008). Moreover, recent reports revealed that the external aboveground parts or phyllosphere features an exceptional diversity of microbes. The recruitment of microbes to the plant's phyllosphere is driven mainly by the adaptability of microbes to tough conditions like UV exposure, desiccation and oligotrophy. Besides, the release of methanol by plant leaves may also act as the driving force for the recruitment of methylotrophic bacteria to the leaves (Galbally and Kirstine 2002). Collection of *Methylobacterium* assemblages from the leaves of different plants and their subsequent comparative analysis reflected that environmental factors are more crucial than genotype in determining microbial community structure (Knief et al. 2010).

5 Techniques Used in the Studies of Core Plant Microbiomes

In order to identify and quantify core plant microbiomes, various cutting-edge techniques are currently being used. These techniques include amplicon sequencing, metagenomics, metatranscriptomics, metaproteomics, and metabolomics. While the amplicon sequencing and metagenomics approaches are used to unveil the compositions of the core microbiome, metatranscriptomics and metaproteomics are employed to discern the functional roles of the core micromicrobiomes. Furthermore, the type of interactions occurring between plants and their core microbiota are unveiled by utilizing metabolomics approach. When all of these techniques are applied combinedly to explore the structural and functional attributes of plant's core microbiome, it is referred to as multiomics approach. Applications of these techniques in core plant microbiome-related studies imparted significant knowledge on the mechanisms employed by single microbial taxa and microbial communities to initiate plant-microbiome interactions and to provide protection to the host plant under stressful conditions. The scientific advantages of these techniques and their contributions to the field of core plant microbiome research are described below.

5.1 Amplicon Sequencing

Amplicon sequencing is one of the most familiar methods presently used for analyzing microbial community structures in the environment. In this technique, a highly conserved region of microbial DNA or a marker gene (16S rDNA for bacteria/archaea and 18S rDNA for fungus) is amplified using polymerase chain reaction (PCR) with the help of specific primers and subsequently sequenced through next-generation sequencing (NGS). The obtained sequence is further matched with sequences available in a suitable database to unveil its organismal identity. Amplicon sequencing is a highly specific and sensitive method capable of targeting functional genes and individual microbial lineages like archaea, bacteria, etc. (Herbold et al. 2015). Due to this high specificity, amplicon sequencing is also used to efficiently recognize rare organisms. The major disadvantage of amplicon sequencing is its high sensitivity-driven proneness to contamination (Glassing et al. 2016). As a result, use of negative and positive controls is indispensable during amplicon sequencing. Amplicon sequencing-based researches have furnished copious amount of data related to the plant-associated core microbiota. The plant species whose microbiomes have been studied through amplicon sequencing approach include rice (Edwards et al. 2015), wheat (Donn et al. 2015; Kumar et al. 2023), corn (Walters et al. 2018), millet (Jin et al. 2017), barley (Bulgarelli et al. 2012), soybean (Mendes et al. 2013), sugarcane (Hamonts et al. 2018), pea (Tkacz et al. 2020), cucumber (Ofek-Lalzar et al. 2014), grapevine (Marasco et al. 2018), and citrus (Xu et al. 2018).

5.2 Shotgun Metagenomics Sequencing

In contrast to amplicon sequencing, shotgun metagenomics sequencing works on different principle. Instead of amplifying and sequencing a specific marker gene, whole genome shotgun sequencing technique analyzes the total microbial DNA pool present in an environmental sample without isolating and culturing the microbes (Sharpton 2014). Although not popular, this technique is frequently used in the study of plant microbiome analysis (Perez-Mon et al. 2021; Bartuv et al. 2023; Chouhan et al. 2023; Masuda et al. 2024). Shotgun metagenomics provide whole genome sequences of all microbes present in a community and also outlines the structural and functional aspects of that microbial community (Eichinger et al. 2016). Although the shotgun metagenomics display low sensitivity than amplicon sequencing and completely fails to identify rare taxa in a community, the biasness for microbial abundances is significantly less (Poretsky et al. 2014). In addition, the obtained data may be presented as draft metagenome sequences. The main advantage of shotgun metagenomics is that one can easily acquire information about the taxonomic as well as functional characteristics of microbial communities present in a given sample. Shotgun metagenomics allows easy identification of genomics signatures of microbes crucial for colonizing plant parts (Ofek-Lalzar et al. 2014; Bulgarelli et al. 2012; Xu et al. 2018). However, the patterns of plant-microbe interaction cannot be revealed through only metagenomics studies; instead, there will be the requirement of other advanced techniques like metatranscriptomics, metaproteomics, and metabolomics.

5.3 Metatranscriptomics, Metaproteomics, and Metabolomics

Metatranscriptomics is an advance molecular technique to study the pattern of gene expression in different microbial assemblages based on high-throughput environmental mRNA sequencing. Like metagenomics, metatranscriptomics also doesn’t require the culturing of microbes. Moreover, it provides information related to the active microbial communities of a given environmental sample, gene expression patterns of these communities, and functions associated with the expressed genes (Zhang et al. 2021; Shakya et al. 2019; Filiatrault 2011). Hence, metatranscriptomics studies yield biological data that are not generated by amplicon sequencing and shotgun metagenomics. Metatranscriptomics is such a powerful technique that it is widely used to study microbiomes associated with humans as well as plants. Its uses in plant microbiome studies are rapidly increasing. Using these techniques, Turner et al. (2013) unveiled the differences in active rhizospheric microbial communities of various plants including pea, wheat, and oat. Apart from this, metatranscriptomics study has also helped in discerning the drought induced alterations in actinobacterial activities in the rhizosphere of sorghum and transcriptional upregulation of carbohydrate as well as amino acid metabolism-related genes of the rhizosphere microbiome (Xu et al. 2018). Despite being a powerful technique, metatranscriptomics studies are needed to be complemented by other advanced approaches like metaproteomics and metabolomics because the transcriptomics data are less relatable to the translational data.

Metaproteomics is a highly advanced technique which makes use of liquid chromatography-tandem mass spectrometry (LC-MS/MS) and performs qualitative and quantitative analyses of proteins synthesized by microbial communities in the environment (https://www.creative-proteomics.com/blog/index.php/a-brief-introduction-to-metaproteomics). As a result, it generates information related to the phenotypes of microbial communities present in a given environmental sample. In addition, metaproteomics also allows spatial and temporal characterization of microbial populations. So far, the rhizosphere and phyllosphere microbiomes of some agricultural crops have been investigated using metaproteomics approach which revealed the molecular phenotypes of those microbial communities (Wang et al. 2011; Moretti et al. 2012; Knief et al. 2012; Bona et al. 2019; Delmotte et al. 2009). Although the studies on metaproteome-based plant microbiome analysis are very scarce, they effectively identified the prevalent taxa and their expressed proteins in different plant compartments (Bona et al. 2019; Knief et al. 2012; Delmotte et al. 2009). Besides, these studies also provided insights into the physiological and metabolic signatures of microbiomes required for colonizing rhizosphere as well as phyllosphere and establishing themselves in those plant-associated habitats (Bona et al. 2019; Knief et al. 2012). Therefore, it is apparent from the above reports that metaproteomics approach can efficiently disclose the molecular and functional traits of microbiomes.

Metabolomics is an emerging technique in the field of plant microbiome research as only a single study is performed till date (Adeniji et al. 2020). It analyses the whole cell extracts of different organisms and identifies the present metabolites based on different mass spectrometry-based techniques like GC-MS, LC-MS, capillary-MS, etc. Since metabolomics studies give information on primary and secondary metabolites, application of this technique in plant microbiome studies may provide information on how microbial populations present in the rhizosphere, endosphere, and phyllosphere interact with the host plant (van Dam and Bouwmeester 2016). Moreover, the results obtained from metabolomics studies can also be used to complement the metatranscriptome and metaproteome data to better understand the functional traits of plant-associated microbiomes.

6 Functional Importance and Roles of Core Plant Microbiome

Plant-associated microbial communities play a wide array of roles in boosting plant growth. The direct methods through which these microbes help their host plants include nutrient mobilization and cycling (phosphorus, nitrogen, potassium, iron, zinc, etc.) (Calvo and Friston 2017), synthesis of different phytohormones (auxins, gibberellins, and cytokinins), and alteration of hormone levels in different plant parts (Bhattacharyya and Jha 2012; Egamberdieva et al. 2017) and stress alleviation of plants by secreting ethylene producing enzyme 1-aminocyclopropane-1-carboxylate (ACC) deaminase (Glick 2005, 2012). However, the indirect strategies employed by plant-associated microbes include reduction of pathogen-induced damages in plants through the production of myriads of bioactive compounds like volatile metabolites, antibiotics, siderophores, lipolytic and proteolytic enzymes, bacteriocins and small peptides (Compant et al. 2005; Ryan et al. 2008), competitive exclusion of niche, intra or inter species competition, and induction of plant immune responses (Vannier et al. 2019). These microbiome-driven protection strategies have been studied rigorously and many of these reports reflected the efficiency of individual taxa to inhibit pathogen growth and enhance plant health.

It was reported earlier that despite being the smallest subset of the whole plant microbiome, core plant microbiota plays various critical roles including nutrient mobilization and protection of host plant from diseases and abiotic stresses (Compant et al. 2019). Moreover, active core microbiomes are also known to regulate various metabolic pathways of plants, nutrient metabolism, and leaf biochemistry (Turnbaugh et al. 2009; Mendes et al. 2013; Kembel et al. 2014). It is generally considered that the intrinsic plant growth-promoting traits of core microbiome resulting from the presence of a specific set of replicators influence the plant health. A recent report on *Arabidopsis* demonstrated that different clonal varieties of this species precisely accompany a set of heritable core microbiome which directly promotes the plant health (Brachi et al. 2017). Moreover, the plant growth-promoting traits like ammonia oxidation, polyhydroxybutyrate production, and nitrogen fixation have recently been reported for the core microbiota of a medicinal plant *Eucommia ulmoides* (Dong et al. 2021). Similarly, the abundance of nitrogen-fixing core bacterial assemblages in the xylem sap of Maize has also been experimentally demonstrated (Zhang et al. 2022). The core root microbiome analysis of coffee plants (*Coffea arabica*) belonging to Central America reflected the presence of many microbes that have previously been reported to possess various plant growth-promoting properties like nitrogen fixation, auxin production, ACC deaminase production, and siderophore release (Fulthorpe et al. 2019).

The resistance of different plant varieties against pathogenic diseases is directly linked to their microbiomes (Kwak et al. 2018). A study delving into the core microbiome analysis of two cotton varieties, i.e., resistant and susceptible to cotton leaf curl disease (CLCuD), clearly showed that the core microbiome composition of cotton plant is correlated with the disease resistance (Aqueel et al. 2024). Apart from this, the roles of core microbiome in improving plant tolerance to heavy metal stress and promoting plant growth have also been enumerated using RNA-based stable isotope probing. In this study, rhizospheric core microbiomes of one cadmium accumulator (*Brassica juncea* L. Czern, Belimov et al. 2005) and four cadmium hyperaccumulators (*Sedum alfredii*, Yang et al. 2004; *Solanum nigrum* L. Wei et al. 2005; the Ganges and Prayon ecotype of *Noccaea caerulescens*, Milner et al. 2014) were compared with that of a non-accumulator (*Sedum alfredii*, Luo et al. 2017), and it was observed that the hyperaccumulators contained same core microbial taxa, but the non-accumulators harboured different set of core microbiota. Moreover, when the shared microbial taxa of cadmium hyperaccumulators were applied in the form of synthetic consortium, significant increase in the plant growth and cadmium tolerance was noticed (Luo et al. 2022). The above reports thus suggest that core microbiomes are crucial for host health and performance under normal as well as stressful conditions. The recruitment of core microbiota to different plant parts and the functional traits of core microbial taxa are illustrated in Fig. 2.

7 Conclusion

Conclusively, the core microbiome of a given plant species or population is comprised of a small fraction of persistent microbial taxa that play crucial eco-physiological functions and improve plant health. Plants recruit these core microbes to their different parts from the microbial pools present in the local soil. The increase in abundance of genes implicated in beneficial functions governs the recruitment of active core microbiota. Since the functions of core microbiota solely depend on

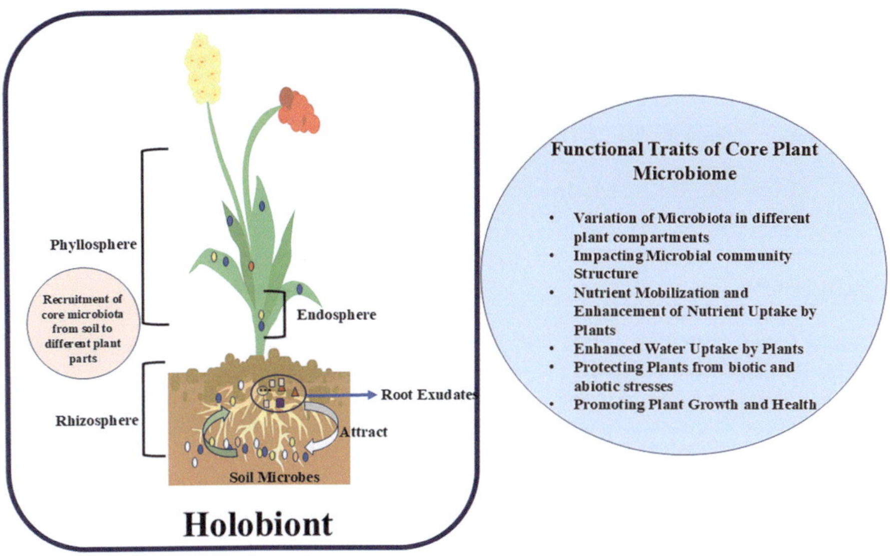

Fig. 2 Recruitment of core microbiota to different plant compartments and functional traits of these microbes

their genetic set up, the subsistence of plant-core microbiome associations relies on the transfer of these genetic traits from one generation to another. Therefore, identification of core microbiome of a specific plant species or population and acquiring knowledge on the factors responsible for their recruitment may explore new perspectives in the field of plant-breeding researches which in turn help increase the soil properties and reduce the use of chemical fertilizers.

8 Future Prospects

Although the existing reports on core plant microbiome have suggested immense agricultural potential of these microbial assemblages, their direct application in agricultural fields is still a big issue because the agricultural application of core microbiota requires accurate identification of functional members harbouring genes related to plant growth promotion or protection. The major questions that are needed to be answered first include the following: (1) Which methods can be used to explore new microbial genes imparting essential plant functions (e.g., nutrient and water uptake, biotic and abiotic stress tolerance)? (2) How can the roles of core microbiota in increasing plant fitness be explained? (3) Which individual functions are specifically played by plants and their microbial partners? (4) Whether the distribution of labour between plants and their core microbiota are impacted by environmental stresses and organismal evolution? (5) Which traits of the host plants are responsible for the recruitment of core microbiome, and how these traits can be applied in plant-breeding projects to enhance plant health and resilience: However, these questions can only be answered after investing significant research efforts in the relevant field.

Acknowledgement The Director of the ICAR-National Bureau of Agriculturally Important Microorganisms Kushmaur, Mau Nath Bhanjan, Uttar Pradesh, 275101, is gratefully acknowledged for providing necessary laboratory facilities.

Conflict of Interest The authors declare no conflict of interest.

References

Adeniji AA, Babalola OO, Loots DT (2020) Metabolomic applications for understanding complex tripartite plant–microbes interactions: strategies and perspectives. Biotechnol Rep 25:00425

Agler MT, Ruhe J, Kroll S, Morhenn C, Kim S, Weigel D, Kemen EM (2016a) Microbial hub taxa link host and abiotic factors to plant microbiome variation. PLoS Biol 14:1002352

Agler MT, Ruhe J, Kroll S, Morhenn C, Kim ST, Weigel D et al (2016b) Microbial hub taxa link host and abiotic factors to plant microbiome variation. PLoS Biol 14:1002352

Aqueel R, Badar A, Roy N et al (2024) Disease resistance correlates with core microbiome diversity in cotton. Curr Microbiol 81:302

Badri DV, Vivanco JM (2009) Regulation and function of root exudates. Plant Cell Environ 32:666–681

Bakker P, Pieterse C, de Jonge R, Berendsen RL (2018) The soilborne legacy. Cell 172:1178–1180

Banerjee S, Schlaeppi K, van der Heijden M (2018) Keystone taxa as drivers of microbiome structure and functioning. Nat Rev Microbiol 16:567–576

Bartuv R, Berihu M, Medina S, Salim S, Feygenberg O, Faigenboim-Doron A, Zhimo VY, Abdelfattah A, Piombo E, Wisniewski M, Freilich S, Droby S (2023) Functional analysis of the apple fruit microbiome based on shotgun metagenomic sequencing of conventional and organic orchard samples. Environ Microbiol 25(9):1728–1746

Belimov AA, Hontzeas N, Safronova VI, Demchinskaya SV, Piluzza G, Bullitta S, Glick BR (2005) Cadmium-tolerant plant growth-promoting bacteria associated with the roots of Indian mustard (*Brassica juncea* L. Czern.). Soil Biol Biochem 37:241–250

Bhattacharyya PN, Jha DK (2012) Plant growth-promoting rhizobacteria (PGPR): emergence in agriculture. World J Microbiol Biotechnol 28:1327–1350

Bona E, Massa N, Novello G, Boatti L, Cesaro P, Todeschini V, Magnelli V, Manfredi M, Marengo E, Mignone F et al (2019) Metaproteomic characterization of the Vitis vinifera rhizosphere. FEMS Microbiol Ecol 95:204

Bonito G, Reynolds H, Robeson MS, Nelson J, Hodkinson BP, Tuskan G, Schadt CW, Vilgalys R (2014) Plant host and soil origin influence fungal and bacterial assemblages in the roots of woody plants. Molecular Ecolo 23:3356–3370

Bouffaud ML, Poirier MA, Muller D, Moënne-Loccoz Y (2014) Root microbiome relates to plant host evolution in maize and other Poaceae. Environ Microbiol 16:2804–2814

Brachi B, Filiault D, Darme P, Le Mentec M, Kerdaffrec E, Rabanal F, Anastasio A, Box M, Duncan S, Morton T, Novikova P, Perisin TT, Woolley R, Yu M, Dean C, Nordborg M, Holme S, Bergelson J (2017) Plant genes influence microbial hubs that shape leaf communities. Biorxiv

Bulgarelli D, Rott M, Schlaeppi K, Ver Loren Van Themaat EVL et al (2012) Revealing structure and assembly cues for *Arabidopsis* root-inhabiting bacterial microbiota. Nature 488:91–95

Cairney J (2011) Ectomycorrhizal fungi: the symbiotic route to the root for phosphorus in forest soils. Plant Soil 344:51–71

Calvo P, Friston K (2017) Predicting green: really radical (plant) predictive processing. J R Soc Interface 14:20170096

Carrion VJ, Perez-Jaramillo J, Cordovez V, Tracanna V et al (2019) Pathogen-induced activation of disease-suppressive functions in the endophytic root microbiome. Science 366:606–612

Chouhan U, Gamad U, Choudhari JK (2023) Metagenomic analysis of soybean endosphere microbiome to reveal signatures of microbes for health and disease. J Genetic Eng Biotechnol 21(1):84

Compant S, Duffy B, Nowak J, Clement C, Barka EA (2005) Use of plant growthpromoting bacteria for biocontrol of plant diseases: principles, mechanisms of action, and future prospects. AEM 71:4951–4959

Compant S, Samad A, Faist H, Sessitsch A (2019) A review on the plant microbiome: ecology, functions, and emerging trends in microbial application. J Adv Res 19:2–37

Courty PE, Buee M, Diedhiou AG, Frey-Klett P (2010) The role of ectomycorrhizal communities in forest ecosystem processes: new perspectives and emerging concepts. Soil Biol Biochem 42:679–698

Dehler CE, Secombes CJ, Martin SAM (2017) Seawater transfer alters the intestinal microbiota profiles of Atlantic salmon (*Salmo salar* L.). Sci Rep 7:13877

Delmotte N, Knief C, Chaffron S, Innerebner G, Roschitzki B, Schlapbach R, von Mering C, Vorholt JA (2009) Community proteogenomics reveals insights into the physiology of phyllosphere bacteria. PNAS, USA 106:16428–16433

Dong C, Zhang Z, Shao Q, Yao T, Liang Z, Han Y (2021) Mycobiota of *Eucommia ulmoides* bark: diversity, rare biosphere and core taxa. Fungal Ecol 53:101090

Donn S, Kirkegaard JA, Perera G, Richardson AE, Watt M (2015) Evolution of bacterial communities in the wheat crop rhizosphere. Environ Microbiol 17:610–621

Edwards J, Johnson C, Santos-Medellin C, Lurie E, Podishetty NK, Bhatnagar S, Eisen JA, Sundaresan V (2015) Structure, variation, and assembly of the rootassociated microbiomes of rice. PNAS USA 112:911–920

Egamberdieva D, Wirth SJ, Alqarawi AA, Abd Allah EF, Hashem A (2017) Phytohormones and beneficial microbes: essential components for plants to balance stress and fitness. Front Microbiol 8:2104

Eichinger V, Nussbaumer T, Platzer A, Jehl MA, Arnold R, Rattei T (2016) Effective DB–updates and novel features for a better annotation of bacterial secreted proteins and Type III, IV, VI secretion systems. Nucleic Acids Res 44:669–674

Engel P, Moran NA (2013) Functional and evolutionary insights into the simple yet specific gut microbiota of the honey bee from metagenomic analysis. Gut Microbes 4:60–65

Eren AM, Borisy GG, Huse SM, Mark Welch JL (2014) Oligotyping analysis of the human oral microbiome. Proc Natl Acad Sci USA 111:2875–2884

Faith JJ et al (2013) The long-term stability of the human gut microbiota. Science 341:1237439

Filiatrault MJ (2011) Progress in prokaryotic transcriptomics. Curr Opin Microbiol 14:579–586

Fulthorpe R, Martin AR, Isaac ME (2019) Root endophytes of coffee (*Coffea arabica*): variation across climatic gradients and relationships with functions. Phytobiomes J 4:27–39

Galbally IE, Kirstine W (2002) The production of methanol by flowering plants and the global cycle of methanol. Jour Atmos Chem 43:195–229

Glassing A, Dowd SE, Galandiuk S, Davis B, Chiodini RJ (2016) Inherent bacterial DNA contamination of extraction and sequencing reagents may affect interpretation of microbiota in low bacterial biomass samples. Gut Pathog 8:24

Glick BR (2012) Plant growth-promoting bacteria: mechanisms and applications. Scientifica 963401:1

Glick BR (2005) Modulation of plant ethylene levels by the bacterial enzyme ACC deaminase. FEMS Microbiol Lett 251:1–7

Hamonts K, Trivedi P, Garg A, Janitz C, Grinyer J, Holford P, Botha FC, Anderson IC, Singh BK (2018) Field study reveals core plant microbiota and relative importance of their drivers. Environ Microbiol 20:124–140

Herbold CW, Pelikan C, Kuzyk O, Hausmann B, Angel R, Berry D et al (2015) A flexible and economical barcoding approach for highly multiplexed amplicon sequencing of diverse target genes. Front Microbiol 6:731

Jin T, Wang Y, Huang Y, Xu J, Zhang P, Wang N, Liu X, Chu H, Liu G, Jiang H, Li Y, Xu J, Kristiansen K, Xiao L, Zhang Y, Zhang G, Du G, Zhang H, Zou H, Zhang H, Jie Z, Liang S, Jia H, Wan J, Lin D, Li J, Fan G, Yang H, Wang J, Bai Y, Xu X (2017) Taxonomic structure and functional association of foxtail millet root microbiome. Gigascience 6(10):1–12

Kembel SW, O'Connor TK, Arnold HK, Hubbell SP, Wright SJ, Green JL (2014) Relationships between phyllosphere bacterial communities and plant functional traits in a neotropical forest. PNAS USA 111:13715–13720

Knief C, Ramette A, Frances L, Alonso-Blanco C, Vorholt JA (2010) Site and plant species are important determinants of the *Methylobacterium* community composition in the plant phyllosphere. ISME J 4:719–728

Knief C, Delmotte N, Chaffron S, Stark M, Innerebner G, Wassmann R, von Mering C, Vorholt JA (2012) Metaproteogenomic analysis of microbial communities in the phyllosphere and rhizosphere of rice. ISME J 6(7):1378–1390

Kumar M, Kumar AA, Sahu KP, Patel A, Reddy B et al (2021) Deciphering core-microbiome of rice leaf endosphere: revelation by metagenomic and microbiological analysis of aromatic and non-aromatic genotypes grown in three geographical zones. Microbiol Res 246:126704

Kumar M, Ansari WA, Zeyad MT, Singh A, Chakdar H, Kumar A, Farooqi MS, Sharma A, Srivastava S, Srivastava AK (2023) Core microbiota of wheat rhizosphere under Upper Indo-Gangetic plains and their response to soil physicochemical properties. Front Plant Sci 14:1186162

Kwak J, Park J (2018) What we can see from very small size sample of metagenomic sequences. BMC Bioinformatic 19:399

Lemanceau P, Blouin M, Muller D, Moenne-Loccoz Y (2017) Let the core microbiota be functional. Trends Plant Sci 22:583–595

Li J, Zhou L, Lin W (2019) Calla lily intercropping in rubber tree plantations changes the nutrient content, microbial abundance, and enzyme activity of both rhizosphere and non-rhizosphere soil and calla lily growth. Ind Crop Prod 132:344–351

Lundberg DS, Lebeis SL, Paredes SH, Yourstone S et al (2012) Defining the core Arabidopsis thaliana root microbiome. Nature 488:86–90

Luo J, Tao Q, Wu K, Li J, Qian J, Liang Y, Yang X, Li T (2017) Structural and functional variability in root-associated bacterial microbiomes of Cd/Zn hyperaccumulator *Sedum alfredii*. Appl Microbiol Biotechnol 101:7961–7976

Luo J, Gu S, Guo X, Liu Y, Tao Q, Zhao H, Liang Y, Banerjee S, Li T (2022) Core microbiota in the rhizosphere of heavy metal accumulators and its contribution to plant performance. Environ Sci Technol 56:12975–12987

Manter DK, Delgado JA, Holm DG, Stong RA (2010) Pyrosequencing reveals a highly diverse and cultivar-specific bacterial endophyte community in potato roots. Micro Ecol 60:157–166

Marasco R, Rolli E, Fusi M, Michoud G, Daffonchio D (2018) Grapevine rootstocks shape underground bacterial microbiome and networking but not potential functionality. Microbiome 6:3

Masuda S, Gan P, Kiguchi Y et al (2024) Uncovering microbiomes of the rice phyllosphere using long-read metagenomic sequencing. Commun Biol 7:357

Mendes R, Garbeva P, Raaijmakers JM (2013) The rhizosphere microbiome: significance of plant beneficial, plant pathogenic, and human pathogenic microorganisms. FEMS Microbiol Rev 37:634–663

Milner MJ, Mitani-Ueno N, Yamaji N, Yokosho K, Craft E, Fei Z, Ebbs S, Clemencia ZM, Ma JF, Kochian LV (2014) Root and shoot transcriptome analysis of two ecotypes of *Noccaea caerulescensun* covers the role ofNcNramp1in Cd hyperaccumulation. Plant J 78:398–310

Mofini MT, Diedhiou AG, Simonin M et al (2022) Cultivated and wild pearl millet display contrasting patterns of abundance and co-occurrence in their root mycobiome. Sci Rep 12:207

Moretti M, Minerdi D, Gehrig P, Garibaldi A, Gullino ML, Riedel K (2012) A bacterial–fungal metaproteomic analysis enlightens an intriguing multicomponent interaction in the rhizosphere of Lactuca sativa. J Proteome Res 11:2061–2077

Naeem S, Loreau M, Inchausti P (2002) Biodiversity and ecosystem functioning: the emergence of a synthetic ecological framework. In: Loreau M, Naeem S, Inchausti P (eds) Biodiversity and ecosystem functioning: synthesis and perspectives, 1st edn. Oxford University Press, Oxford, pp 3–11

Ofek M, Voronov-Goldman M, Hadar Y, Minz D (2013) Host signature effect on plant root-associated microbiomes revealed through analyses of resident vs. active communities. Environ Microbiol 16:2157–2167

Ofek-Lalzar M, Sela N, Goldman-Voronov M, Green SJ, Hadar Y, Minz D (2014) Niche and host-associated functional signatures of the root surface microbiome. Nat Com 5:4950

Paine RT (1969) A note on trophic complexity and community stability. Am Nat 103:91–93

Peiffer JA, Spor A, Koren O, Jin Z, Green Tringe S, Dangl JL, Buckler ES, Ley RE (2013) Diversity and heritability of the maize rhizosphere microbiome under field conditions. Proceed Nat Acadm Sci, USA 16:6548–6553

Perez-Mon C, Qi W, Vikram S, Frossard A, Makhalanyane T, Cowan D, Frey B (2021) Shotgun metagenomics reveals distinct functional diversity and metabolic capabilities between 12000-year-old permafrost and active layers on Muot da Barba Peider (Swiss Alps). Microb Genom 7(4):000558

Poretsky R, Rodriguez RL, Luo C, Tsementzi D, Konstantinidis KT (2014) Strengths and limitations of 16S rRNA gene amplicon sequencing in revealing temporal microbial community dynamics. PLoS One 9:93827

Reinhold-Hurek B, Hurek T (2011) Living inside plants: bacterial endophytes. Curr Opi Plant Biol 14:435–443

Risely A (2019) Applying the core microbiome to understand host–microbe systems. J Anim Ecol 89(7):1549–1558

Ryan RP, Germaine K, Franks A, Ryan DJ, Dowling DN (2008) Bacterial endophytes: recent developments and applications. FEMS Microbiol Lett 278:1–9

Sabree ZL, Hansen AK, Moran NA (2012) Independent studies using deep sequencing resolve the same set of core bacterial species dominating gut communities of honey bees. PLoS One 7:e41250

Sahu KP, Kumar A, Sakthivel K, Reddy B et al (2022) Deciphering core phyllomicrobiome assemblage on rice genotypes grown in contrasting agroclimatic zones: implications for phyllomicrobiome engineering against blast disease. Environ Microbio 17:28

Schlaeppi K, Dombrowski N, Oter RG, Ver Loren van Themaat E, Schulze-Lefert P (2014) Quantitative divergence of the bacterial root microbiota in *Arabidopsis thaliana* relatives. Proceed Nat Acadm Sci, USA 111:585–592

Schlatter DC, Yin C, Hulbert S, Paulitz TC (2020) Core rhizosphere microbiomes of dryland wheat are influenced by location and land use history. Appl Environ Microbiol 86:2119–2135

Schreiter S, Ding G-C, Heuer H, Neumann G, Sandmann M, Grosch R, Kropf S, Smalla K (2014) Effect of the soil type on the microbiome in the rhizosphere of field-grown lettuce. Front Microbiol 5:144

Shade A, Handelsman J (2012) Beyond the Venn diagram: the hunt for a core microbiome. Environ Microbiol 14:4–12

Shakya M, Gottel N, Castro H, Yang ZK, Gunter L, Labb EJ, Muchero W, Bonito G, Vilgalys R, Tuskan G et al (2013) A multifactor analysis of fungal and bacterial community structure in the root microbiome of mature *Populus deltoides* trees. PLoS One 8:76382

Shakya M et al (2019) Advances and challenges in metatranscriptomic analysis. Front Genet 10:904

Sharpton TJ (2014) An introduction to the analysis of shotgun metagenomic data. Front Plant Sci 16(5):209

Shetty SA, Hugenholtz F, Lahti L, Smidt H, de Vos WM (2017) Intestinal microbiome landscaping: insight in community assemblage and implications for microbial modulation strategies. FEMS Microbiol Rev 41:182–199

Smith SE, Read DJ (2008) Mycorrhizal symbiosis. New Phytol 137:563–568

Stopnisek N, Shade A (2021) Persistent microbiome members in the common bean rhizosphere: an integrated analysis of space, time, and plant genotype. ISME J 15:2708–2722

Sun L, Qiu F, Zhang X, Dai X, Dong X, Song W (2008) Endophytic bacterial diversity in rice (*Oryza sativa* L.) roots estimated by 16S rDNA sequence analysis. Microb Ecol 55:415–424

Tkacz A, Bestion E, Bo Z, Hortala M, Poole PS (2020) Influence of plant fraction, soil, and plant species on microbiota: a multikingdom comparison. MBio 11:02785–02719

Toju H, Peay KG, Yamamichi M, Narisawa K, Hiruma K et al (2018) Core microbiomes for sustainable agroecosystems. Nat Plants 4:247–257

Trivedi P, Leach JE, Tringe SG, Sa T, Singh BK (2020) Plant-microbiome interactions: from community assembly to plant health. Nat Rev Microbiol 18:607–621

Turnbaugh PJ et al (2007) The human microbiome project. Nature 449:804–810

Turnbaugh PJ, Hamady M, Yatsunenko T, Cantarel BL, Duncan A, Ley RE et al (2009) A core gut microbiome in obese and lean twins. Nature 457:480–484

Turner TR, Ramakrishnan K, Walshaw J, Heavens D, Alston M, Swarbreck D, Osbourn A, Grant A, Poole PS (2013) Comparative metatranscriptomics reveals kingdom level changes in the rhizosphere microbiome of plants. ISME J 7:2248–2258

van Dam NM, Bouwmeester HJ (2016) Metabolomics in the rhizosphere: tapping into belowground chemical communication. Trends Plant Sci 21:256–265

Vandenkoornhuyse P, Baldauf SL, Leyval C, Straczek J, Young JPW (2002) Extensive fungal diversity in plant roots. Science 295:2051

Vandenkoornhuyse P, Quaiser A, Duhamel M, Le Van A, Dufresne A (2015) The importance of the microbiome of the plant holobiont. New Phytol 206:1196–1206

Vannier N, Agler M, Hacquard S (2019) Microbiota-mediated disease resistance in plants. PLoS Patholo 15:1007740

Walker BH (1992) Biodiversity and ecological redundancy. Conserv Biol 6:18–23

Walters WA, Jin Z, Youngblut N, Wallace JG, Sutter J, Zhang W, Gonzalez-Pena A, Peiffer J, Koren O, Shi Q et al (2018) Large-scale replicated field study of maize rhizosphere identifies heritable microbes. PNAS USA 115:7368–7373

Wang HB, Zhang ZX, Li H, He HB, Fang CX, Zhang AJ, Li QS, Chen RS, Guo XK, Lin HF et al (2011) Characterization of metaproteomics in crop rhizospheric soil. J Proteome Res 10:932–940

Wei S, Zhou Q, Wang X, Zhang K, Guo G, Lena QMA (2005) Newly-discovered Cd-hyperaccumulator Solanum nigrum L. Chin Sci Bull 50:33–38

Willing BP et al (2010) A pyrosequencing study in twins shows that gastrointestinal microbial profiles vary with inflammatory bowel disease phenotypes. Gastroenterology 139:1844–1854

Xu J, Zhang Y, Zhang P, Trivedi P, Riera N, Wang Y et al (2018) The structure and function of the global citrus rhizosphere microbiome. Nat Commun 9:4894

Yang XE, Long XX, Ye HB, He ZL, Calvert DV, Stoffella PJ (2004) Cadmium tolerance and hyperaccumulation in a new Zn-hyperaccumulating plant species (*Sedum alfredii* Hance). Plant Soil 259:181–189

Yeoh YK, Dennis PG, Paungfoo-Lonhienne C, Weber L, Brackin R, Ragan MA, Schmidt S, Hugenholtz P (2017) Evolutionary conservation of a core root microbiome across plant phyla along a tropical soil chronosequence. Nat Commun 8:215

Zaura E, Keijser BJF, Huse SM, Crielaard W (2009) Defining the healthy "core microbiome" of oral microbial communities. BMC Microbiol 9:259

Zhang Y et al (2021) Metatranscriptomics for the human microbiome and microbial community functional profiling. Annu Rev Biomed Data Sci 4:279–311

Zhang L, Zhang M, Huang S, Li L, Gao Q, Wang Y, Zhang S, Huang S, Yuan L, Wen Y, Liu K, Yu X, Li D, Zhang L, Xu X, Wei H, He P, Zhou W, Philippot L, Ai C (2022) A highly conserved core bacterial microbiota with nitrogen-fixation capacity inhabits the xylem sap in maize plants. Nat Commun, 13(1):3361

Microbiomes: Exploring their Potentials and Developing Strategies for Sustainable Legume Production

N. S. Nysanth, Koj Haniya, M. Senthilkumar, B. Ramakrishnan, and K. Swarnalakshmi

1 Introduction

Legumes are well-known for their unique capability to fix atmospheric nitrogen directly into the plant's usable form in their root nodules through the symbiotic relationship with rhizobia. The benefits of integrating legumes in agricultural systems lie in their nitrogen-fixing ability, as they contribute a significant amount of nitrogen to the soil ecosystem (Peoples et al. 2009). With unique biological capabilities to improve soil fertility, reduce dependence on chemical inputs, and enhance biodiversity, legumes contribute more to sustainable crop production (Plaza-Bonilla et al. 2017).

Legume interaction with specific rhizobial symbionts brings a competitive advantage to the host under nitrogen-deficient environments. In addition to rhizobia, potential rhizobacteria and endophytes were recognized as promising inoculants for crop production (Swarnalakshmi et al. 2013a, 2016b; Timmusk et al. 2017; O'Callaghan et al. 2022; da Silva et al. 2023). The intricate relationship between the plant-microbe interactions provides exciting opportunities to improve crop resilience and productivity.

Furthermore, in natural conditions, rhizobia constantly interact with complex soil and plant-associated microbial communities/microbiomes. Growth and survival are the key determinants of plant fitness, influenced by the interaction between the plant host and its associated microbiome, collectively known as the holobiont or the

N. S. Nysanth · K. Haniya · B. Ramakrishnan · K. Swarnalakshmi (✉)
Division of Microbiology, ICAR - Indian Agricultural Research Institute (IARI), New Delhi, India
e-mail: swarna_micro@iari.res.in

M. Senthilkumar
Division of Basic Sciences, ICAR - Indian Institute of Pulses Research (IIPR), Kanpur, Uttar Pradesh, India

Renu et al. (eds.), *Meta-omics in Crop Improvement*,
https://doi.org/10.1007/978-981-95-2294-1_10

holomicrobiome. The holomicrobiome provides additional genes and functions that improve plant fitness and symbiotic interaction (Vandenkoornhuyse et al. 2015). Rhizosphere microbiome is crucial to soil nutrient cycling and ecosystem functioning (Falkowski et al. 2008; Van Der Heijden et al. 2008). They interact with the plants in different manners, viz., good (beneficial microbiome), bad (plant pathogenic microbiome), and ugly (human pathogen travel via food chain) (Ali et al. 2017). The beneficial microbiomes in toto can enhance nitrogen fixation, nutrient uptake, plant health, and productivity. Sequencing technology advancements have revealed the microbiome as a highly dynamic and responsive system that can be strategically and scientifically managed to enhance agricultural sustainability. These understandings pave the way to farming without depending much on synthetic chemical fertilizers. The microbiome-based strategy is considered a possible game changer in agriculture and offers a more natural and eco-friendly approach to tackle challenges in crop cultivation. Hence, the need for microbiome research is significant in sustainable legume production under global environmental challenges as these crops are grown under marginal lands. The present chapter comprehensively explores the interactions between legumes and their associated microbiomes, explaining scientific understanding, technological innovations, and future prospects. We aim to provide insights into how microbiome manipulation can contribute to more sustainable and productive legume cultivation strategies.

2 Legume-Associated Microbes and Their Beneficial Roles

The fascinating story of legume-microbial interactions began in the mid-nineteenth century with one of the most intriguing discoveries on biological nitrogen fixation (BNF) by Boussingault (1838). Five decades later, Hellriegel and Wilfarth, in 1888, reported that legume plants form a specialized root nodule structure that converts atmospheric di-nitrogen (N_2) gas to ammonium (NH_3) by symbiotic nitrogen fixation (SNF). This groundbreaking discovery further facilitated the successful isolation of root nodule bacteria named *Bacillus radicicola* (Beijerinck 1888), later renamed *Rhizobium leguminosarum* (Frank 1889), which is capable of nodulating *Pisum* and *Vicia*. The early twentieth century witnessed the specificity of complex legume-rhizobia interaction, leading to cross-inoculation groups where different legume species selectively established associations with particular rhizobial strains (Fred et al. 1932). This finding marked the birth of rhizobiology, the study of plant-*Rhizobia* interactions, as a distinct field of study. Rhizobial species colonizing legume crops are distributed across several families (Berrada and Fikri-Benbrahim 2014), viz., *Rhizobiaceae* (*Allorhizobium*, *Neorhizobium*, *Pararhizobium*, *Rhizobium*, *Shinella*, and *Sinorhizobium*), *Phyllobacteriaceae* (*Aminobacter*, *Mesorhizobium*, *Phyllobacterium*), *Methylobacteriaceae* (*Methylobacterium*, *Microvirga*), *Brucellaceae* (*Ochrobactrum*), *Bradyrhizobiaceae* (*Bradyrhizobium*), *Hyphomicrobiaceae* (*Devosia*), and *Xanthobacteraceae* (*Azorhizobium*). These milestone scientific discoveries facilitated the development of impactful inoculants for agricultural use. Nitragin (*Rhizobium* sp.), the first commercial rhizobial biofertilizer for crop growth enhancement, was patented in the USA (Nobbe and Hiltner 1896). Subsequently, rapid progress in molecular biology revolutionized our perspectives of the legume-*Rhizobium* symbiotic relationship. The nodulation genes, elucidating Nod factor structures, and the process of infection thread formation provided deep insights into the molecular mechanism underlying legume-rhizobium symbiosis (Peters et al. 1986; Lerouge et al. 1990). Beyond that, several studies meticulously examined the complex signaling cascades that orchestrate the rhizobial colonization within root nodules of diverse legume plants (Dénarié et al. 1996; Freiberg et al. 1997; Haag et al. 2013; Denison 2000). A remarkable advancement was noticed in transforming bacteria into nitrogen-fixing bacteroids within root nodules (Paau et al. 1978, 1979; Oke and Long 1999).

The rhizobial interactions were the major focus of early research; scientists gradually recognized that legumes accommodate a broad spectrum of microorganisms. Though BNF within root nodules provides an additional nitrogen input, this process requires large amounts of energy and phosphorus. Hence, legumes have evolved with adaptation strategies for improved P uptake. Root colonization with mycorrhiza is considered one of the most successful strategies (Smith and Smith 2011; Vance et al. 2003). Not only legume-rhizobial symbiosis but also legumes-arbuscular mycorrhizal fungi (AMF) association is also highly specific, where the hosts prefer specific AMF that are efficient suppliers of phosphorous (Valdenegro et al. 2001). Most legumes were found to form beneficial partnerships with both AMF and rhizobia, where AMF reside in the root cortex region and rhizobia within root nodules (Chalk et al. 2006; Scheublin and Van Der Heijden 2006; Soumare et al. 2015; Loo et al. 2022). While mycorrhizal associations play a crucial role in phosphorous acquisition, identification of plant growth-promoting rhizobacteria (PGPR) expanded our understanding of the diversity of legume-associated microorganisms and their functions. Kloepper and Schroth (1978) coined "rhizobacteria" to describe a beneficial soil bacterial community that aggressively colonizes plant root systems. In 1981, Kloepper and Schroth renamed these bacteria as plant growth-promoting rhizobacteria (PGPR), primarily functioning in the rhizosphere and root surface. Accumulating evidence suggests the potential role of diverse PGPRs in sustainable legume production (Glick 1995; Panwar and Swarnalakshmi 2005; Pérez-Montano et al. 2014; Swarnalakshmi et al. 2020, 2023; Shah et al. 2021; Kaschuk et al. 2022). They benefit the host plant through a multitude of mechanisms. The direct and indirect mechanisms of PGPR include the production of phytohormones (Ortíz-Castro et al. 2008; Malik and Sindhu 2011; Gamalero and Glick 2015; Uzma et al. 2022), increasing nutrient availability (Elkoca et al. 2010; Sharma et al. 2017; Swarnalakshmi et al. 2017; Matse et al. 2020) and biotic and abiotic stress alleviation (Ryu et al. 2003; Santoro et al. 2015; Anand et al. 2021a; Swarnalakshmi et al. 2023).

Analogous to PGPR, another group of plant-associated microbes live within plant tissues such as seed, root, shoot, leaves, or root nodules and exert several beneficial effects on host plants, such as stimulation of plant growth, health, and protection (Reinhold-Hurek and Hurek 2011; Swarnalakshmi et al. 2013b, 2019; Hardoim et al. 2015; Verma et al. 2019; Compant et al. 2021). Such microorganisms reside within the plant tissues without causing apparent disease symptoms, commonly known as endophytes (Kado 1992; Schulz and Boyle 2006). The term "endophyte" ("endon"-within; and "phyte"-plant) was introduced by de Bary (1866). Though the endophytes are known to be derived from the rhizosphere soil (Castro-Sowinski et al. 2007; Compant et al. 2010), these bacterial populations are better protected from environmental stresses and microbial competition compared to rhizosphere bacteria (Quadt-Hallmann et al. 1997). The common endophytic bacteria in legumes include members of the major phyla like *Actinobacteria*, *Proteobacteria*, and *Firmicutes* (Gyaneshwar et al. 2011; Swarnalakshmi et al. 2016a). They can improve the nodule formation and N_2 fixation of legumes when co-inoculated with potential rhizobial strains (Swarnalakshmi et al. 2017, 2021; Debnath et al. 2023). A large number of non-rhizobial endophytes (NREs) are reported to reside within the root nodules along with rhizobia (Muresu et al. 2008; De Meyer et al. 2015; Peix et al. 2015; Martinez-Hidalgo and Hirsch 2017; Trabelsi et al. 2017; Ferchichi et al. 2019). The endophyte colonization of root nodules is a host-controlled event, and they infect via rhizobial-initiated infection thread. Rhizobial-driven Nod factor signal and exopolysaccharide selectively guide the NREs toward nodule primordia (Zgadzaj et al. 2015). The co-occurrence of NREs in root nodules was first reported by Beijerinck and Van Delden (1902), and they improve, either directly or indirectly, to nodule development, nitrogen fixation, and plant growth promotion when co-inoculated with rhizobia (Queiroz Rego et al. 2018; Musyoka et al. 2020).

3 The Plant-Microbiome Interactions

The term microbiome, derived by combining two words, "micro" and "biome," was first defined by Whipps et al. (1988) as a specific group of microbes living in a well-defined environment with unique physical and chemical characteristics. The "Microbiome" definition emphasizes the unique properties, functions, and interactions of microbial communities with their surroundings, leading to the creation of specific ecological niches. Later, in 2001, Lederberg and McCray redefined in an ecological context as the "collection of resident microorganisms that live in a specific environment." This definition is the most accepted as it comprehensively covers the ecological roles of microorganisms, including beneficial, neutral, and pathogenic. The microbiome is considered the host plants' secondary genome. Plants' characteristics and on-field performance are primarily affected by the diversity and composition of soil and plant-associated microbiomes, which are as-yet-uncultured (Nichols 2007). The plant microbiomes result from inherited and externally adapted microbial populations (Huitzil et al. 2023). Their beneficial function includes improved nutrient uptake, plant hormone production, disease resistance, and stress tolerance. They can alter the plant physiology, including regulation of osmotic pressure, changes in stomatal responses, improved root morphology, N accumulation, metabolism, and uptake of minerals (Hussain et al. 2018). The microbiome interactions in rice, wheat, maize, soybean, and many other crops are well documented (Peiffer et al. 2013; Edwards et al. 2015; Rascovan et al. 2016). The nodule microbiome structure of clover, lotus, and alfalfa suggested a direct connection between microbial community and symbiotic performance of rhizobia (Martinez-Hidalgo and Hirsch 2017). The bacterial community coexisting in the legume root nodules with specialized metabolic pathways exhibits a dynamic interaction for shaping the overall microbiome structure and function (Hansen et al. 2020). The core microbiome of plants enriches genes involved in C-metabolism, unlike noncore microbiome strains (Toju et al. 2018; de Souza et al. 2019). It has been known that microbial diversity across various metabolic functions often displays significant functional redundancy and plays a central role in plant growth and development. It implies that multiple taxonomically distinct organisms coexist, each capable of performing the same metabolic functions (Allison and Martiny 2008; Nelson et al. 2016; Louca et al. 2018). Metagenomics and metatranscriptome approaches are the key tools to identify and quantify multifunctional redundancy or overlapping functional roles of microbial communities and to expose the mechanisms shaping microbial community composition and function (Miki et al. 2014; Souza et al. 2015; Beier et al. 2017). Growing evidence shows that functional redundancy of plant-associated microbiomes can help to sustain ecosystem functions and enhance crop resilience to environmental challenges.

4 Microbiome Assembly in Legumes

The rhizosphere, a nutrient hotspot, is highly regulated by the host and supports many microorganisms (Bais et al. 2006; Fierer et al. 2012). Plant-derived metabolites are known to play a definite role in the microbiome assemblage in the rhizosphere (Hacquard et al. 2017; Yang et al. 2017; Zhang et al. 2017). The composition of rhizosphere communities significantly impacts host fitness (Mendes et al. 2013). Rhizosphere communities are known to be more diverse than those in other plant compartments (Zhalnina et al. 2018; Rolfe et al. 2019). The knowledge of their structural diversity, functional potential, and specific interactions can be translated into effective agricultural strategies. The host recruits beneficial root-associated microbes from the rhizosphere region to other specific plant compartments (Tkacz et al. 2020). Only the selected rhizosphere microbes enter root tissues and form a unique root endophytic microbiome, different from the rhizosphere microbiome (Lundberg et al. 2012). Legumes allow the entry of only the subset of root-associated microbial communities that attribute nodule development and functioning. Thus, compartment-specific microbiome assembly in legumes is a highly selective process driven by the host's ability to recruit and sustain microbial symbionts that contribute to nodule development, nitrogen fixation, and overall plant health (Zgadzaj et al. 2016; Tkacz et al. 2022). Recent evidence on the plant compartment-specific microbiome of different legume plants by employing metagenome sequencing and efficient bioinformatic tools further strengthens these findings.

The amplicon (16S) and shotgun metagenome-based study on *Medicago truncatula* found soil origin as one of the key determinants in structuring rhizosphere microbiome (Brown et al. 2020). The bacterial diversity of cowpea root nodules primarily depends on soil type, with plant genotype playing a relatively minor role (Leite et al. 2017; Sneha et al. 2021). A clear difference in bacterial community composition was noted when rhizocompartments (nodule endosphere, root endosphere, rhizosphere, and root zone) of soybean and alfalfa were subjected to microbiome analysis. This study also suggested that soil type significantly influences the shaping of the rhizosphere and root zone microbiome. At the same time, plant species are the primary determinant of nodule and root endophytic composition (Xiao et al. 2017). In addition, comparable findings have been documented in other studies (Trabelsi et al. 2017; Bender et al. 2022). Cultivar-specific rhizosphere microbiome changes were observed when whole genome analysis of two different lentil cultivars was evaluated. In addition, a drastic difference in the diversity and function of genes related to nitrogen and phosphorus cycling suggests the benefits of exploring the microbiome to hasten the nutrient cycling processes in soil (Pramanik et al. 2020). Hamel et al. (2018) reported that pulse crops significantly affect the microbial diversity and composition of wheat rhizosphere soil if the frequency of pulse crops in cropping systems is increased. Sharaf et al. (2019) observed that soybean's root nodule bacteriome richness partly depends on cultivar and water status.

Higher productivity and drought tolerance were observed in plants grown with greater microbial diversity. Prudent et al. (2020) signified the importance of restoring soil microbial biodiversity for stress-resilient crop production. The experimental results showed that pea yield stability under drought stress mainly relied on the soil microbial communities. A stable and active soil microbial community was found to improve soil structure and increase the water availability to the roots during drought stress conditions. Based on rhizosphere soil metagenome analysis of *Mimosa tenuiflora*, Taketani et al. (2015) observed greater stability in the functional potential of microbial communities across dry and rainy seasons. Dai et al. (2019) revealed how drought stress conditions reshape the rhizosphere community of peanut. *Acidobacteria*, *Actinobacteria*, *Cyanobacteria*, *Chloroflexi*, *Proteobacteria*, and *Saccharibacteria* were the predominant phyla in the rhizosphere soil during normal conditions. Under drought conditions, *Actinobacteria* and *Acidobacteria* showed relatively high abundance during seedling and podding stages, whereas *Cyanobacteria* and *Gemmatimonadetes* predominated during the flowering stage. Moreover, enrichment in metabolic activity, defense mechanism, and signal transduction-related sequences were also noticed during drought. Various factors influencing microbiome assembly in legumes are summarized in Table 1.

The assembly of root nodule microbiome of legumes has been extensively explored, and many fascinating findings have been made. Other than rhizobia, diverse bacterial species with nodule-forming ability have been reported in Lucerne (*Medicago sativa*) root nodules (Wigley et al. 2017). Hakim et al. (2018) described mung bean root nodule was found to occupy several NREs belonging to different phyla such as *Actinobacteria*, *Bacteroides*, *Firmicutes*, and *Proteobacteria*. Furthermore, the absence of *Rhizobium*, very low abundance of *Mesorhizobium,* and codominance of both *Bradyrhizobium* and *Ensifer* strains in the nodules confirms the host-mediated selection of specific rhizobial symbionts in nodules. Based on the metabarcoding of root nodules of three lotus genotypes: *Lotus burttii*, *Lotus*

Table 1 Factors influencing legume microbiome assembly

Legume crop	Factors	Salient finding(s)	Reference
Chickpea (*Cicer arietinum* L.)	Plant genotype	Genotype-specific changes in the root microbial community structure with consistency in structure across years	Bazghaleh et al. (2015)
Cowpea (*Vigna unguiculata* L.)	Soil type and host genotype	Soil type strongly shapes root nodule microbiome composition; plant genotype plays minor role	Leite et al. (2017)
	Salinity stress	Native soil community significantly alter the nodule microbiome under salinity stress conditions	Mukhtar et al. (2020)
Soybean (*Glycine max*)	Soil type and host genotype	Rhizosphere microbial composition is cooperatively modulated by both soil type and genotype	Liu et al. (2019)
Pea (*Pisum sativum*)	Host plant and soil management practices	Rhizosphere enrichment of genes related to plant growth promotion and stress tolerance under different management practices	Chaudhari et al. (2020)
Common bean (*Phaseolus vulgaris* L.)	Soil	Rhizobacterial diversity of rhizosphere soil increased from native soil to agricultural soil, however co-occurrence network in agricultural soil exhibited lower complexity than in native soil	Pérez-Jaramillo et al. (2019)
	Drought stress	Stable transmission of bacterial members within the plant microbiome across generations persists, even in the face of environmental stress	Sulesky-Grieb et al. (2024)
	Drought stress	Parental plants exposed to abiotic stress influence its seed microbiome	Bintarti et al. (2022)
Alfalfa (*Medicago sativa*) and Red clover (*Trifolium pratense*)	Drought stress	Deleterious impact of drought on nodulation, plant growth, and carbon and nitrogen cycling enzymes and positive effect on soil nitrogen availability and certain microbial taxa on soil	Dollete et al. (2024)
Medicago truncatula	Soil origin and host genotype	Root and rhizosphere microbial community is strongly influenced by plant genotype and soil origin, respectively	Brown et al. (2020)
Clover (*Trifolium repens* and *T. subterraneum*)	Soil pH	Cooperative functioning of rhizobia and non-rhizobial endophytes in the root nodule improves the persistence of clover under pH stress	Shah et al. (2021)
Bambara groundnut (*Vigna subterranea*)	Plant growth stage	Rhizosphere microbial community structure is similar during vegetative, flowering and pod filling stage	Ajilogba et al. (2022)
Lentil (*Lens culinaris* Medik.)	Landrace seed origin and host genotype	Landrace possess higher microbial diversity in their rhizosphere compared to modern improved cultivars	Gleridou et al. (2023)

japonicus, and *Lotus corniculatus*, Crosbie et al. (2021) reported that the nodule bacteriome of *Lotus* species differ based on soil inputs. Still, the host genotype specifically influences variations linked to plant health. Mayhood and Mirza (2021) found that soybean plants recruit specific bacterial genera of rhizobia and NREs to the nodule, irrespective of the age, location, or size of the root nodules present on a root system. Prominent nodules of field-grown peanut plants treated with a commercial inoculum were found to be dominated by *Bradyrhizobium* (more than 99%). In contrast, small nodules occupied a diverse set of NREs along with *Bradyrhizobium* (Hossain et al. 2023). Several extensive investigations have further enriched our understanding of nodule microbiomes (Zheng et al. 2020; Crosbie et al. 2022; Favero et al. 2021; Shah et al. 2021; Jesus et al. 2023; Adan et al. 2024).

Many researchers have monitored the complex interactions in the legume microbiome in the presence of introduced microbe(s). Metagenome studies showed similar bacterial community structures in legume plants grown with and without bacterial inoculation. For example, Fan et al. (2018) reported that inoculating *Mesorhizobium loti* in *Robinia pseudoacacia* did not affect the root microbial diversity. A similar result was later forwarded by Dos Santos et al. (2022), who studied the changes in soybean plants upon inoculation with plant growth-promoting bacteria *Bacillus subtilis* by employing 16S amplicon analysis. Inoculation did not change the root microbiome; however, fertilizer application altered the bacterial diversity. In contrast, our recent study revealed that inoculation and fertilizer application can differentially alter microbiome structure in chickpea (Nysanth et al. 2024). Table 2 summarizes the impact of microbial inoculation on the microbiome associated with different legumes. This growing volume of data on legume-associated microbiomes suggests effective strategies to employ newly designed inoculants for biofertilizer programs to support sustainable legume production.

5 Methodologies Used in Microbiome Study

5.1 Microbial Diversity Through Metagenomes

Earlier studies on the diversity and function of plant-associated microbes solely relied on culture-dependent approaches. However, a significant fraction of the microbial diversity reservoir, called microbial dark matter, is largely inaccessible using conventional methods (Whitman et al. 1998; Marcy et al. 2007). Many unculturable bacteria differ substantially, especially nutritionally, from typical eubacteria, making it difficult for them to grow using standard culture media. For instance, oligotrophs grow very slowly under laboratory conditions, and visible turbidity in liquid media or colonies on solid media is rarely observed. Their slow growth indicates that employing conventional techniques in their studies is particularly challenging (Staley and Konopka 1985). Cultivation-based methods provide foundational insights but pose significant limitations, as a low prevalent microbial community remained unculturable (Stewart et al. 2012). Only 1–5% of the total microbial community observed in environmental samples can be cultured in the laboratory, and

Table 2 Impact of microbial inoculants on legume microbiome

Host	Microbial inoculant(s)	Salient findings	Reference
Chickpea *(Cicer arietinum L.)*	*Mesorhizobium ciceri Mesorhizobium muleiense*	Bacterial community structure varies with distinct rhizobial inoculant	Zhang et al. (2023)
	Mesorhizobium ciceri	Inoculation leads to a distinct shift in structure and composition of native microbiome compared to fertilizer application	Nysanth et al. (2024)
Peanut *(Arachis hypogea)*	*Bradyrhizobium* (*strain Lb8*)	Inoculation of *Bradyrhizobium* strain in both organic and inorganic practices significantly alters the rhizosphere microbiome composition with dominant bacterial taxa, largely similar in both cultivation types	Paudel et al. (2023)
Soybean (*Glycine max*)	*Rhizobium* sp.	Inoculant shapes rhizobacterial composition and co-occurrence network in genotype-dependent manner	Zhong et al. (2019)
	Bradyrhizobium sp.	Genotype and inoculant together shape rhizosphere microbial communities	Xu et al. (2020)
	Bacillus subtilis	Transient changes in the root microbial diversity Inoculum concentrations and their interaction with mineral fertilizer level has no effect on root endophytic microbial structure	Dos Santos et al. (2022)
Alfalfa (*Medicago sativa* L.)	*Lactobacillus* (L.) *plantarum PS-8*	Significant shift in silage microbiota composition with higher abundance of beneficial lactic acid bacteria and lower levels of harmful microorganisms	Cao et al. (2021)
	Bacillus thuringiensis NL-11	Inoculation decreased bacterial alpha diversity and the stability and complexity of fungal co-occurrence networks, while improving the stability and complexity of bacterial co-occurrence networks	Nie et al. (2024)
Medicago truncatula	*Sinorhizobium (ensifer) meliloti 1021, Sinorhizobium (ensifer) meliloti 1022 and Sinorhizobium (ensifer) medicae 419*	Inoculation has significantly altered bacterial community in the rhizosphere and endosphere with no changes in fungal community structure	Lagunas et al. (2023)
Mimosa pudica	*Paraburkholderia phymatum*	Inoculation has significant changes in rhizoplane bacterial structure, and *Streptomyces* spp. was identified as keystone taxa involved in nodulation by *P. phymatum*	Welmillage et al. (2021)

this phenomenon is known as "great plate count anomaly" (Staley and Konopka 1985). Since the advancement of bioinformatic tools and metagenomics sequencing, the disparity of "great plate count anomaly" became increasingly evident. Sequence-based or culture-independent approaches could expose this hidden information to a great extent; however, a significant portion remains to be explored. These methods follow extracting DNA directly from samples of interest, followed by high-throughput sequencing to identify microbial taxa and analyze their functional diversity. Studies consistently report the presence of vast microbial diversity identifiable through their DNA but not yet isolated in culture (Hugenholtz 2002; Parks et al. 2018). Most of the bacteria identified through their DNA have never been successfully cultured. Although there have been notable successes in isolating previously unculturable microbes, the rate at which new unculturable microorganisms are discovered overtakes the rate at which new species are cultivated (Gutleben et al. 2018).

Culture-dependent studies may overrepresent certain bacterial or fungal groups during isolation using appropriate media. In contrast, culture-independent studies could decipher the highly diverse and dynamic microbial community residing in root nodules. The tremendous progress of sequencing technologies coupled with bioinformatics tools largely accelerated the scope for exploring the composition and functions of culturable and unculturable microbes in diverse environments. Various culture-independent approaches have been introduced to analyze diverse microbial communities (Su et al. 2012). Among them, taxonomy based on 16S ribosomal RNA (16S rRNA)-sequence has been widely explored and remains an efficient tool in present microbiome study (Woese 1987; Johnson et al. 2019). New sequencing technologies offer cost-effective and comprehensive gene surveys, especially for studying microbial communities based on 16S rRNA gene and whole genome analysis (Cardenas and Tiedje 2008). The term "metagenomics" was proposed by Handelsman et al. (1998) for the studies investigating the sets of genomics data obtained from a microbial consortium. Subsequently, metagenomics studies helped the scientific community to realize the diversity and functional characteristics of microorganisms in diverse, complex environments. Moreover, it can also tell us how adaptive microorganisms impact their environment (Mendes et al. 2017). Sequencing the full length of the rRNA gene is most appropriate for an accurate metagenomics analysis targeting environmental bacteria. However, considering the limitation in sequencing technologies to cover the entire length, researchers mainly aim at selected regions of 100–350 bases in the 16S rRNA gene, known as hyper-variable regions, flanked by more conserved sequence regions (Lane et al. 1985). With the help of primers targeting hypervariable regions specifically, we can survey environments containing thousands of bacterial sequences (Cardenas and Tiedje 2008). Metagenomics studies are of two types, targeted or amplicon and shotgun metagenome sequencing, and they are chosen depending upon the experimental objectives. Targeted metagenomics could be employed when the study explores the microbial community diversity. It targets genome markers that are common in almost all organisms of interest and is a strong representative of phylogeny and taxonomy. For amplicon sequencing of bacteria and archaea, researchers target the RNA (16S) locus of small subunit of ribosomes (Pace et al. 1986; Hugenholtz and Pace 1996). However, in the case of fungus and eukaryotes, intergenic transcribed spacer (ITS) regions, the small subunit (SSU) rRNA (18S) and large subunit (LSU) rRNA (26S) are widely used (Mota-Gutierrez et al. 2019; McNichol et al. 2021; Martin et al. 2022).

In comparison, shotgun approaches are used when the research aims to examine functional genes and metabolic diversity. Unlike amplicon sequencing targeting specific DNA regions, the environmental DNA extraction is followed by shearing into small fragments, which are then sequenced individually (Quince et al. 2017). It resolves many of the limitations, such as biases on polymerase chain reaction (PCR), variations in diversity estimates, sequencing errors, and, more importantly, the lack of functional information associated with the amplicon metagenome approach (Liu et al. 2008; Brooks et al. 2015; Chiu and Chao 2016). Targeted and shotgun metagenomics analyses led to the characterization of the microbial diversity and functional potential associated with large numbers of agriculturally important crops, such as rice (Sun et al. 2008; Oliveira et al. 2022), sugarcane (Pisa et al. 2011), tomato (Ottesen et al. 2013), banana (Thomas and Sekhar 2017), ginger (Liu et al. 2017), oil seed (Gkarmiri et al. 2017), rape seed (Bazghaleh et al. 2020), wheat (Rossmann et al. 2020), maize (Ali et al. 2021), pearl millet (Ndour et al. 2021), sorghum (Wipf et al. 2021), and sugarcane (Wang et al. 2024).

Upon completion of sequencing, it is crucial to preprocess the raw sequence data to ensure the reads are of high quality for further analysis. Preprocessing metagenomics sequences involves removing linkers and adapters, eliminating duplicate as well as chimeric sequences, demultiplexing barcoded samples, and performing quality control to ensure high-quality reads for downstream analysis. Over the years, many software tools and platforms have been developed for this task. Some of these very frequently explored tools include FASTX-Toolkit (Su et al. 2012), Trimmomatic (Bolgee et al. 2014), and FastQC for summary statistics (Huson et al. 2007). Chimeras present in the sequence data can be removed using UCHIME (Edgar et al. 2011), while the data can be denoised and analyzed using MOTHUR (Schloss et al. 2009), QIIME (Caporaso et al. 2010), AmpliconNoise (Quince et al. 2011), or MEGAN (Huson et al. 2007). Metagenomics studies have significantly expanded our insights on plant-microbe interactions by exploring developing bioinformatic tools and platforms. In recent years, numerous hypothesis-driven plant microbiome research have benefited from metagenomics approaches. Besides, many real-time genome databases for plant-associated microbial communities available in the public domain, such as Greengenes (DeSantis et al. 2006), Ribosomal Database Project (RDP) (Cole et al. 2008), SILVA (Pruesse et al. 2007), and UNITE (Abarenkov et al. 2010) boost the progress of metagenomics studies. Metagenomics tools and methodologies have been critically reviewed by Oulas et al. (2015). Figure 1 represents a simplified illustration of the targeted and shotgun metagenomics workflow.

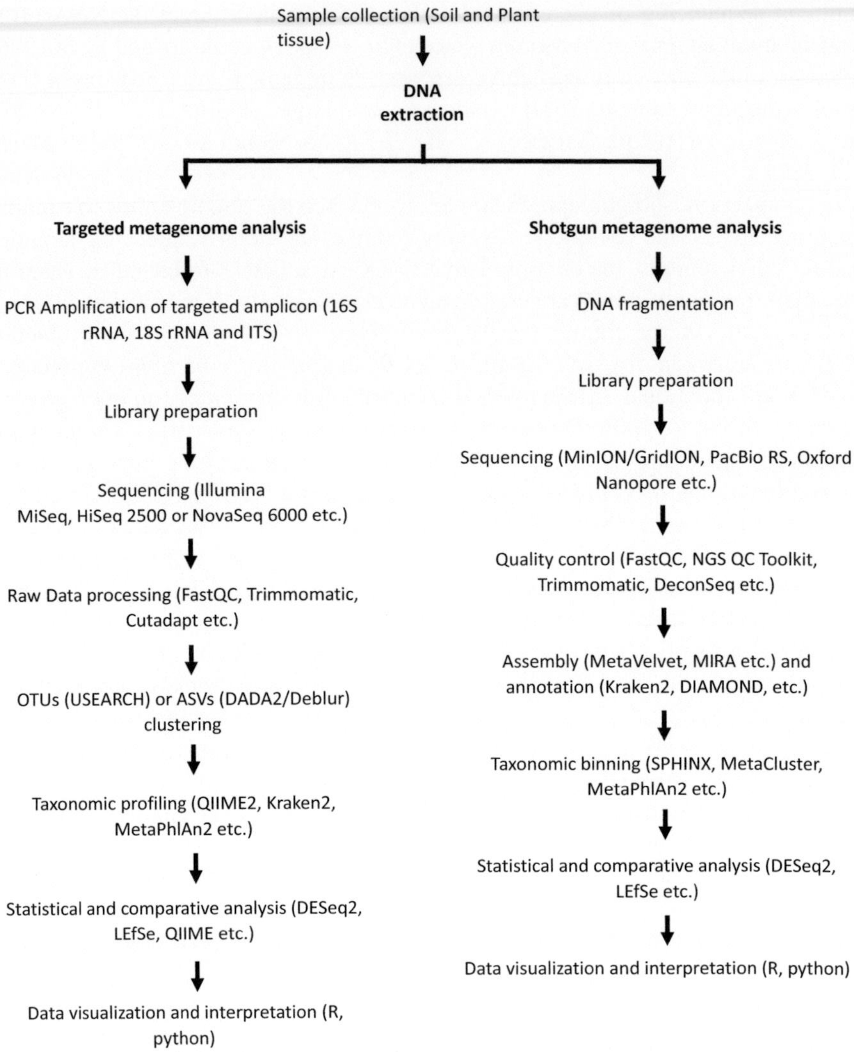

Fig. 1 Bioinformatic workflow diagram for targeted amplicon and shotgun metagenome analysis

5.2 Microbial Function Through Metatranscriptomes

Characterization of complex microbial communities can be achieved with the help of metagenomics approaches. However, metagenomics techniques do not provide information on the actual function of microorganisms, their genes in a particular environment, or their changes in response to various factors (Moran 2009). Metatranscriptomics analysis involves sequencing the total RNA in an environment to identify the active microbial members, genes, and metabolic pathways (Urich et al. 2008). It enables the comparison of relative abundance and functional potential of microorganisms across different samples and spanning all domains of life without polymerase chain reaction bias (Turner et al. 2013). The primer designing step of metagenome studies is one of the potential sources of bias during PCR amplification of genomics DNA (Pinto and Raskin 2012). Conversely, metatranscriptomes rely on either oligo(dT) primers or random hexamers. Oligo(dT) primers specifically bind to the poly(A) tails of eukaryotic mRNAs, enabling the selective synthesis of cDNA from eukaryotic mRNA populations (Nam et al. 2002). Random primers, on the other hand, short oligodeoxyribonucleotides bind nonspecifically to RNA and synthesize cDNA from RNA, including mRNA, rRNA, and noncoding RNAs (Armour et al. 2009). The metatranscriptomics procedure involving these primers is a nontarget approach to synthesize cDNA from the total RNA, avoiding PCR biases caused by specific primers in the metagenomics approach (Urich et al. 2008; Turner et al. 2013). This powerful approach provides a real-time snapshot of which genes are actively expressed and which metabolic pathways function at any given moment. These metatranscriptomics studies are generally performed in a complementary deoxyribonucleic acid (cDNA), oligonucleotide, or ribonucleic acid (RNA) dependent manner. cDNA-based microarrays, which study gene expression at large scale in a DNA-dependent manner (Parro et al. 2007; Garg et al. 2010), have been employed in legume-microbe interaction studies mainly to capture transcriptome dynamics during seed, flower, and nodule developmental stages (He et al. 2009; Garg and Jain 2013; Verdier et al. 2013). The workflow begins with RNA extraction from samples that require immediate preservation to maintain RNA integrity. Afterward, abundant ribosomal RNA is removed to concentrate the messenger RNA fraction. This enriched mRNA is further reverse-transcribed to cDNA and sequenced. This procedure yields numerous mRNA-derived cDNA sequence reads that reflect actively expressed genes in the plant and microbial community (Poretsky et al. 2005; Urich et al. 2008; Cabeza et al. 2014; Kudapa et al. 2014; Alves-Carvalho et al. 2015). It is a hybridization-based technique where cDNA is directly or indirectly labeled with fluorescence dyes (Do and Choi 2007). Modern high-throughput sequencing technologies have made sequencing ribonucleic acid (RNA-seq) possible, making metatranscriptomics studies relatively less challenging. RNA-seq technology revolutionized transcriptomics analysis by sequencing all RNA molecules in an environment in a high-throughput manner. This approach possesses several advantages over DNA-based microarrays regarding transcript coverage, sensitivity, and identifying novel transcripts (Ozsolak and Milos 2011). Stable isotope probing (SIP) is an excellent strategy in metatranscriptomics studies that facilitates the identification of active microbial communities by monitoring the assimilation of labeled substrates in near-natural conditions (Dumont and Murrell 2005). In this technique, metabolically active microorganisms incorporate the labeled substances into their biomass, enabling differentiation from inactive microbial populations. Subsequently, specific microorganisms that exhibited metabolic activity are tracked using various molecular techniques (Neufeld et al. 2007). Integrating RNA-SIP with metatranscriptomics studies helps linking microbial community structure and function by associating actively expressed genes with particular microbial groups (Dumont et al. 2013; Lueders et al. 2016). However,

RNA-seq offers only partial characterization, especially for genes and transcripts with low expression levels. This problem was resolved with the advent of targeted RNA-seq (CaptureSeq) (Mercer et al. 2014) that can capture a greater number of samples with high sequencing depth (Ding et al. 2020). The unknown role of diverse plant and bacterial genes involved in legume developmental processes has been brought to light with the help of these metatranscriptomics approaches. For example, using a dual plant-bacteria RNA-seq, an approach to simultaneously analyze the transcriptome of two organisms in a shared environment, Sauviac et al. (2022) identified hundreds of differentially expressed plant and bacterial genes during nodule senescence of *Medicago truncatula*, forming the symbiotic association with *Sinorhizobium meliloti.* RNA-seq aids in understanding the functional potential of legume microbiome, its changes under varying conditions, and how it differs from other crops. Turner et al. (2013) reported that the metabolically active rhizosphere microbiome of pea (*Pisum sativum*) differed significantly from that of cereal grains such as oats and wheat, with the pea microbiome showing greater strength and distinctiveness. In a comparative transcriptome study using the RNA sequencing approach (Nasr Esfahani et al. 2017), phosphorous deficiency was found to induce distinct changes in the nodule gene expression profile in *Mesorhizobium mediterraneum* SWRI9-(MmSWRI9)-chickpea and *M. ciceri* CP-31-(McCP-31)-chickpea association, of which the McCP-31-chickpea association has recorded relatively higher SNF capacity compared to MmSWRI9-chickpea during Pi-deficient conditions. The workflow for metatranscriptome analysis is provided in Fig. 2.

5.3 Strategies to Cultivate the Uncultivable Microorganisms

Microbiome data have shown that a substantial amount of microbial diversity remains unexplored due to cultivability limitations (Lundberg et al. 2013). Traditionally, bacterial inoculants have been developed based on plant growth-promoting traits, and screening for a single or few traits has failed under field conditions. In contrast to the traditional approach, synthetic microbials (SynComs) can be designed using metagenomics data sets with multiple traits, viz., robust colonization (high abundant bacteria), persistence (consistent across the developmental stages) with specific functions (Nichols et al. 2010; Vorholt et al. 2017; Kong et al. 2018). The dominant core microbiome, which efficiently colonizes plants, can be included in SynComs to ensure inoculum stability under natural conditions (Niu et al. 2017). Yin et al. (2022) showed the beneficial effect of SynCom on disease control and plant resilience. Improving the cultivable efficiency of these uncultured microbes is crucial for designing SynComs. Novel strategies have been proposed to cultivate viable but yet to cultivate microbial strains (Kaeberlein et al. 2002; Kehe et al. 2019). The microfluidic platform is a promising method for cultivating uncultured microorganisms (Aleklett et al. 2018). By replicating natural conditions under the microenvironment, microbe-microbe interactions and their survival can be improved (Nichols et al. 2010). The community-based culture collection (CBC) method for picking non-confluent colonies instead of targeting axenic pure culture

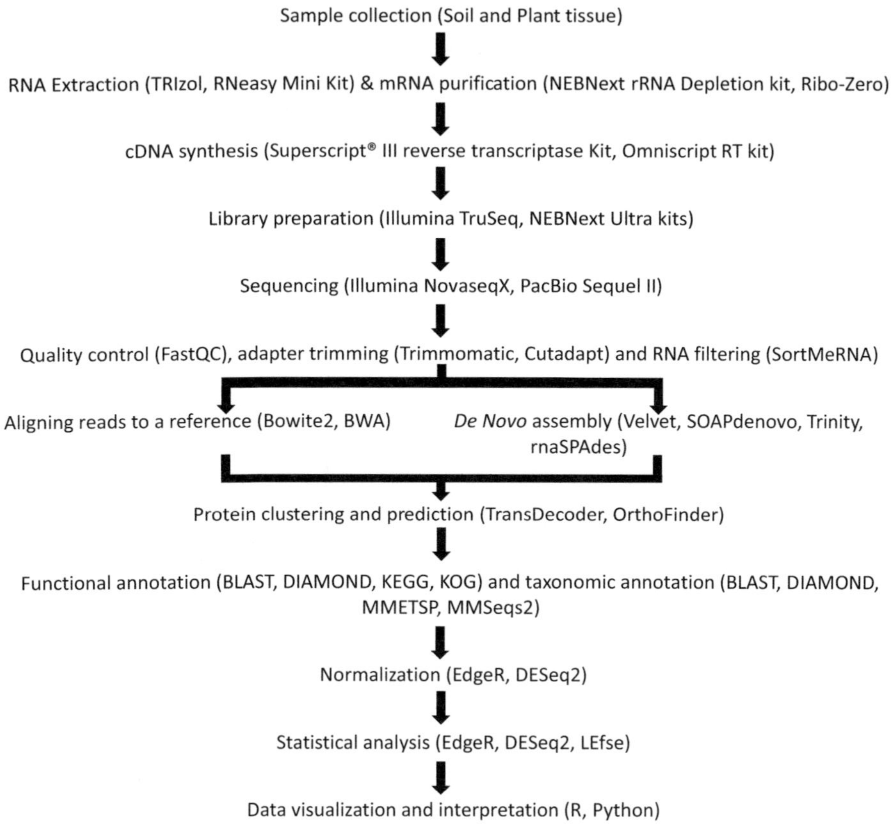

Fig. 2 Bioinformatic workflow diagram for metatranscriptome analysis

greatly improves large-scale plant-microbiome isolation. The coculture method and supplementing isolation media with soil/plant extract have improved the retrieval of microbiomes in natural conditions (Stewart 2012).

5.4 Microbiome Engineering in Legumes

Accumulating knowledge of the legume microbiome offers significant opportunities for strategic interventions such as microbiome engineering to enhance crop growth, improve soil health, and sustainably increase legume productivity. Manipulating the legume microbiome can enhance nitrogen fixation, alleviate biotic and abiotic stresses, and even discover novel bioactive compounds. Recent advancements now focus on engineering legume-associated microbiomes to optimize beneficial interactions for improved plant health and productivity. The microbiome associated with a specific host can significantly influence the host's fitness (Foo et al. 2017). Gut microbiome perturbation affects digestion, immune function, and the host's mental health in humans and animals. Similarly, the rhizosphere microbiome in plants impacts the growth, nutrient uptake, and resistance to pathogens (Burz et al. 2023). Even subtle changes in microbiome composition can lead to dysbiosis, disrupting the host's overall functioning (Mueller and Sachs 2015). To prevent disease or stress, and increase host fitness, the host microbiome can be modulated with the desired outcome using the microbiome engineering technique, which is used in agriculture to develop more sustainable farming practices. By manipulating the root-associated and rhizosphere microbiome, it is possible to promote plant growth, increase crop yields, reduce the need for agrochemicals, and increase crop resilience to environmental stressors (Qiu et al. 2019). Hence, microbiome engineering practice can serve as a valuable strategy for enhancing crop productivity sustainably. Advances in "multi-omics" techniques have provided sound evidence of intricate interactions within microbial ecosystems and between microbiomes and their host environments. These advancements have allowed engineering microbial communities to perform specific functions (Lawson et al. 2019). However, successful microbiome engineering also requires considering factors such as soil environment, plant host genetics, and cultivation methods, which are often overlooked (Batool et al. 2024). Microbiome engineering involves intentional manipulation of the microbial community composition either by modifying the environment to promote the growth of specific Indigenous microbes (top-down approach) or by introducing synthetic microbial strains or consortia into the indigenous community (bottom-up approach) (Hu et al. 2022). Figure 3 illustrates the two approaches to microbiome engineering in legumes:

In the top-down approach, the environmental factors are modulated to select/optimize beneficial indigenous microbial communities with desirable functions from the naturally occurring microbiome (Lawson et al. 2019). Host-mediated artificial selection is a top-down approach that is often employed for microbiome engineering in rhizosphere and root-associated microbiome. The plant per se selectively shapes the associated microbial community (Berruto and Demirer 2024). This is achieved by growing genotypically identical crops in replicates in the starter soil microbiome, and with every generational cycle, the desired phenotypes are observed. Further, the soil from the best-performing crop is selected for the next cycle, followed by reinoculation of genotypically identical crops with the selected soil. After every cycle, the diversity and composition of the soil microbial community will be shaped toward the required microbiome. A minimum of ten growth cycles is required to establish a suitable microbial community in the rhizosphere soil to enhance crop health and fitness (Mueller and Sachs 2015). The microbiome-engineered *Brachypodium distachyon*, a model grass plant, for the salinity stress through host-mediated artificial selection could increase the seed yield to 55–205% (Mueller et al. 2021). Host-mediated artificial selection is an eco-friendly approach for abiotic stress management in legumes. A study was conducted on *Vigna radiata* to mitigate salinity stress by microbiome engineering of the host rhizosphere by Anand et al. (2021b), where a multigenerational plant growth experiment was carried out to engineer the rhizosphere microbiome using the rhizosphere soil of the best-performing plant as a growing media for the next generation. With every

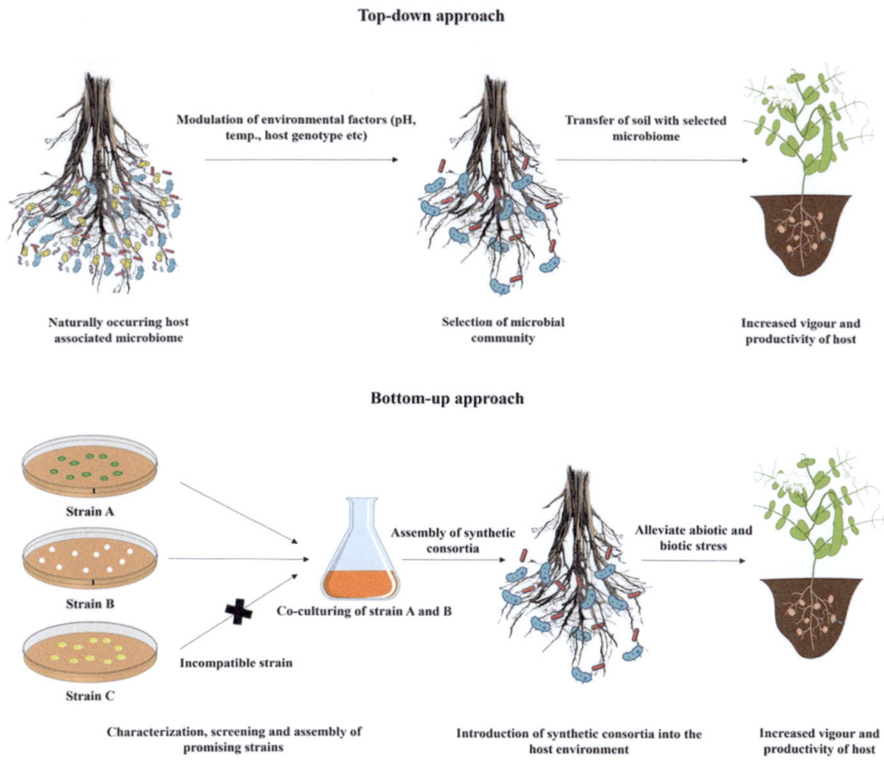

Fig. 3 Illustration of two approaches in microbiome engineering. The upper panel and lower panel describe the top-down approach and bottom-up approaches, respectively

growth cycle, the plant biometrics improved, and there was a decrease in plant biochemical salinity-induced stress markers. The bacterial diversity decreased with every cycle, and the richness of the engineered rhizosphere of the microbial species was similar to that of the control. In another study, *Crotalaria juncea* and *Canavalia ensiformis* have been used as trap crops to attract beneficial microbes from the soil microbiome into their nodule and rhizosphere. Effective AMF and bacteria were identified, and the microbial consortium was formulated as a biofertilizer using the substrate and the roots of the host plant. The metagenomics analysis of the formulated biofertilizer reveals a diverse spectrum of plant growth-promoting microbes that can help the crop mitigate abiotic and biotic stresses (Hernandez-Alvarez et al. 2023). Hansen et al. (2020) derived an inoculum by macerating root nodules of field-grown *Medicago sativa* and serially passed it to the next generations in a gnotobiotic system that maintained specific microbial entophytes only. Microbial community profiling showed that *Brevibacillus brevis*, *Paenibacillus* sp., *Pantoea agglomerans*, and *Pseudomonas* sp. were the key endophytic species spatially segregated in the nodules. However, *B. brevis* produces antimicrobial molecules, which may protect against plant pathogens.

The top-down approach has several advantages as it doesn't require prior details on how microbes interact in the microbial community and with the plant host. Besides, functional redundancy in mixed microbial populations yields a stable microbiome. However, there are limitations as it depends on the indigenous microbial community, which may not always possess the functional traits or diversity required for a particular application. The results of top-down approaches can be less predictable due to their reliance on the intricate dynamics of natural communities, which remain only partially understood (Hu et al. 2022).

In contrast, the bottom-up approach involves constructing a microbiome de novo by introducing well-characterized microbial strains or synthetic consortia into the host-associated microbiome. It employs constructing a microbiome with specific, well-defined mechanisms for more precise regulation and control of its functional attributes (Lyu et al. 2024). Subsequently, the synthetically assembled strains or consortia can be introduced into the host or host environment to manipulate the host's microbiome. It can be done directly through inoculation into the rhizosphere or seeds and spraying microparticles of synthetically assembled strains or consortia onto the plant tissues (Orozco-Mosqueda et al. 2018). The bottom-up approach begins with the selection and screening of beneficial microbes in vitro, and the consortium is constructed by examining the compatibility of the selected strains. These synthetically made consortiums can be directly inoculated into the host environment, where it interacts with the host and other microbes in the host environment. These synthetic consortiums can improve crop's resilience against abiotic and biotic stresses. *Pseudomonas aeruginosa* has been reported to have the ability to alleviate drought stress in *Vigna radiata* (Uzma et al. 2022). Five drought-tolerant strains of *Pseudomonas aeruginosa* were screened for IAA (indole acetic acid), siderophore, and ACC (1-aminocyclopropane-1-carboxylate) deaminase production. When inoculated to *Vigna radiata*, these strains successfully established themselves within the plant and improved plant growth under drought-stress conditions. Inoculation of multiple microbes as a consortium can also be done, as Laranjeira et al. (2021) described in chickpea. The plant growth-promoting bacteria (*Mesorhizobium* sp., *Burkholderia* sp., and *Pseudomonas* sp.) and AMF (*Rhizophagus irregularis*, *Funneliformis geosporum*, *Claroideoglomus claroideum*) were inoculated to mitigate the drought stress in chickpea. The bottom-up approach has several advantages over the top-down approach (Table 3). It focuses more on individual organisms' metabolic pathways, which allows for constructing precise microbiomes with proper optimization. It takes account of the predictive computational models such as flux balance analysis, which helps to simulate and understand the metabolic fluxes and interactions within the microbial community and guide the selection of specific strains that can affect the results (Lawson et al. 2019). Finally, this approach facilitates the exploration of novel microbial combinations and functions (Lyu et al. 2024). Even though the bottom-up approach has several advantages over the top-down approach, it has certain limitations, as many microorganisms with crucial functional roles have yet to be cultured. Understanding the metabolic pathway and microbial interaction within the microbiome is a daunting task, and there is always

Table 3 Key differences between the Top-down and Bottom-up approach in microbiome engineering

Aspects	Top-down approach	Bottom-up approach
Manipulation at the ecosystem level	Yes	No
Manipulation of the individual organism	No	Yes
Requirement of prior knowledge of metabolic pathways in the microbiome	No	Yes
Selection by host and host environment manipulation	Yes	No
In vitro selection and assembling of microbes	No	Yes
Precise manipulation of microbial interaction	No	Yes
Ecologically safe and easy to implement	Yes	No
Easy to customize	No	Yes

uncertainty about the disruption of the natural microbial ecosystem due to the introduction of synthetically engineered strains (Hu et al. 2022).

The advancement of multi-omics technologies can construct better strategies for legume microbiome engineering to enhance crop yield and nutrient use efficiency with resistance and resilience to biotic and abiotic stresses, respectively. The use of synthetic communities for improving crop yield and productivity is well-known; however, host-mediated artificial selection in legumes still needs to be explored (Petrushin et al. 2023). Despite the increasing popularity of microbiome engineering, attempts to engineer microbiome in legumes are scarce (Quides and Atamian 2021). Nevertheless, employing microbiome engineering in legume production will contribute to achieving sustainable global food and nutritional security.

6 Conclusion

Considering the great importance of legumes, understanding their associated microbiome establishes a foundation for designing and developing innovative strategies to enhance productivity. A coupling of culture-dependent and culture-independent techniques in recent studies can tell us the exact changes in bacterial community structure and composition, their specific interactions, and the definite mechanism behind growth enhancement or decline at a deeper level. By deciphering microbial communities' complex interactions, dynamics, and genetic potential for legumes growth and development, we can design new synthetic communities that can improve crop performance in diverse environmental conditions. This vast knowledge holds great promise for creating refined microbial inoculants that promote sustainable legume production. Additionally, integrating modern biotechnological strategies such as gene editing to engineer both plant-associated microbes and host plants can further optimize beneficial interactions and enhance the plants resilience to various environmental stresses.

Conflict of Interest Authors declare no conflict of interest.

References

Abarenkov K, Henrik Nilsson R, Larsson KH, Alexander IJ, Eberhardt U, Erland S, Pennanen T (2010) The UNITE database for molecular identification of fungi–recent updates and future perspectives. New Phytol 186(2):281–285. https://www.jstor.org/stable/27797548

Adan IH, Asudi GO, Niassy S, Jalloh AA, Mutua JM, Chidawanyika F, Khamis F, Khan Z, Subramanian S, Dubois T, Mutyambai DM (2024) Comparative microbiome diversity in root-nodules of three *Desmodium* species used in push-pull cropping system. Front Microbiol 15:1395811. https://doi.org/10.3389/fmicb.2024.1395811

Ajilogba CF, Olanrewaju OS, Babalola OO (2022) Plant growth stage drives the temporal and spatial dynamics of the bacterial microbiome in the rhizosphere of *Vigna subterranea*. Front Microbiol 13:825377. https://doi.org/10.3389/fmicb.2022.825377

Aleklett K, Kiers ET, Ohlsson P, Shimizu TS, Caldas VEA, Hammer EC (2018) Build your own soil: exploring microfluidics to create microbial habitat structures. ISME J 12(2):312–319. https://doi.org/10.1038/ismej.2017.184

Ali MA, Naveed M, Mustafa A, Abbas A (2017) The good, the bad, and the ugly of rhizosphere microbiome. In: Kumar V, Kumar M, Sharma S, Prasad R (eds) Probiotics and plant health. Springer, Singapore, pp 253–290. https://doi.org/10.1007/978-981-10-3473-2_11

Ali M, Ahmad Z, Ashraf MF, Dong W (2021) Maize endophytic microbial-communities revealed by removing PCR and 16S rRNA sequencing and their synthetic applications to suppress maize banded leaf and sheath blight. Microbiol Res 242:126639. https://doi.org/10.1016/j.micres.2020.126639

Allison SD, Martiny JB (2008) Resistance, resilience, and redundancy in microbial communities. Proc Natl Acad Sci 105:11512–11519. https://doi.org/10.1073/pnas.0801925105

Alves-Carvalho S, Aubert G, Carrere S, Cruaud C, Brochot AL, Jacquin F, Klein A, Martin C, Boucherot K, Kreplak J, Da Silva C (2015) Full-length de novo assembly of RNA-seq data in pea (*Pisum sativum* L.) provides a gene expression atlas and gives insights into root nodulation in this species. Plant J 84(1):1–9. https://doi.org/10.1111/tpj.12967

Anand G, Bhattacharjee A, Shrivas VL, Dubey S, Sharma S (2021a) ACC deaminase positive *Enterobacter*-mediated mitigation of salinity stress, and plant growth promotion of *Cajanus cajan*: a lab to field study. Physiol Mol Biol Plants 27(7):1547–1557. https://doi.org/10.1007/s12298-021-01031-0

Anand G, Goel V, Dubey S, Sharma S (2021b) Tailoring the rhizospheric microbiome of *Vigna radiata* by adaptation to salt stress. Plant Growth Regul 93:79–88. https://doi.org/10.1007/s10725-020-00667-4

Armour CD, Castle JC, Chen R, Babak T, Loerch P, Jackson S, Shah JK, Dey J, Rohl CA, Johnson JM, Raymond CK (2009) Digital transcriptome profiling using selective hexamer priming for cDNA synthesis. Nat Methods 6(9):647–649. https://doi.org/10.1038/nmeth.1360

Bais HP, Weir TL, Perry LG, Gilroy S, Vivanco JM (2006) The role of root exudates in rhizosphere interactions with plants and other organisms. Annu Rev Plant Biol 57(1):233–266. https://doi.org/10.1146/annurev.arplant.57.032905.105159

de Bary A (1866) Morphologie und physiologie der pilze, flechten und myxomyceten. Engelmann

Batool M, Carvalhais LC, Fu B, Schenk PM (2024) Customized plant microbiome engineering for food security. Trends Plant Sci 29(4):482–494. https://doi.org/10.1016/j.tplants.2023.10.012

Bazghaleh N, Hamel C, Gan Y, Tar'an B, Knight JD (2015) Genotype-specific variation in the structure of root fungal communities is related to chickpea plant productivity. Appl Environ Microbiol 81(7):2368–2377. https://doi.org/10.1128/AEM.03692-14

Bazghaleh N, Bell JK, Mamet SD, Moreira ZM, Taye ZM, Williams S, Norris C, Dowhy T, Arcand M, Lamb EG, Links M (2020) An intensive multilocation temporal dataset of fungal and bacterial communities in the root and rhizosphere of *Brassica napus*. Data Brief 31:106143. https://doi.org/10.1016/j.dib.2020.106143

Beier S, Shen D, Schott T, Jürgens K (2017) Metatranscriptomic data reveal the effect of different community properties on multifunctional redundancy. Mol Ecol 26(24):6813–6826. https://doi.org/10.1111/mec.14409

Beijerinck MW (1888) Cultur des *Bacillus radicicola* aus den Knöllchen. Bot Ztg 46:740–750

Beijerinck MW, Van Delden A (1902) Ueber die Assimilation des freien Stickstoffs durch Bakterien. Centralbl Bakt Abt II 9:3–43

Bender FR, Alves LC, da Silva JF, Ribeiro RA, Pauli G, Nogueira MA, Hungria M (2022) Microbiome of nodules and roots of soybean and common bean: searching for differences associated with contrasting performances in symbiotic nitrogen fixation. Int J Mol Sci 23(19):12035. https://doi.org/10.3390/ijms231912035

Berrada H, Fikri-Benbahim K (2014) Taxonomy of the rhizobia: current perspectives. Microbiol Res J Int 4:616–639

Berruto CA, Demirer GS (2024) Engineering agricultural soil microbiomes and predicting plant phenotypes. Trends Microbiol 32(9):858–873. https://doi.org/10.1016/j.tim.2024.02.003

Bintarti AF, Kearns PJ, Sulesky-Grieb A, Shade A (2022) Abiotic treatment to common bean plants results in an altered endophytic seed microbiome. Microbiol Spect 10(2):e00210–e00221. https://doi.org/10.1128/spectrum.00210-21

Bolger AM, Lohse M, Usadel B (2014) Trimmomatic: a flexible trimmer for Illumina sequence data. Bioinfomatics 15:2114–2120. https://doi.org/10.1093/bioinformatics/btu170

Boussingault J (1838) Recherches chimiques sur la vegetation enterprises dans le but d'examiner si les plantes prennent de I'atmosphere. Annales de Chimie et de Physique 67:1–54

Brooks JP, Edwards DJ, Harwich MD, Rivera MC, Fettweis JM, Serrano MG, Reris RA, Sheth NU, Huang B, Girerd P (2015) Vaginal microbiome consortium (additional members). The truth about metagenomics: quantifying and counteracting bias in 16S rRNA studies. BMC Microbiol 15:1–14. https://doi.org/10.1186/s12866-015-0351-6

Brown SP, Grillo MA, Podowski JC, Heath KD (2020) Soil origin and plant genotype structure distinct microbiome compartments in the model legume *Medicago truncatula*. Microbiome 8:1–17. https://doi.org/10.1186/s40168-020-00915-9

Burz SD, Causevic S, Dal Co A, Dmitrijeva M, Engel P, Garrido-Sanz D, Greub G, Hapfelmeier S, Hardt W, Hatzimanikatis V, Heiman CM, Herzog MK, Hockenberry A, Keel C, Keppler A, Lee S, Luneau J (2023) From microbiome composition to functional engineering, one step at a time. Microbiol Mol Biol Rev 87(4):e00063–e00023. https://doi.org/10.1128/mmbr.00063-23

Cabeza RA, Liese R, Lingner A, von Stieglitz I, Neumann J, Salinas-Riester G, Pommerenke C, Dittert K, Schulze J (2014) RNA-seq transcriptome profiling reveals that *Medicago truncatula* nodules acclimate N_2 fixation before emerging P deficiency reaches the nodules. J Exp Bot 65(20):6035–6048. https://doi.org/10.1093/jxb/eru341

Cao C, Bao W, Li W, Zhao F, Kwok LY, Zhang W, Zhang H (2021) Changes in physico-chemical characteristics and viable bacterial communities during fermentation of alfalfa silages inoculated with *Lactobacillus plantarum*. World J Microbiol Biotechnol 37:1–4. https://doi.org/10.1007/s11274-021-03095-2

Caporaso JG, Kuczynski J, Stombaugh J, Bittinger K, Bushman FD, Costello EK, Fierer N, Peña AG, Goodrich JK, Gordon JI, Huttley GA (2010) QIIME allows analysis of high-throughput community sequencing data. Nat Methods 7(5):335–336. https://doi.org/10.1038/nmeth.f.303

Cardenas E, Tiedje JM (2008) New tools for discovering and characterizing microbial diversity. Curr Opin Biotechnol 19(6):544–549. https://doi.org/10.1016/j.copbio.2008.10.010

Castro-Sowinski S, Herschkovitz Y, Okon Y, Jurkevitch E (2007) Effects of inoculation with plant growth-promoting rhizobacteria on resident rhizosphere microorganisms. FEMS Microbiol Lett 276(1):1–1. https://doi.org/10.1111/j.1574-6968.2007.00878.x

Chalk PM, Souza RD, Urquiaga S, Alves BJ, Boddey RM (2006) The role of arbuscular mycorrhiza in legume symbiotic performance. Soil Biol Biochem 38(9):2944–2951. https://doi.org/10.1016/j.soilbio.2006.05.005

Chaudhari D, Rangappa K, Das A, Layek J, Basavaraj S, Kandpal BK, Shouche Y, Rahi P (2020) Pea (*Pisum sativum* L.) plant shapes its rhizosphere microbiome for nutrient uptake and stress amelioration in acidic soils of the North-East region of India. Front Microbiol 11:968. https://doi.org/10.3389/fmicb.2020.00968

Chiu CH, Chao A (2016) Estimating and comparing microbial diversity in the presence of sequencing errors. PeerJ 4:e1634. https://doi.org/10.7717/peerj.1634

Cole JR, Wang Q, Cardenas E, Fish J, Chai B, Farris RJ, Garrity GM (2008) The ribosomal database project: improved alignments and new tools for rRNA analysis. Nucleic Acids Res 37(suppl_1):D141–D145. https://doi.org/10.1093/nar/gkn879

Compant S, Clément C, Sessitsch A (2010) Plant growth-promoting bacteria in the rhizo-and endosphere of plants: their role, colonization, mechanisms involved and prospects for utilization. Soil Biol Biochem 42(5):669–678. https://doi.org/10.1016/j.soilbio.2009.11.024

Compant S, Cambon MC, Vacher C, Mitter B, Samad A, Sessitsch A (2021) The plant endosphere world–bacterial life within plants. Environ Microbiol 23(4):1812–1829. https://doi.org/10.1111/1462-2920.15240

Crosbie DB, Mahmoudi M, Radl V, Brachmann A, Schloter M, Kemen E, Marín M (2021) The microbiome of *Lotus* nodules varies with plant health in a species-specific manner. BioRxiv 19:2021–2005. https://doi.org/10.1101/2021.05.19.441130

Crosbie DB, Mahmoudi M, Radl V, Brachmann A, Schloter M, Kemen E, Marín M (2022) Microbiome profiling reveals that *Pseudomonas* antagonises parasitic nodule colonisation of cheater rhizobia in *Lotus*. New Phytol 234(1):242–255. https://doi.org/10.1111/nph.17988

Dai L, Zhang G, Yu Z, Ding H, Xu Y, Zhang Z (2019) Effect of drought stress and developmental stages on microbial community structure and diversity in peanut rhizosphere soil. Int J Mol Sci 20(9):2265. https://doi.org/10.1016/S2095-3119(19)62608-0

De Meyer SE, De Beuf K, Vekeman B, Willems A (2015) A large diversity of non-rhizobial endophytes found in legume root nodules in Flanders (Belgium). Soil Biol Biochem 83:1–11. https://doi.org/10.1016/j.soilbio.2015.01.002

Debnath S, Chakraborty S, Langthasa M, Choure K, Agnihotri V, Srivastava A, Rai PK, Tilwari A, Maheshwari DK, Pandey P (2023) Non-rhizobial nodule endophytes improve nodulation, change root exudation pattern and promote the growth of lentil, for prospective application in fallow soil. Front Plant Sci 14:1152875. https://doi.org/10.3389/fpls.2023.1152875

Dénarié J, Debellé F, Promé JC (1996) *Rhizobium* lipo-chitooligosaccharide nodulation factors: signaling molecules mediating recognition and morphogenesis. Annu Rev Biochem 65(1):503–535. https://doi.org/10.1146/annurev.bi.65.070196.002443

Denison RF (2000) Legume sanctions and the evolution of symbiotic cooperation by rhizobia. Am Nat 156(6):567–576. https://doi.org/10.1086/316994

DeSantis TZ, Hugenholtz P, Larsen N, Rojas M, Brodie EL, Keller K, Andersen GL (2006) Greengenes, a chimera-checked 16S rRNA gene database and workbench compatible with ARB. Appl Environ Microbiol 72(7):5069–5072. https://doi.org/10.1128/AEM.03006-05

Ding P, Ngou BP, Furzer OJ, Sakai T, Shrestha RK, MacLean D, Jones JD (2020) High-resolution expression profiling of selected gene sets during plant immune activation. Plant Biotechnol J 18(7):1610–1619. https://doi.org/10.1111/pbi.13327

Do JH, Choi DK (2007) cDNA labeling strategies for microarrays using fluorescent dyes. Eng Life Sci 7(1):26–34. https://doi.org/10.1002/elsc.200620169

Dollete D, Lumactud RA, Carlyle CN, Szczyglowski K, Hill B, Thilakarathna MS (2024) Effect of drought stress on symbiotic nitrogen fixation, soil nitrogen availability and soil microbial diversity in forage legumes. Plant Soil 495(1):445–467. https://doi.org/10.1007/s11104-023-06348-1

Dos Santos RM, Cueva-Yesquén LG, Garboggini FF, Desoignies N, Rigobelo EC (2022) Inoculum concentration and mineral fertilization: effects on the endophytic microbiome of soybean. Front Microbiol 13:900980. https://doi.org/10.3389/fmicb.2022.900980

Dumont MG, Murrell JC (2005) Stable isotope probing—linking microbial identity to function. Nat Rev Microbiol 3(6):499–504. https://doi.org/10.1038/nrmicro1162

Dumont MG, Pommerenke B, Casper P (2013) Using stable isotope probing to obtain a targeted metatranscriptome of aerobic methanotrophs in lake sediment. Environ Microbiol Rep 5(5):757–764. https://doi.org/10.1111/1758-2229.12078

Edgar RC, Haas BJ, Clemente JC, Quince C, Knight R (2011) UCHIME improves sensitivity and speed of chimera detection. Bioinformatics 27(16):2194–2200. https://doi.org/10.1093/bioinformatics/btr381

Edwards J, Johnson C, Santos-Medellín C, Lurie E, Podishetty NK, Bhatnagar S, Eisen JA, Sundaresan V (2015) Structure, variation, and assembly of the root-associated microbiomes of rice. Proc Natl Acad Sci 112(8):E911–E920. https://doi.org/10.1073/pnas.1414592112

Elkoca E, Turan M, Donmez MF (2010) Effects of single, dual and triple inoculations with *Bacillus subtilis*, *Bacillus megaterium* and *Rhizobium leguminosarum* bv. *phaseoli* on nodulation, nutrient uptake, yield and yield parameters of common bean (*Phaseolus vulgaris* l. cv.'elkoca-05'). J Plant Nutr 33(14):2104–2019. https://doi.org/10.1080/01904167.2010.519084

Falkowski PG, Fenchel T, Delong EF (2008) The microbial engines that drive Earth's biogeochemical cycles. Science 320(5879):1034–1039. https://doi.org/10.1126/science.1153213

Fan M, Xiao X, Guo Y, Zhang J, Wang E, Chen W, Lin Y, Wei G (2018) Enhanced phytoremdiation of *Robinia pseudoacacia* in heavy metal-contaminated soils with rhizobia and the associated bacterial community structure and function. Chemosphere 197:729–740. https://doi.org/10.1016/j.chemosphere.2018.01.102

Favero VO, Carvalho RH, Motta VM, Leite AB, Coelho MR, Xavier GR, Rumjanek NG, Urquiaga S (2021) *Bradyrhizobium* as the only rhizobial inhabitant of mung bean (*Vigna radiata*) nodules in tropical soils: a strategy based on microbiome for improving biological nitrogen fixation using bio-products. Front Plant Sci 11:602645. https://doi.org/10.3389/fpls.2020.602645

Ferchichi N, Toukabri W, Boularess M, Smaoui A, Mhamdi R, Trabelsi D (2019) Isolation, identification and plant growth promotion ability of endophytic bacteria associated with lupine root nodule grown in Tunisian soil. Arch Microbiol 201:1333–1349. https://doi.org/10.1007/s00203-019-01702-3

Fierer N, Leff JW, Adams BJ, Nielsen UN, Bates ST, Lauber CL, Owens S, Gilbert JA, Wall DH, Caporaso JG (2012) Cross-biome metagenomic analyses of soil microbial communities and their functional attributes. Proc Natl Acad Sci 109(52):21390–21395. https://doi.org/10.1073/pnas.1215210110

Foo JL, Ling H, Lee YS, Chang MW (2017) Microbiome engineering: current applications and its future. Biotechnol J 12(3):28133942. https://doi.org/10.1002/biot.201600099

Frank B (1889) Uber die Pilzsymbiose der leguminosen. Berichte der Deutschen Beru Deut Bot Ges 7:332–346

Fred EB, Baldwin IL, McCoy E (1932) Root nodule bacteria and leguminous plants. University of Wiscosin Press, p 343

Freiberg C, Fellay R, Bairoch A, Broughton WJ, Rosenthal A, Perret X (1997) Molecular basis of symbiosis between *Rhizobium* and legumes. Nature 387(6631):394–401. https://doi.org/10.1038/387394a0

Gamalero E, Glick BR (2015) Bacterial modulation of plant ethylene levels. Plant Physiol 169(1):13–22. https://doi.org/10.1104/pp.15.00284

Garg R, Jain M (2013) Transcriptome analyses in legumes: a resource for functional genomics. Plant Genome 6(3):2013–2004. https://doi.org/10.3835/plantgenome2013.04.0011

Garg R, Jhanwar S, Tyagi AK, Jain M (2010) Genome-wide survey and expression analysis suggest diverse roles of glutaredoxin gene family members during development and response to various stimuli in rice. DNA Res 17(6):353–367. https://doi.org/10.1093/dnares/dsq023

Gkarmiri K, Mahmood S, Ekblad A, Alström S, Högberg N, Finlay R (2017) Identifying the active microbiome associated with roots and rhizosphere soil of oilseed rape. Appl Environ Microbiol 83(22):e01938–e01917. https://doi.org/10.1128/AEM.01938-17

Gleridou A, Giannopoulos G, Polidoros AN, Mylona PV (2023) Lentil landrace seed origin and genotype affects rhizosphere microbiome. Agronomy 13(12):2910. https://doi.org/10.3390/agronomy13122910

Glick BR (1995) The enhancement of plant growth by free living bacteria. Can J Microbiol 41(2):109–117. https://doi.org/10.1139/m95-015

Gutleben J, Chaib De Mares M, Van Elsas JD, Smidt H, Overmann J, Sipkema D (2018) The multi-omics promise in context: from sequence to microbial isolate. Crit Rev Microbiol 44(2):212–229. https://doi.org/10.1080/1040841X.2017.1332003

Gyaneshwar P, Hirsch AM, Moulin L, Chen WM, Elliott GN, Bontemps C, Estrada-de Los Santos P, Gross E, dos Reis Jr FB (2011) Legume-nodulating betaproteobacteria: diversity, host range, and future prospects. Mol Plant Microbe Interact 24(11):1276–1288. pmid:21830951

Haag AF, Arnold MF, Myka KK, Kerscher B, Dall'Angelo S, Zanda M, Mergaert P, Ferguson GP (2013) Molecular insights into bacteroid development during *Rhizobium*–legume symbiosis. FEMS Microbiol Rev 37(3):364–383. https://doi.org/10.1111/1574-6976.12003

Hacquard S, Spaepen S, Garrido-Oter R, Schulze-Lefert P (2017) Interplay between innate immunity and the plant microbiota. Annu Rev Phytopathol 55:565–589. https://doi.org/10.1146/annurev-phyto-080516-035623

Hakim S, Mirza BS, Zaheer A, Mclean JE, Imran A, Yasmin S, Sajjad Mirza M (2018) Retrieved 16S rRNA and *nif*H sequences reveal co-dominance of *Bradyrhizobium* and *Ensifer* (*Sinorhizobium)* strains in field-collected root nodules of the promiscuous host *Vigna radiata* (L.) R. wilczek. Appl Microbiol Biotechnol 102:485–497. https://doi.org/10.1007/s00253-017-8609-6

Hamel C, Gan Y, Bainard LD (2018) High frequency cropping of pulses modifies soil nitrogen level and the rhizosphere bacterial microbiome in 4-year rotation systems of the semiarid prairie. Appl Soil Ecol 126:47–56. https://doi.org/10.1016/j.apsoil.2018.01.003

Handelsman J, Rondon MR, Brady SF, Clardy J, Goodman RM (1998) Molecular biological access to the chemistry of unknown soil microbes: a new frontier for natural products. Chem Biol 5:R245–R249. https://doi.org/10.1016/s1074-5521(98)90108-9

Hansen BL, Pessotti RD, Fischer MS, Collins A, El-Hifnawi L, Liu MD, Traxler MF (2020) Cooperation, competition, and specialized metabolism in a simplified root nodule microbiome. MBio 11(4):100–128. https://doi.org/10.1128/mbio.01917-20

Hardoim PR, van Overbeek LS, Berg G, Pirttila AM, Compant S, Campisano A, Doring M, Sessitch A (2015) The hidden world within plants: ecological and evolutionary considerations for defining functioning of microbial endophytes. Microbiol Mol Biol Rev 79:293–320. https://doi.org/10.1128/mmbr.00050-14

He J, Benedito VA, Wang M, Murray JD, Zhao PX, Tang Y, Udvardi MK (2009) The *Medicago truncatula* gene expression atlas web server. BMC Bioinform 10:1–9. https://doi.org/10.1186/1471-2105-10-441

Hellriegel H, Wilfarth H (1888) Untersuchungen über die Stickstoffnahrung der Graminen und Leguminosen. Beilageheft zu der Zeitschrift des Vereins für Rübenzucker-Industrie Deutschen Reichs, 234 pp.

Hernandez-Alvarez C, Peimbert M, Rodrıguez-Martin P, Trejo-Aguilar D, Alcaraz LD (2023) A study of microbial diversity in a biofertilizer consortium. PLoS One 18(8):e0286285. https://doi.org/10.1371/journal.pone.0286285

Hossain MS, DeLaune PB, Gentry TJ (2023) Microbiome analysis revealed distinct microbial communities occupying different sized nodules in field-grown peanut. Front Microbiol 14:1075575. https://doi.org/10.3389/fmicb.2023.1075575

Hu H, Wang M, Huang Y, Xu Z, Xu P, Nie Y, Tang H (2022) Guided by the principles of microbiome engineering: accomplishments and perspectives for environmental use. mLife 1(4):382–398. https://doi.org/10.1002/mlf2.12043

Hugenholtz P (2002) Exploring prokaryotic diversity in the genomic era. Genome Biol 3:1–8. https://doi.org/10.1186/gb-2002-3-2-reviews0003

Hugenholtz P, Pace NR (1996) Identifying microbial diversity in the natural environment: a molecular phylogenetic approach. Trends Biotechnol 14:190–197. https://doi.org/10.1016/0167-7799(96)10025-1

Huitzil S, Huepe C, Aldana M, Frank A (2023) The missing link: how the holobiont concept provides a genetic framework for rapid evolution and the inheritance of acquired characteristics. Front Ecol Evol 11:1279938. https://doi.org/10.3389/fevo.2023.1279938

Huson DH, Auch AF, Qi J, Schuster SC (2007) MEGAN analysis of metagenomic data. Genome Res 17(3):377–386. https://doi.org/10.1101/gr.5969107

Hussain SS, Mehnaz S, Siddique KHM (2018) Harnessing the plant microbiome for improved abiotic stress tolerance. In: Egamberdieva D, Ahmad P (eds) Plant microbiome: stress response. Microorganisms for sustainability, vol 5. Springer, Singapore. https://doi.org/10.1007/978-981-10-5514-0_2

Jesus JG, Máguas C, Dias R, Nunes M, Pascoal P, Pereira M, Trindade H (2023) What if root nodules are a guesthouse for a microbiome? The case study of Acacia longifolia. Biology 12(9):1168. https://doi.org/10.3390/biology12091168

Johnson JS, Spakowicz DJ, Hong BY, Petersen LM, Demkowicz P, Chen L, Leopold SR, Hanson BM, Agresta HO, Gerstein M, Sodergren E (2019) Evaluation of 16S rRNA gene sequencing for species and strain-level microbiome analysis. Nat Commun 10(1):5029. https://doi.org/10.1038/s41467-019-13036-1

Kado CI (1992) Plant pathogenic bacteria. In: Balows A, Truper HG, Dworkin M, Harder W, Schleifer KH (eds) The prokaryotes. Springer, New York, pp 660–662

Kaeberlein T, Lewis K, Epstein SS (2002) Isolating "uncultivable" microorganisms in pure culture in a simulated natural environment. Science 296(5570):1127–1129. https://doi.org/10.1126/science.1070633

Kaschuk G, Auler AC, Vieira CE, Dakora FD, Jaiswal SK, da Cruz SP (2022) Coinoculation impact on plant growth promotion: a review and meta-analysis on coinoculation of rhizobia and plant growth-promoting bacilli in grain legumes. Braz J Microbiol 4:2027–2037. https://doi.org/10.1007/s42770-022-00800-7

Kehe J, Kulesa A, Ortiz A, Ackerman CM, Thakku SG, Sellers D, Kuehn S, Gore J, Friedman J, Blainey PC (2019) Massively parallel screening of synthetic microbial communities. Proc Natl Acad Sci 116(26):12804–12809. https://doi.org/10.1073/pnas.1900102116

Kloepper JW, Schroth MN (1978) Plant growth-promoting rhizobacteria on radishes. In: Proceedings of the 4th international conference on plant pathogenic bacteria, vol 2. Gilbert-Clarey, Tours, pp 879–882

Kloepper JW, Schroth MN (1981) Relationship of *in vitro* antibiosis of plant growth promoting rhizobacteria to plant growth and the displacement of root microflora. Phytopathology 71:1020–1024. https://doi.org/10.1094/phyto-71-1020

Kong Z, Hart M, Liu H (2018) Paving the way from the lab to the field: using synthetic microbial consortia to produce high-quality crops. Front Plant Sci 9:1467. https://doi.org/10.3389/fpls.2018.01467

Kudapa H, Azam S, Sharpe AG, Taran B, Li R, Deonovic B, Cameron C, Farmer AD, Cannon SB, Varshney RK (2014) Comprehensive transcriptome assembly of chickpea (*Cicer arietinum* L.) using Sanger and next generation sequencing platforms: development and applications. PLoS One 9(1):e86039. https://doi.org/10.1371/journal.pone.0086039

Lagunas B, Richards L, Sergaki C, Burgess J, Pardal AJ, Hussain RM, Richmond BL, Baxter L, Roy P, Pakidi A, Stovold G (2023) Rhizobial nitrogen fixation efficiency shapes endosphere bacterial communities and *Medicago truncatula* host growth. Microbiome 11(1):146. https://doi.org/10.1186/s40168-023-01592-0

Lane DJ, Pace B, Olsen GJ, Stahl DA, Sogin ML, Pace NR (1985) Rapid determination of 16S ribosomal RNA sequences for phylogenetic analyses. Proc Natl Acad Sci 82(20):6955–6959. https://doi.org/10.1073/pnas.82.20.6955

Laranjeira S, Fernandes-Silva A, Reis S, Torcato C, Raimundo F, Ferreira L, Carnide V, Marques G (2021) Inoculation of plant growth promoting bacteria and arbuscular mycorrhizal fungi improve chickpea performance under water deficit conditions. Appl Soil Ecol 164:103927. https://doi.org/10.1016/j.apsoil.2021.103927

Lawson CE, Harcombe WR, Hatzenpichler R, Lindemann SR, Loffler FE, O'Malley MA, Martin HG, Pfleger BF, Raskin L, Venturelli OS, Weissbrodt DG, Noguera DR, McMohan KD (2019) Common principles and best practices for engineering microbiomes. Nat Rev Microbiol 17:725–741. https://doi.org/10.1038/s41579-019-0255-9

Lederberg J, McCray AT (2001) Ome SweetOmics--A genealogical treasury of words. Scientist 15(7):8. http://lhncbc.nlm.nih.gov/publication/lhncbc-2001-047

Leite J, Fischer D, Rouws LF, Fernandes-Júnior PI, Hofmann A, Kublik S, Schloter M, Xavier GR, Radl V (2017) Cowpea nodules harbor non-rhizobial bacterial communities that are shaped by soil type rather than plant genotype. Front Plant Sci 7:2064. https://doi.org/10.3389/fpls.2016.02064

Lerouge P, Roche P, Faucher C, Maillet F, Truchet G, Promé JC, Dénarié J (1990) Symbiotic host-specificity of *Rhizobium meliloti* is determined by a sulphated and acylated glucosamine oligosaccharide signal. Nature 344:781–784. https://doi.org/10.1038/344781a0

Liu Z, DeSantis TZ, Andersen GL, Knight R (2008) Accurate taxonomy assignments from 16S rRNA sequences produced by highly parallel pyrosequencers. Nucleic Acids Res 36(18):e120. https://doi.org/10.1093/nar/gkn491

Liu Y, Wu L, Wu X, Li H, Liao Q, Zhang X, Sun Z, Li W (2017) Analysis of microbial diversity in soil under ginger cultivation. Scientifica 1:8256865. https://doi.org/10.1155/2017/8256865

Liu F, Hewezi T, Lebeis SL, Pantalone V, Grewal PS, Staton ME (2019) Soil indigenous microbiome and plant genotypes cooperatively modify soybean rhizosphere microbiome assembly. BMC Microbiol 19:1–9. https://doi.org/10.1186/s12866-019-1572-x

Loo WT, Chua KO, Mazumdar P, Cheng A, Osman N, Harikrishna JA (2022) Arbuscular mycorrhizal symbiosis: a strategy for mitigating the impacts of climate change on tropical legume crops. Plants 11(21):2875. https://doi.org/10.3390/plants11212875

Louca S, Polz MF, Mazel F, Albright MB, Huber JA, O'Connor MI, Ackermann M, Hahn AS, Srivastava DS, Crowe SA, Doebeli M (2018) Function and functional redundancy in microbial systems. Nat Ecol Evol 2(6):936–943. https://doi.org/10.1038/s41559-018-0519-1

Lueders T, Dumont MG, Bradford L, Manefield M (2016) RNA-stable isotope probing: from carbon flow within key microbiota to targeted transcriptomes. Curr Opin Biotechnol 41:83–89. https://doi.org/10.1016/j.copbio.2016.05.001

Lundberg DS, Lebeis SL, Paredes SH, Yourstone S, Gehring J, Malfatti S, Tremblay J, Engelbrektson A, Kunin V, Rio TG, Edgar RC (2012) Defining the core *Arabidopsis thaliana* root microbiome. Nature 488(7409):86–90. https://doi.org/10.1038/nature11237

Lundberg DS, Yourstone S, Mieczkowski P, Jones CD, Dangl JL (2013) Practical innovations for high-throughput amplicon sequencing. Nat Methods 10(10):999–1002. https://doi.org/10.1038/nmeth.2634

Lyu X, Nuhu M, Candry P, Wolfanger J, Betenbaugh M, Saldivar A, Zuniga C, Wang Y, Shrestha S (2024) Top-down and bottom-up microbiome engineering approaches to enable biomanufacturing from waste biomass. J Ind Microbiol Biotechnol 51:kuae025. https://doi.org/10.1093/jimb/kuae025

Malik DK, Sindhu SS (2011) Production of indole acetic acid by *Pseudomonas* sp.: effect of coinoculation with *Mesorhizobium* sp. *cicer* on nodulation and plant growth of chickpea (*Cicer arietinum*). Physiol Mol Biol Plants 17:25–32. https://doi.org/10.1007/s12298-010-0041-7

Marcy Y, Ouverney C, Bik EM, Lösekann T, Ivanova N, Martin HG, Szeto E, Platt D, Hugenholtz P, Relman DA, Quake SR (2007) Dissecting biological "dark matter" with single-cell genetic analysis of rare and uncultivated TM7 microbes from the human mouth. Proc Natl Acad Sci 104(29):11889–11894. https://doi.org/10.1073/pnas.0704662104

Martin JL, Santi I, Pitta P, John U, Gypens N (2022) Towards quantitative metabarcoding of eukaryotic plankton: an approach to improve 18S rRNA gene copy number bias. Metabarcod Metagenom 6:e85794. https://doi.org/10.3897/mbmg.6.85794

Martinez-Hidalgo P, Hirsch AM (2017) The nodule microbiome: N_2 -fixing rhizobia do not live alone. Phytobiomes 1(2):70–82. https://doi.org/10.1094/PBIOMES-12-16-0019-RVW

Matse DT, Huang CH, Huang YM, Yen MY (2020) Effects of coinoculation of *Rhizobium* with plant growth promoting rhizobacteria on the nitrogen fixation and nutrient uptake of *Trifolium repens* in low phosphorus soil. J Plant Nutr 43(5):739–752. https://doi.org/10.1080/01904167.2019.1702205

Mayhood P, Mirza BS (2021) Soybean root nodule and rhizosphere microbiome: distribution of rhizobial and nonrhizobial endophytes. Appl Environ Microbiol 87(10):e02884–e02820. https://doi.org/10.1128/AEM.02884-20

McNichol J, Berube PM, Biller SJ, Fuhrman JA (2021) Evaluating and improving small subunit rRNA PCR primer coverage for bacteria, archaea, and eukaryotes using metagenomes from global ocean surveys. Msystems 6(3):110–128. https://doi.org/10.1128/msystems.00565-21

Mendes R, Garbeva P, Raaijmakers JM (2013) The rhizosphere microbiome: significance of plant beneficial, plant pathogenic, and human pathogenic microorganisms. FEMS Microbiol Rev 37(5):634–663. https://doi.org/10.1111/1574-6976.12028

Mendes LW, Braga LP, Navarrete AA, de Souza DG, Silva GG, Tsai SM (2017) Using metagenomics to connect microbial community biodiversity and functions. Curr Issues Mol Biol 24(1):103–118. https://doi.org/10.21775/cimb.024.103

Mercer TR, Clark MB, Crawford J, Brunck ME, Gerhardt DJ, Taft RJ, Nielsen LK, Dinger ME, Mattick JS (2014) Targeted sequencing for gene discovery and quantification using RNA CaptureSeq. Nat Protoc 9(5):989–1009. https://doi.org/10.1038/nprot.2014.058

Miki T, Yokokawa T, Matsui K (2014) Biodiversity and multifunctionality in a microbial community: a novel theoretical approach to quantify functional redundancy. Proc R Soc B Biol Sci 281(1776):20132498. https://doi.org/10.1098/rspb.2013.2498

Moran MA (2009) Metatranscriptomics: eavesdropping on complex microbial communities. Microbe 4(7):7. https://doi.org/10.1128/microbe.4.329.1

Mota-Gutierrez J, Ferrocino I, Rantsiou K, Cocolin L (2019) Metataxonomic comparison between internal transcribed spacer and 26S ribosomal large subunit (LSU) rDNA gene. Int J Food Microbiol 290:132–140. https://doi.org/10.1016/j.ijfoodmicro.2018.10.010

Mueller UG, Sachs JL (2015) Engineering microbiomes to improve plant and animal health. Trends Microbiol 23(10):606–617. https://doi.org/10.1016/j.tim.2015.07.009

Mueller UG, Juenger TE, Kardish MR, Carlson AL, Burns KM, Edwards JA, Smith CC, Fang CC, Des Marais DL (2021) Artificial selection on microbiomes to breed microbiomes that confer salt tolerance to plants. MSystem 6(6):e0112521. https://doi.org/10.1128/mSystems.01125-21

Mukhtar S, Hirsch AM, Khan N, Malik KA, Humm EA, Pellegrini M, Shi B, Briscoe L, Huntemann M, Clum A, Foster B (2020) Impact of soil salinity on the cowpea nodule-microbiome and the isolation of halotolerant PGPR strains to promote plant growth under salinity stress. Phytobiomes J 4(4):364–374. https://doi.org/10.1094/PBIOMES-09-19-0057-R

Muresu R, Polone E, Sulas L, Baldan B, Tondello A, Delogu G, Cappuccinelli P, Alberghini S, Benhizia Y, Benhizia H, Benguedouar A (2008) Coexistence of predominantly nonculturable rhizobia with diverse, endophytic bacterial taxa within nodules of wild legumes. FEMS Microbiol Ecol 63(3):383–400. https://doi.org/10.1111/j.1574-6941.2007.00424.x

Musyoka DM, Njeru EM, Nyamwange MM, Maingi JM (2020) Arbuscular mycorrhizal fungi and *Bradyrhizobium* co-inoculation enhances nitrogen fixation and growth of green grams (*Vigna radiata* L.) under water stress. J Plant Nutr 43(7):1036–1047. https://doi.org/10.1080/01904167.2020.1711940

Nam DK, Lee S, Zhou G, Cao X, Wang C, Clark T, Chen J, Rowley JD, Wang SM (2002) Oligo (dT) primer generates a high frequency of truncated cDNAs through internal poly (A) priming during reverse transcription. Proc Natl Acad Sci 99(9):6152–6156. https://doi.org/10.1073/pnas.092140899

Nasr Esfahani M, Inoue K, Chu HD, Nguyen KH, Van Ha C, Watanabe Y, Burritt DJ, Herrera-Estrella L, Mochida K, Tran LS (2017) Comparative transcriptome analysis of nodules of two *Mesorhizobium*–chickpea associations with differential symbiotic efficiency under phosphate deficiency. Plant J 91(5):911–926. https://doi.org/10.1111/tpj.13616

Ndour PM, Barry CM, Tine D, De la Fuente CC, Gueye M, Barakat M, Ortet P, Achouak W, Ndoye I, Sine B, Laplaze L (2021) Pearl millet genotype impacts microbial diversity and enzymatic activities in relation to root-adhering soil aggregation. Plant Soil 464:109–129. https://doi.org/10.1007/s11104-021-04917-w

Nelson MB, Martiny AC, Martiny JB (2016) Global biogeography of microbial nitrogen-cycling traits in soil. Proc Natl Acad Sci 113(29):8033–8040. https://doi.org/10.1073/pnas.1601070113

Neufeld JD, Dumont MG, Vohra J, Murrell JC (2007) Methodological considerations for the use of stable isotope probing in microbial ecology. Microb Ecol 53:435–442. https://doi.org/10.1007/s00248-006-9125-x

Nichols D (2007) Cultivation gives context to the microbial ecologist. FEMS Microbiol Ecol 60(3):351–357. https://doi.org/10.1111/j.1574-6941.2007.00332.x

Nichols D, Cahoon N, Trakhtenberg EM, Pham L, Mehta A, Belanger A, Kanigan T, Lewis K, Epstein S (2010) Use of ichip for high-throughput in situ cultivation of "uncultivable" microbial species. Appl Environ Microbiol 76(8):2445–2450. https://doi.org/10.1128/AEM.01754-09

Nie H, Shi Y, Yang X, Zeng J, Tang Y, Liu X, Sun L, Zhou Y, Xu X, Liu M, Li C (2024) Microbial inoculant-induced modifications of rhizospheric metabolites and microbial communities enhance plant growth. Plant Soil 22:1–9. https://doi.org/10.1007/s11104-024-07102-x

Niu B, Paulson JN, Zheng X, Kolter R (2017) Simplified and representative bacterial community of maize roots. Proc Natl Acad Sci 114(12):E2450–E2459. https://doi.org/10.1073/pnas.1616148114

Nobbe F, Hiltner L (1896) Inoculation of the soil for cultivating leguminous plants. US patent 570:813

Nysanth NS, Senthilkumar M, Pooniya V, Viswanathan C, Swarnalakshmi K (2024) Mesorhizobial inoculation and fertilizer application differentially influence native bacterial community structure associated with chickpea. J Food Legumes 37(4):410–419. https://doi.org/10.59797/jfl.v37.i4.226

O'Callaghan M, Ballard RA, Wright D (2022) Soil microbial inoculants for sustainable agriculture: limitations and opportunities. Soil Use Manag 38(3):1340–1369. https://doi.org/10.1111/sum.12811

Oke V, Long SR (1999) Bacteroid formation in the *Rhizobium*–legume symbiosis. Curr Opin Microbiol 2(6):641–646. https://doi.org/10.1016/S1369-5274(99)00035-1

Oliveira C, Shakiba E, North D, McGraw M, Ballard E, Barrett-D'Amico M, Glazko G, Rahmatallah Y (2022) 16S rRNA gene-based metagenomic analysis of rhizosphere soil bacteria in arkansas rice crop fields. Agronomy 12(1):222. https://doi.org/10.3390/agronomy12010222

Orozco-Mosqueda MDC, Rocha-Granados MDC, Glick BR, Santoyo G (2018) Microbiome engineering to improve biocontrol and plant growth-promoting mechanisms. Microbiol Res 208:25–31. https://doi.org/10.1016/j.micres.2018.01.005

Ortíz-Castro R, Valencia-Cantero E, López-Bucio J (2008) Plant growth promotion by *Bacillus megaterium* involves cytokinin signaling. Plant Signal Behav 3(4):263–265. https://doi.org/10.4161/psb.3.4.5204

Ottesen AR, González Peña A, White JR, Pettengill JB, Li C, Allard S, Rideout S, Allard M, Hill T, Evans P, Strain E (2013) Baseline survey of the anatomical microbial ecology of an important food plant: *Solanum lycopersicum* (tomato). BMC Microbiol 13:1–2. https://doi.org/10.1186/1471-2180-13-114

Oulas A, Pavloudi C, Polymenakou P, Pavlopoulos GA, Papanikolaou N, Kotoulas G, Arvanitidis C, Iliopoulos L (2015) Metagenomics: tools and insights for analyzing next-generation sequencing data derived from biodiversity studies. Bioinform Biol Insights 9:75. https://doi.org/10.4137/BBI.S12462

Ozsolak F, Milos PM (2011) RNA sequencing: advances, challenges and opportunities. Nat Rev Genet 12(2):87–98. https://doi.org/10.1038/nrg2934

Paau AS, Cowles JR, Raveed D (1978) Development of bacteroids in alfalfa (*Medicago sativa*) nodules. Plant Physiol 62(4):526–530. https://doi.org/10.1104/pp.62.4.526

Paau AS, Oro J, Cowles JR (1979) DNA content of free-living rhizobia and bacteroids of various *rhizobium*-legume associations. Plant Physiol 63(2):402–405. https://doi.org/10.1104/pp.63.2.402

Pace NR, Stahl DA, Lane DJ, Olsen GJ (1986) The analysis of natural microbial populations by ribosomal rna sequences. In: Marshall KC (ed) Advances in microbial ecology, vol 9. Springer, Boston. https://doi.org/10.1007/978-1-4757-0611-6_1

Panwar JDS, Swarnalakshmi K (2005) PGPR: a new group of biofertilizers for sustainable crop productivity. In: Developments in physiology, biochemistry and molecular biology of plants, vol 1. New India Publishing Agency, New Delhi, pp 159–180

Parks DH, Chuvochina M, Waite DW, Rinke C, Skarshewski A, Chaumeil PA, Hugenholtz P (2018) A standardized bacterial taxonomy based on genome phylogeny substantially revises the tree of life. Nat Biotechnol 36(10):996–1004. https://doi.org/10.1038/nbt.4229

Parro V, Moreno-Paz M, González-Toril E (2007) Analysis of environmental transcriptomes by DNA microarrays. Environ Microbiology 9(2):453–464. https://doi.org/10.1111/j.1462-2920.2006.01162.x

Paudel D, Wang L, Poudel R, Acharya JP, Victores S, de Souza CH, Rios E, Wang J (2023) Elucidating the effects of organic vs. conventional cropping practice and rhizobia inoculation on rhizosphere microbial diversity and yield of peanut. Environ Microbiome 18(1):60. https://doi.org/10.1186/s40793-023-00517-6

Peiffer JA, Spor A, Koren O, Jin Z, Tringe SG, Dangl JL, Buckler ES, Ley RE (2013) Diversity and heritability of the maize rhizosphere microbiome under field conditions. Proc Natl Acad Sci 110(16):6548–6553. https://doi.org/10.1073/pnas.1302837110

Peix A, Ramírez-Bahena MH, Velázquez E, Bedmar EJ (2015) Bacterial associations with legumes. Crit Rev Plant Sci 34(1–3):17–42. https://doi.org/10.1080/07352689.2014.897899

Peoples MB, Brockwell J, Herridge DF, Rochester IJ, Alves BJ, Urquiaga S, Boddey RM, Dakora FD, Bhattarai S, Maskey SL, Sampet C (2009) The contributions of nitrogen-fixing crop legumes to the productivity of agricultural systems. Symbiosis 48:1–7

Pérez-Jaramillo JE, De Hollander M, Ramírez CA, Mendes R, Raaijmakers JM, Carrión VJ (2019) Deciphering rhizosphere microbiome assembly of wild and modern common bean (*Phaseolus vulgaris*) in native and agricultural soils from Colombia. Microbiome 7:1–6. https://doi.org/10.1186/s40168-019-0727-1

Pérez-Montaño F, Alías-Villegas C, Bellogín RA, Del Cerro P, Espuny MR, Jiménez-Guerrero I, López-Baena FJ, Ollero FJ, Cubo T (2014) Plant growth promotion in cereal and leguminous agricultural important plants: from microorganism capacities to crop production. Microbiol Res 169(5–6):325–336. https://doi.org/10.1016/j.micres.2013.09.011

Peters NK, Frost JW, Long SR (1986) A plant flavone, luteolin, induces expression of *Rhizobium meliloti* nodulation genes. Science 233:977–980

Petrushin IS, Vasilev IA, Markova YA (2023) Drought tolerance of legumes: physiology and the role of the microbiome. Curr Issues Mol Biol 45(8):6311–6324. https://doi.org/10.3390/cimb45080398

Pinto AJ, Raskin L (2012) PCR biases distort bacterial and archaeal community structure in pyrosequencing datasets. PLoS One 7(8):e43093. https://doi.org/10.1371/journal.pone.0043093

Pisa G, Magnani GS, Weber H, Souza EM, Faoro H, Monteiro RA, Daros E, Baura V, Bespalhok JP, Pedrosa FO, Cruz LM (2011) Diversity of 16S rRNA genes from bacteria of sugarcane rhizosphere soil. Braz J Med Biol Res 44:1215–1221. https://doi.org/10.1590/S0100-879X2011007500148

Plaza-Bonilla D, Nolot JM, Raffaillac D, Justes E (2017) Innovative cropping systems to reduce N inputs and maintain wheat yields by inserting grain legumes and cover crops in southwestern France. Eur J Agron 82:331–341. https://doi.org/10.1016/j.eja.2016.05.010

Poretsky RS, Bano N, Buchan A, LeCleir G, Kleikemper J, Pickering M, Pate WM, Moran MA, Hollibaugh JT (2005) Analysis of microbial gene transcripts in environmental samples. Appl Environ Microbiol 71(7):4121–4126. https://doi.org/10.1128/AEM.71.7.4121-4126.2005

Pramanik K, Das A, Banerjee J, Das A, Chatterjee S, Sharma R, Kumar S, Gupta S (2020) Metagenomic insights into rhizospheric microbiome profiling in lentil cultivars unveils differential microbial nitrogen and phosphorus metabolism under rice-fallow ecology. Int J Mol Sci 21(23):8895. https://doi.org/10.3390/ijms21238895

Prudent M, Dequiedt S, Sorin C, Girodet S, Nowak V, Duc G, Salon C, Maron PA (2020) The diversity of soil microbial communities matters when legumes face drought. Plant Cell Environ 43(4):1023–1035. https://doi.org/10.1111/pce.13712

Pruesse E, Quast C, Knittel K, Fuchs BM, Ludwig W, Peplies J, Glöckner FO (2007) SILVA: a comprehensive online resource for quality checked and aligned ribosomal RNA sequence data compatible with ARB. Nucleic Acids Res 35(21):7188–7196. https://doi.org/10.1093/nar/gkm864

Qiu Z, Egidi E, Liu H, Kaur S, Singh BK (2019) New frontiers in agriculture productivity: optimised microbial inoculants and in situ microbiome engineering. Biotechnol Adv 37(6):107371. https://doi.org/10.1016/j.biotechadv.2019.03.010

Quadt-Hallmann A, Hallmann J, Kloepper JW (1997) Bacterial endophytes in cotton: location and interaction with other plant-associated bacteria. Can J Microbiol 43(3):254–259. https://doi.org/10.1139/m97-035

Queiroz Rego CH, Cardoso FB, da Silva Cândido AC, Teodoro PE, Alves CZ (2018) Co-inoculation with *Bradyrhizobium* and *Azospirillum* increases yield and quality of soybean seeds. Agron J 110(6):2302–2309. https://doi.org/10.2134/agronj2018.04.0278

Quides KW, Atamian HS (2021) A microbiome engineering framework to evaluate rhizobial symbionts of legumes. Plant Soil 463:631–642. https://doi.org/10.1007/s11104-021-04892-2

Quince C, Lanzen A, Davenport RJ, Turnbaugh PJ (2011) Removing noise from pyrosequenced amplicons. BMC Bioinform 12:38. https://doi.org/10.1186/1471-2105-12-38

Quince C, Walker AW, Simpson JT, Loman NJ, Segata N (2017) Shotgun metagenomics, from sampling to analysis. Nat Biotechnol 35(9):833–844. https://doi.org/10.1038/nbt.3935

Rascovan N, Carbonetto B, Perrig D, Díaz M, Canciani W, Abalo M, Alloati J, González-Anta G, Vazquez MP (2016) Integrated analysis of root microbiomes of soybean and wheat from agricultural fields. Sci Rep 6(1):28084. https://doi.org/10.1038/srep28084

Reinhold-Hurek B, Hurek T (2011) Living inside plants: bacterial endophytes. Curr Opin Plant Biol 14(4):435–443. https://doi.org/10.1016/j.pbi.2011.04.004

Rolfe SA, Griffiths J, Ton J (2019) Crying out for help with root exudates: adaptive mechanisms by which stressed plants assemble health-promoting soil microbiomes. Curr Opin Microbiol 49:73–82. https://doi.org/10.1016/j.mib.2019.10.003

Rossmann M, Perez-Jaramillo JE, Kavamura VN, Chiaramonte JB, Dumack K, Fiore-Donno AM, Mendes LW, Ferreira MM, Bonkowski M, Raaijmakers JM, Mauchline TH (2020) Multitrophic interactions in the rhizosphere microbiome of wheat: from bacteria and fungi to protists. FEMS Microbiol Ecol 96(4):fiaa032. https://doi.org/10.1093/femsec/fiaa032

Ryu CM, Farag MA, Hu CH, Reddy MS, Wei HX, Paré PW, Kloepper JW (2003) Bacterial volatiles promote growth in *Arabidopsis*. Proc Natl Acad Sci 100(8):4927–4932. https://doi.org/10.1073/pnas.0730845100

Santoro M, Cappellari L, Giordano W, Banchio E (2015) Production of volatile organic compounds in PGPR. In: Handbook for *Azospirillum*: technical issues and protocols, pp 307–317. https://doi.org/10.1007/978-3-319-06542-7_17

Sauviac L, Rémy A, Huault E, Dalmasso M, Kazmierczak T, Jardinaud MF, Legrand L, Moreau C, Ruiz B, Cazalé AC, Valière S (2022) A dual legume-*Rhizobium* transcriptome of symbiotic nodule senescence reveals coordinated plant and bacterial responses. Plant Cell Environ 45(10):3100–3121. https://doi.org/10.1111/pce.14389

Scheublin TR, Van Der Heijden MG (2006) Arbuscular mycorrhizal fungi colonize nonfixing root nodules of several legume species. New Phytol 172(4):732–738. https://doi.org/10.1111/j.1469-8137.2006.01858.x

Schloss PD, Westcott SL, Ryabin T, Hall JR, Hartmann M, Hollister EB, Lesniewski RA, Oakley BB, Parks DH, Robinson CJ, Sahl JW (2009) Introducing mothur: open-source, platform-independent, community-supported software for describing and comparing microbial communities. Appl Environ Microbiol 75(23):7537–7541. https://doi.org/10.1128/AEM.01541-09

Schulz B, Boyle C (2006) What are endophytes? In: Schulz BJE, Boyle CJC, Sieber TN (eds) Microbial root endophytes. Soil biology, vol 9. Springer, Berlin, Heidelberg. https://doi.org/10.1007/3-540-33526-9_1

Shah AS, Wakelin SA, Moot DJ, Blond C, Laugraud A, Ridgway HJ (2021) *Trifolium repens* and *T. subterraneum* modify their nodule microbiome in response to soil pH. J Appl Microbiol 131(4):1858–1869. https://doi.org/10.1111/jam.15050

Sharaf H, Rodrigues RR, Moon J, Zhang B, Mills K, Williams MA (2019) Unprecedented bacterial community richness in soybean nodules vary with cultivar and water status. Microbiome 7:1–8. https://doi.org/10.1186/s40168-019-0676-8

Sharma R, Paliwal JS, Chopra P, Dogra D, Pooniya V, Bisaria VS, Swarnalakshmi K, Sharma S (2017) Survival, efficacy and rhizospheric effects of bacterial inoculants on *Cajanus cajan*. Agric Ecosyst Environ 240:244–252. https://doi.org/10.1016/j.agee.2017.02.01

da Silva TR, Rodrigues RT, Jovino RS, Carvalho JRDS, Leite J, Hoffman A, Fischer D, Ribeiro PRDA, Rouws LFM, Radl V, Fernandes-Júnior PI (2023) Not just passengers, but co-pilots! Non-rhizobial nodule-associated bacteria promote cowpea growth and symbiosis with (brady) rhizobia. J Appl Microbiol 134(1):lxac013. https://doi.org/10.1093/jambio/lxac013

Smith SE, Smith FA (2011) Roles of arbuscular mycorrhizas in plant nutrition and growth: new paradigms from cellular to ecosystem scales. Annu Rev Plant Biol 62(1):227–250. https://doi.org/10.1146/annurev-arplant-042110-103846

Sneha GR, Swarnalakshmi K, Sharma M, Reddy K, Bhoumik A, Suman A, Kannepalli A (2021) Soil type influence nutrient availability, microbial metabolic diversity, eubacterial and diazotroph abundance in chickpea rhizosphere. World J Microbiol Biotechnol 37:167. https://doi.org/10.1007/s11274-021-03132

Soumare A, Diop T, Manga A, Ndoye I (2015) Role of arbuscular mycorrhizal fungi and nitrogen fixing bacteria on legume growth under various environmental stresses. Int J Biosci 7(4):31–46. https://doi.org/10.12692/ijb/7.4.31-46

Souza RC, Hungria M, Cantão ME, Vasconcelos AT, Nogueira MA, Vicente VA (2015) Metagenomic analysis reveals microbial functional redundancies and specificities in a soil under different tillage and crop-management regimes. Appl Soil Ecol 86:106–112. https://doi.org/10.1016/j.apsoil.2014.10.010

de Souza RSC, Armanhi JSL, de Damasceno NB, Imperial J, Arruda P (2019) Genome sequences of a plant beneficial synthetic bacterial community reveal genetic features for successful plant colonization. Front Microbiol 10:1779. https://doi.org/10.3389/fmicb.2019.01779

Staley JT, Konopka A (1985) Measurement of in situ activities of non-photosynthetic microorganisms in aquatic and terrestrial habitats. Ann Rev Microbiol 39(1):321–346. https://doi.org/10.1146/annurev.micro.39.1.321

Stewart EJ (2012) Growing unculturable bacteria. J Bacteriol 194:4151–4160. https://doi.org/10.1128/jb.00345-12

Su C, Lei L, Duan Y, Zhang KQ, Yang J (2012) Culture-independent methods for studying environmental microorganisms: methods, application, and perspective. Appl Microbiol Biotechnol 93:993–1003. https://doi.org/10.1007/s00253-011-3800-7

Sulesky-Grieb A, Simonin M, Bintarti AF, Marolleau B, Barret M, Shade A (2024) Stable, multigenerational transmission of the bean seed microbiome despite abiotic stress. Msystems 9(11):e00951–e00924. https://doi.org/10.1128/msystems.00951-24

Sun L, Qiu F, Zhang X, Dai X, Dong X, Song W (2008) Endophytic bacterial diversity in rice (*Oryza sativa* L.) roots estimated by 16S rDNA sequence analysis. Microb Ecol 55:415–424. https://doi.org/10.1007/s00248-007-9287-1

Swarnalakshmi K, Senthilkumar M, Singh M (2013a) Potential and prospect of bio-fertilizers in pulse production under resource conservation technologies. In: Ghosh PK, Kumar N, Venkatesh MS, Hazra KK, Nadarajan N (eds) Resource conservation technology in pulses. Scientific Publishers, pp 251–271

Swarnalakshmi K, Senthilkumar M, Sachan S (2013b) Microbial inoculation through endophytic bacteria for sustainable agriculture. In: Kumar S, Prasad B (eds) Modern technologies for sustainable agriculture. New India Publishing Agency, New Delhi, pp 55–76

Swarnalakshmi K, Senthilkumar M, Ramakrishnan B (2016a) Endophytic actinobacteria: nitrogen fixation, phytohormone production and antibiosis. In: Subramaniam G, Arumugam S, Rajendran V (eds) Plant growth promoting actinobacteria. Springer Science and Business Media, Singapore. https://doi.org/10.1007/978-981-10-0707-1_8

Swarnalakshmi K, Yadav V, Senthilkumar M, Dhar DW (2016b) Biofertilizers for higher pulse production in India: scope, accessibility and challenges. Indian J Agron 61:S173–S181

Swarnalakshmi K, Pooniya V, Paul S (2017) Synergistic interaction of *Piriformospora indica* and microbial inoculants on symbiotic potential, plant nutrition and productivity of chickpea (*Cicer arietinum*). Indian J Agron 62(4):481–488. https://doi.org/10.59797/ija.v62i4.5555

Swarnalakshmi K, Rajkhowa S, Senthilkumar M, Dhar DW (2019) Influence of endophytic bacteria on growth promotion and protection against diseases in associated plants. In: Singh DP, Prabha R (eds) Microbial interventions in agriculture and environment, vol 3. Springer Nature Singapore Pte Ltd, pp 263–287. https://doi.org/10.1007/978-981-32-9084-6_12

Swarnalakshmi K, Yadav V, Tyagi D, Dhar DW, Kannepalli A, Senthilkumar M (2020) Significance of plant growth promoting rhizobacteria in grain legumes: growth promotion and crop production. Plants 9(11):1596. https://doi.org/10.3390/plants9111596

Swarnalakshmi K, Rajkhowa S, Pooniya V, Kannepalli A (2021) Endophytic bacteria improve mesorhizobial nodulation, plant growth and yield in chickpea (*Cicer arietinum*). Indian J Agric Sci 91(7):22–26. https://doi.org/10.56093/ijas.v91i7.115027

Swarnalakshmi K, Yadav V, Tyagi D, Dhar DW, Annapurna K (2023) Mechanisms of plant growth promoting rhizobacteria in grain legumes for biotic and abiotic stress management. In: Kumar N, Nath CP, Sah U, Singh NP (eds) Sustainable production of pulses in diverse agroecosystems, Vol. 2. Stress management and livelihood security. Scientific Publisher, Jodhpur, pp 133–152. (ISBN: 978-93-90749-19-5)

Taketani RG, Kavamura VN, Mendes R, Melo IS (2015) Functional congruence of rhizosphere microbial communities associated to leguminous tree from Brazilian semiarid region. Environ Microbiol Rep 7(1):95–101. https://doi.org/10.1111/1758-2229.12187

Thomas P, Sekhar AC (2017) Cultivation versus molecular analysis of banana (*Musa* sp.) shoot-tip tissue reveals enormous diversity of normally uncultivable endophytic bacteria. Microb Ecol 73(4):885–899. https://doi.org/10.1007/s00248-016-0877-7

Timmusk S, Behers L, Muthoni J, Muraya A, Aronsson AC (2017) Perspectives and challenges of microbial application for crop improvement. Front Plant Sci 8:49. https://doi.org/10.3389/fpld.2017.00049

Tkacz A, Bestion E, Bo Z, Hortala M, Poole PS (2020) Influence of plant fraction, soil, and plant species on microbiota: a multikingdom comparison. MBio 11(1):100–128. https://doi.org/10.1128/mbio.02785-19

Tkacz A, Ledermann R, Martyn A, Schornack S, Oldroyd GE, Poole PS (2022) Nodulation and nitrogen fixation in *Medicago truncatula* strongly alters the abundance of its root microbiota and subtly affects its structure. Environ Microbiol 24(11):5524–5533. https://doi.org/10.1111/1462-2920.16164

Toju H, Peay KG, Yamamichi M, Narisawa K, Hiruma K, Naito K, Fukuda S, Ushio M, Nakaoka S, Onoda Y, Yoshida K (2018) Core microbiomes for sustainable agroecosystems. Nat Plants 4(5):247–257. https://doi.org/10.1038/s41477-018-0139-4

Trabelsi D, Chihaoui SA, Mhamdi R (2017) Nodules and roots of *Vicia faba* are inhabited by quite different populations of associated bacteria. Appl Soil Ecol 119:72–79. https://doi.org/10.1016/j.apsoil.2017.06.002

Turner TR, Ramakrishnan K, Walshaw J, Heavens D, Alston M, Swarbreck D, Osbourn A, Grant A, Poole PS (2013) Comparative metatranscriptomics reveals kingdom level changes in the rhizosphere microbiome of plants. ISME J 7:2248–2258. https://doi.org/10.1038/ismej.2013.119

Urich T, Lanzén A, Qi J, Huson DH, Schleper C, Schuster SC (2008) Simultaneous assessment of soil microbial community structure and function through analysis of the meta-transcriptome. PLoS One 3(6):e2527. https://doi.org/10.1371/journal.pone.0002527

Uzma M, Iqbal A, Hasnain S (2022) Drought tolerance induction and growth promotion by indole acetic acid producing *Pseudomonas aeruginosa* in *Vigna radiata*. PLoS One 17(2):e0262932. https://doi.org/10.1371/journal.pone.0262932

Valdenegro M, Barea JM, Azcón R (2001) Influence of arbuscular-mycorrhizal fungi, *Rhizobium meliloti* strains and PGPR inoculation on the growth of *Medicago arborea* used as model legume for re-vegetation and biological reactivation in a semi-arid mediterranean area. Plant Growth Regul 34:233–240. https://doi.org/10.1023/A:1013323529603

Van Der Heijden MG, Bardgett RD, Van Straalen NM (2008) The unseen majority: soil microbes as drivers of plant diversity and productivity in terrestrial ecosystems. Ecol Lett 11(3):296–310. https://doi.org/10.1111/j.1461-0248.2007.01139.x

Vance CP, Uhde-Stone C, Allan DL (2003) Phosphorus acquisition and use: critical adaptations by plants for securing a non-renewable resource. New Phytol 157(3):423–447. https://doi.org/10.1046/j.1469-8137.2003.00695.x

Vandenkoornhuyse P, Quaiser A, Duhamel M, Le Van A, Dufresne A (2015) The importance of the microbiome of the plant holobiont. New Phytol 206(4):1196–1206. https://doi.org/10.1111/nph.13312

Verdier J, Torres-Jerez I, Wang M, Andriankaja A, Allen SN, He J, Tang Y, Murray JD, Udvardi MK (2013) Establishment of the *Lotus japonicus* gene expression stlas (L j GEA) and its use to explore legume seed maturation. Plant J 74(2):351–362. https://doi.org/10.1111/tpj.12119

Verma SK, Kharwar RN, White JF (2019) The role of seed-vectored endophytes in seedling development and establishment. Symbiosis 78:107–113. https://doi.org/10.1007/s13199-019-00619-1

Vorholt JA, Vogel C, Carlström CI, Müller DB (2017) Establishing causality: opportunities of synthetic communities for plant microbiome research. Cell Host Microbe 22:142–155. https://doi.org/10.1016/j.chom.2017.07.004

Wang B, Liu X, Qi Y, Chen Z, Zhu Y, Wang Y, Zheng H, Fang W (2024) Sugarcane rhizosphere microbiota: exploring diversity across varied topographies and growth environments. Plant Soil 6:1–4. https://doi.org/10.1007/s11104-024-06688-6

Welmillage SU, Zhang Q, Sreevidya VS, Sadowsky MJ, Gyaneshwar P (2021) Inoculation of *Mimosa pudica* with *Paraburkholderia phymatum* results in changes to the rhizoplane microbial community structure. Microbes Environ 36(1):ME20153. https://doi.org/10.1264/jsme2.ME20153

Whipps J, Lewis K, Cooke R (1988) Mycoparasitism and plant disease control. In: Burge M (ed) Fungi in biological control systems. Manchester University Press, pp 161–187. https://doi.org/10.1071/app9890106c

Whitman WB, Coleman DC, Wiebe WJ (1998) Prokaryotes: the unseen majority. Proc Natl Acad Sci 95(12):6578–6583. https://doi.org/10.1073/pnas.95.12.6578

Wigley K, Moot D, Wakelin SA, Laugraud A, Blond C, Seth K, Ridgway H (2017) Diverse bacterial taxa inhabit root nodules of lucerne (*Medicago sativa* L.) in New Zealand pastoral soils. Plant Soil 420:253–262. https://doi.org/10.1007/s11104-017-3395-6

Wipf HM, Xu L, Gao C, Spinner HB, Taylor J, Lemaux P, Mitchell J, Coleman-Derr D (2021) Agricultural soil management practices differentially shape the bacterial and fungal microbiomes of *Sorghum bicolor*. Appl Environ Microbiol 87(5):e02345–e02320. https://doi.org/10.1128/AEM.02345-20

Woese CR (1987) Bacterial evolution. Microbiol Rev 51(2):221–271. https://doi.org/10.1128/mmbr.51.2.221-271.1987

Xiao X, Chen W, Zong L, Yang J, Jiao S, Lin Y, Wang E, Wei G (2017) Two cultivated legume plants reveal the enrichment process of the microbiome in the rhizocompartments. Mol Ecol 26(6):1641–1651. https://doi.org/10.1111/mec.14027

Xu H, Yang Y, Tian Y, Xu R, Zhong Y, Liao H (2020) *Rhizobium* inoculation drives the shifting of rhizosphere fungal community in a host genotype dependent manner. Front Microbiol 10:3135. https://doi.org/10.3389/fmicb.2019.03135

Yang Y, Wang N, Guo X, Zhang Y, Ye B (2017) Comparative analysis of bacterial community structure in the rhizosphere of maize by high throughput pyrosequencing. PLoS One 12:e178425. https://doi.org/10.1371/journal.pone.0178425

Yin C, Hagerty CH, Paulitz TC (2022) Synthetic microbial consortia derived from rhizosphere soil protect wheat against a soil borne fungal pathogen. Front Microbiol 13:908981. https://doi.org/10.3389/fmicb.2022.908981

Zgadzaj R, James EK, Kelly S, Kawaharada Y, de Jonge N, Jensen DB, Madsen LH, Radutoiu S (2015) A legume genetic framework controls infection of nodules by symbiotic and endophytic bacteria. PLoS Genet 11(6):e1005280. https://doi.org/10.1371/journal.pgen.1005280

Zgadzaj R, Garrido-Oter R, Jensen DB, Koprivova A, Schulze-Lefert P, Radutoiu S (2016) Root nodule symbiosis in *Lotus japonicus* drives the establishment of distinctive rhizosphere, root, and nodule bacterial communities. Proc Natl Acad Sci 113(49):E7996–E8005. https://doi.org/10.1073/pnas.1616564113

Zhalnina K, Louie KB, Hao Z, Mansoori N, Da Rocha UN, Shi S, Cho H, Karaoz U, Loqué D, Bowen BP, Firestone MK (2018) Dynamic root exudate chemistry and microbial substrate preferences drive patterns in rhizosphere microbial community assembly. Nat Microbiol 3(4):470–480. https://doi.org/10.1038/s41564-018-0129-3

Zhang X, Zhang R, Gao J, Wang X, Fan F, Ma X, Yin H, Zhang C, Feng K, Deng Y (2017) Thirty-one years of rice-rice-green manure rotations shape the rhizosphere microbial community and enrich beneficial bacteria. Soil Biol Biochem 104:208–217. https://doi.org/10.1016/j.soilbio.2016.10.023

Zhang J, Wang N, Li S, Wang J, Feng Y, Wang E, Li Y, Yang T, Chen W (2023) The effect of different rhizobial symbionts on the composition and diversity of rhizosphere microorganisms of chickpea in different soils. Plants 12(19):3421. https://doi.org/10.3390/plants12193421

Zheng Y, Liang J, Zhao DL, Meng C, Xu ZC, Xie ZH, Zhang CS (2020) The root nodule microbiome of cultivated and wild halophytic legumes showed similar diversity but distinct community structure in Yellow River Delta saline soils. Microorganisms 8(2):207. https://doi.org/10.3390/microorganisms8020207

Zhong Y, Yang Y, Liu P, Xu R, Rensing C, Fu X, Liao H (2019) Genotype and rhizobium inoculation modulate the assembly of soybean rhizobacterial communities. Plant Cell Environ 42(6):2028–2044. https://doi.org/10.1111/pce.13519

Deciphering Nodule Microbiome for Improving Symbiotic Efficiency and Plant Health

Amrita Vishwakarma, Sonika Pandey, Madhubala, Priyanka Mishra, Manishi Tripathi, and Murugesan Senthilkumar

1 Introduction

Legumes including pulses establish a tripartite association with nitrogen-fixing microsymbiont called *Rhizobium* and other root nodule endophytes. Understanding the community structure of root and nodule-associated endophytic bacteria along with rhizobacteria helps to assess microbe-mediated plant development. The presence of non-rhizobial endophytic bacteria was identified as *Agrobacterium radiobacter* in 1902. The community structure and abundance of nodule endophytes within the plant are dynamic in nature. Various factors including soil type, plant species, interaction among microbial population, and microbial interaction with plants play a critical role in determining community structure locally as well as on a larger scale. Organic compounds exuded from plant roots can determine bacterial community structure in the rhizosphere which in turn may affect plant-associated microbial assemblies. Plant tissues are being colonized by a variety of microbial communities which are collectively termed as plant microbiomes. The plant microbiomes interact with their hosts and determine the growth behavior and fitness of the crop plants. Members of endophytic microbial assemblies of a plant are recruited

A. Vishwakarma · S. Pandey · Madhubala · M. Senthilkumar (✉)
Division of Basic Sciences, ICAR-Indian Institute of Pulses Research,
Kanpur, Uttar Pradesh, India

P. Mishra
Division of Basic Sciences, ICAR-Indian Institute of Pulses Research,
Kanpur, Uttar Pradesh, India

Department of Microbiology, Chhatrapati Shahu Ji Maharaj University,
Kanpur, Uttar Pradesh, India

M. Tripathi
Department of Microbiology, Chhatrapati Shahu Ji Maharaj University,
Kanpur, Uttar Pradesh, India

Renu et al. (eds.), *Meta-omics in Crop Improvement*,
https://doi.org/10.1007/978-981-95-2294-1_11

from the soil, sometimes transmitted via seeds. The root tissues are most heavily colonized by microorganisms among all other plant tissues. Community structure of endophytic bacteria is considered as a subset of rhizobacterial community. Most of the endophytes seem to originate from soil and air, and some may be transmitted via seed and determine the plant's fitness.

Root nodules of legume plants are specialized ecological niche for the growth and survival of nodule-associated bacteria. The *Rhizobium* cells along with non-rhizobial endophytes enter into the majority of legume plants through the formation of infection thread. These non-rhizobial strains can enter the root nodules of legumes in association with infecting rhizobial bacteria. Non-rhizobial nodule endophytic bacteria were isolated from root nodules of peanut and these endophytes were found to colonize the root nodules during co-inoculation with potential host specific *Bradyrhizobium* strains (Ibanez et al. 2009). A significant number of bacteria belonging to phylum Proteobacteria were found to colonize the root nodules of a wide range of root nodules of leguminous plants. These bacteria were identified as *Bacillus*, *Enterobacter*, *Agrobacterium*, *Acinetobacter*, *Paenibacillus*, *Pantoea*, *Mycobacterium*, *Micromonospora*, and *Pseudomonas*. Culture-dependent approach revealed the presence of 15 different bacterial species as non-rhizobial endophytic bacteria with a population density of 10^4 viable bacteria per gram of red clover nodules (Sturz et al. 1997). The metagenomics approach revealed higher diversity and abundance of non-rhizobial nodule-inhabiting bacteria (Martinez-Hidalgo and Hirsch 2017). Several researchers had reported the abundance of non-rhizobial bacteria in the range of 1–40% (Mayhood and Mirza 2021; Sharaf et al. 2019; Wigley et al. 2017).

2 Root Nodule Microbiome of Legumes

Community structure of root nodule microbiome of several legumes including pulses, oilseeds, and tree legumes were revealed recently. Chickpea root nodules were colonized by a diverse microbial community with the detection of 567 bacterial genera. *Mesorhizobium* is the predominant bacteria colonizing root nodules of chickpea with relative abundance of 0.74–0.82. *Mesorhizobium huakuii*, *M. septentrionale*, *M. camelthorni*, and *M. opportunistum* contribute 11.13% reads of *Mesorhizobium* retrieved from root nodules of chickpea cv. DCP 92–3 (Alok et al. 2020). The remaining unclassified reads (88.87%) at species level indicated higher diversity of *Mesorhizobium* at species level in Indian Soils. Other non-rhizobial endophytic bacteria of chickpea root nodules are *Calothrix*, *Rickettsia*, *Methylobacillus*, *Arthrobacter*, *Bacillus*, *Streptomyces*, *Saccharopolyspora*, *Rhodococcus*, *Ramlibacter*, *Propionivibrio*, *Janthinobacterium*, *Kaistobacter*, *Sphingomonas*, *Ammoniphilus*, and *Rubrobacter*. The sequences belonging to *Actinocatenispora*, *Pseudaminobacter*, *Burkholderia*, and *Shinella* with RA of >0.001 are retrieved only from root nodules of chickpea.

The abundance and distribution of rhizobial and non-rhizobial endophytes of soybean root nodules was studied by Mayhood and Mirza (2021). The most dominant sequences of *Bradyrhizobium* with a relative abundance of 93–99% were retrieved from each of the 193 tested soybean nodules regardless of nodule size or its location (tap or lateral roots). Certain other soybean-compatible rhizobia such as *Mesorhizobium* (Adhikari et al. 2012), *Rhizobium* (Kapembwa et al. 2016), and *Ensifer* (Yan et al. 2014; Pastorino et al. 2015) with low relative abundance were also detected. In general, *Bradyrhizobium* was dominant in soybean nodules under acidic soil conditions (Zhang et al. 2011), while *Ensifer* dominated in root nodules of soybean grown in alkaline soils (Pastorino et al. 2015).

In few cases, the extent of colonization of soybean root nodules by *Bradyrhizobium* is poor (66%), while the majority of non-rhizobial endophytic bacteria belong to *Pseudomonadaceae* (19.8%), *Enterobacteriaceae* (11.8%), and *Paenibacillaceae* (5.3%) (Sharaf et al. 2019). Nodule bacteria belonging to *Nitrobacter* and *Tardiphaga* were detected with relative abundance of 1–5% in almost all tested root nodules of soybean (Mayhood and Mirza 2021). A slightly higher abundance (3–4%) of *Nitrobacter* was found in a subset of nodule samples; however, the variation did not correlate with the size of root nodules, their position on the root system (tap versus lateral roots), and soil depth. *Tardiphaga*-related sequences with abundance up to 1% were retrieved in 97% of total soybean root nodules. *Nitrobacter* and *Tardiphaga* are rarely isolated as plant endophytes (Bromfield et al. 2017). Previously, culture-dependent approach did not produce any evidence for their presence due to slow growth as chemoorganotrophs or lack of nitrite for the growth as chemolithotrophs and growth rate of up to a 10-day doubling time in culture (Safronova et al. 2015). However, culture-independent studies revealed the colonization of soybean as well as alfalfa root nodules by *Nitrobacter* and *Tardiphaga* (Bromfield et al. 2017; Chizhevskaya et al. 2018).

Apart from *Nitrobacter* and *Tardiphaga*, other nodule endophytic bacteria *Novosphingobium*, *Pseudomonas*, *Bacillus*, *Stenotrophomonas*, and *Sphingomonas* were also detected in soybean nodules (Bender et al. 2022; Zhao et al. 2018; Zhang et al. 2011). Recent studies had identified many nodule-endophytic bacteria such as *Pseudomonas* (Sharaf et al. 2019; Zhao et al. 2018), *Variovorax* (Aserse et al. 2013), *Novosphingobium* (Aserse et al. 2013), *Flavobacterium* (Sharaf et al. 2019), *Achromobacter* (Aserse et al. 2013), *Bacillus* (Sharaf et al. 2019; Zhao et al. 2018), etc., in soybean root nodules. Inconsistent occurrence with relatively low abundance of these nodule endophytes in soybean was recently reported (Mayhood and Mirza 2021). Sequences related to *Novosphingobium* and *Pseudomonas* were retrieved from nearly 25%, 16% of the 193 tested soybean nodules, respectively. The above result suggested that *Novosphingobium* and *Pseudomonas* were not preferentially selected as nodule endophytes by host plants (Mayhood and Mirza 2021).

Microbial taxa that are consistently present in a particular habitat are termed as core microbiome of that habitat. These core microbes can be considered as generalist species that are capable of thriving in diverse environments with observed occurrence >95% confidence (Sriswasdi et al. 2017), while the specialists have less than 95% confidence in observed occurrence. Nodule-microbiome of normal nodulation soybean (NNS) and super nodulation soybean (SNS) were analyzed by Wang et al. (2023). A total of 87 generalist amplicon sequence variants (ASV) in NNS soybean and 37 generalists in SNS were detected to present across the soybean plant and soil samples and considered as core general microbiota. The identified core generalists ASVs belonged to SC-I-84, C0119, uncultured-f-Reyranellaceae, uncultured-f-Xanthobacteraceae, uncultured-o-Gaiellales, *Sphingomonas*, *Aeromicrobium*, *Rhodanobacter*, and *Pseudarthrobacter*. Rhizosphere microbiota had higher values of averaged community niche breadth indicated by Levin's index suggested that they are metabolically more flexible at the community level among all other root-associated microbiota. Similarly, the rhizosphere component was reported with the largest prevalence values of generalists, while root nodules have the largest specialist prevalence values (Wang et al. 2023). These core generalist species of nodule habitat of soybean were consistently detected through plant developmental stages. The identified core microbes of soybean tissues via *Sphingomonas*, *Aeromicrobium*, *Pseudarthrobacter*, and *Rhodanobacter* were also detected in other crops like rice (Guo et al. 2021), *Trifolium repens* L. (Wang et al. 2022), sugarcane (Hamonts et al. 2018), and sweet potato (Pepe-Ranney et al. 2019). The occurrence of certain members of core microbes across plant species indicated that the core microbes experienced selective pressures and adapted sufficiently to colonize and survive within plant tissues.

The taxonomic structure of generalist microbiota varies significantly with different microhabitats and host genotypes. Soybean genotypes with contrasting nodulation potential influenced the taxonomic profiles of core microbes. Rhizobiales were enriched in root nodules of the super nodulating genotype and subsequently had lower bacterial diversity and abundance in comparison to the normal nodulating genotype (Wang et al. 2023). This indicated that different strategies were employed for recruiting core microbial assembly of soybean with different phenotypes. Similarly, significant differences in the metabolic profiles of root exudates among host genotypes determine the structure of core microbes that colonize the roots. Development of omics technologies including metabolomics analysis, metagenomics and meta-transcriptome analysis can improve our understanding on the effect of association of core microbiota as well as occasional taxa with plant traits.

Bacterial taxa associated with the root nodules of *Inga punctata* were evaluated (Eaton et al. 2024) at different stages (1-year-old, 2-year-old, 13-year-old, and old trees). *Bradyrhizobium diazoefficiens* was the dominant taxon of youngest nodules from 1-year-old tree saplings with 63.16–85.71% mean percent sequences (MPS) of the total nodule bacterial DNA. MPS of other 26 nodule-associated bacteria (NAB) ranged between 7.4% and 12.2%, and most of the NAB with potential PGP traits suggested their importance on nodule development and plant growth. Simple NAB communities successionally developed into more complex taxonomic communities (Eaton et al. 2024). Root nodules of *Prosopis cineraria* were enriched with nitrogen-fixing bacteria such as *Ensifer* and *Sinorhizobium* species along with many other non-rhizobial endophytes (Ali et al. 2024). *Bradyrhizobium* was dominant in root nodules of *Acacia longifolia* (ca. 77%), followed by *Tardiphaga* (ca. 9%), and

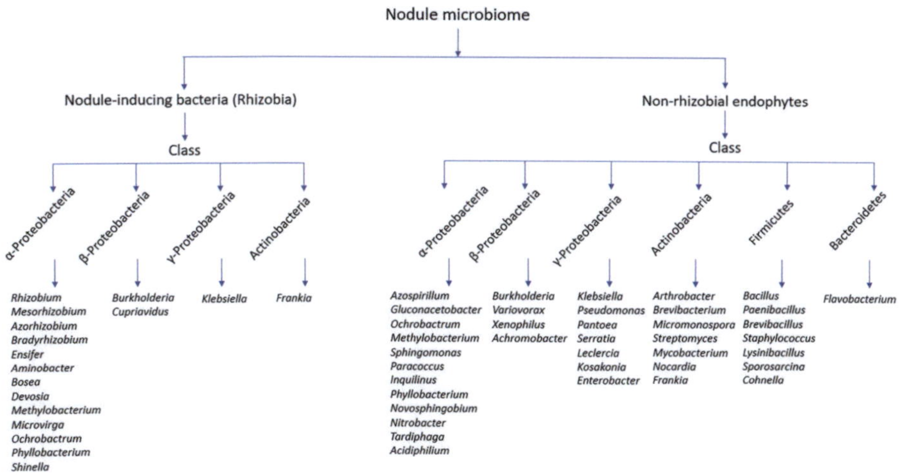

Fig. 1 Root nodule microbiome of legumes

Synechococcus (ca. 5%). The diverse nature of root nodule endophytes suggested their role in facilitating *A. longifolia* invasive success (Jesus et al. 2023).

Vertical and horizontal transfer of plant-associated microbes was reported previously. Nodule habitat received the majority of the bacterial microbiotas with an average of 81.1% from the root habitat and root microhabitats attributable to the rhizosphere (43.0% on average) and contribute to the rhizobacterial community. Depletion and enrichment of certain bacterial genera across the microhabitats suggest that the bacterial colonization of soybean roots is not a passive process and plant can select certain members of bacterial consortia (Wang et al. 2023). Members belonging to *Streptomycetaceae* and *Rhizobiaceae* were enriched in root endosphere, while a larger proportion of ASVs (>500) belonging to *Xanthobacteraceae* and Subgroup-6 were depleted. The lower abundance of *Acidobacteria* in the root endosphere compared to the rhizosphere and the higher abundance of *Proteobacteria* from soil to the endosphere suggested that certain bacterial members are well adapted to colonize specific niches. Evolutionary adaptation of certain endophytic bacterial communities enables them to overcome plant innate immunity and results in successful colonization (Kemen et al. 2015) (Fig. 1).

3 Developmental Stages of Nodule and Legume Plant Influence Nodule Microbiome

Physical and biochemical constituents of root nodules at different developmental stages generally vary. It is important to know whether microbial succession occurs during nodule development. The consistent occurrence and abundance of small nodules (<1.0 mm) were observed across the plant growth stages of peanut (Ellman-Stortz 2021). Small-sized with light red/pink nodules of peanut had less nitrogenase activity than medium-sized root nodules (Tajima et al. 2007). 16S rRNA amplicon strategy can be used to reveal the microbiome of distinct types of nodules formed on the field-grown peanut (Hossain et al. 2023).

Proteobacteria is the dominant phylum (~99.96%) of the bacteria in the big nodules and coinhabited with a single phylum *Actinobacteriota* (~0.03%). The relative abundance of *Proteobacteria* in small nodules was ~68.67% along with 15 different phyla including *Actinobacteriota* that was observed in the small nodules. During the development of small-nodules, non-rhizobial endophytes such as members of *Actinobacteriota*, *Bacteroidota*, *Chloroflexi*, and *Patescibacteria* had become more dominant in small nodules. The phyla *Patescibacteria* and *Chloroflexi* were not at all enriched in the early stage (R2) of plant growth. Temporal variation in the relative abundance of *Bradyrhizobium* in soybean nodule microhabitats was studied. In mature nodules, abundance of *Bradyrhizobium* was significantly decreased, while *Chryseobacterium* along with other saprophytic bacteria recorded significant increase in abundance (Wang et al. 2023).

Nodule microbiome during the development of root nodules is dynamic. The community structure, species richness, relative abundance of few microbiotas may vary during different stages of nodule development (Hansen et al. 2020). The relative abundance of predominant genera *Ensifer* in *Medicago sativa* is decreased from 96% in the early stages of active root nodule to 86% in senescent nodules. Similarly, relative abundance of *Pseudomonas* had reduced from 2.45% in young root nodules to 0.03% in active and 0.05% in senescent nodules. In contrast, relative abundance of *Streptomyces* and *Actinoplanes* of *Actinobacteria* was increased from 0.1% in young nodules to 1.60% in senescent nodules (Hansen et al. 2020). However, their relative abundances did not vary in young or active root nodules. Nodule size also determines the level of abundance of predominant nodule bacteria.

4 Nodule Microbiome of Legumes with Contrasting Symbiotic Efficiency

Legumes including pulses are symbiotically associated with microsymbionts termed rhizobia, and all symbiotic associations are not equally effective in reducing atmospheric nitrogen. Microsymbionts along with NRE may also influence the symbiotic efficiency by changing biochemical constituents of root nodules through their metabolism. Bender et al. (2022) investigated the microbiome structure of two contrasting symbioses, soybean-*Bradyrhizobium* with a high capacity of BNF (Herridge et al. 2022) and common bean-*Rhizobium* with a low capacity of BNF (Costa et al. 2018) under three different conditions via (i) uninoculated control, (ii) *Rhizobium*-inoculated, and (iii) *Azospirillum* along with *Rhizobium*-inoculated condition.

Higher abundance of *Bradyrhizobium* was observed in root nodules of more efficient soybean symbiosis in comparison with common bean. The abundance of *Bradyrhizobium* was slightly increased from 97.6% in uninoculated control plants to 99.2% in *Bradyrhizobium*-inoculated plants. However, co-inoculation of *Azospirillum brasilense* decreased the relative abundance of *Bradyrhizobium* to

95.9%. In contrast, abundance of the *Rhizobium* symbiont in common bean root nodules was not improved upon inoculation with potential strains of *Rhizobium*. The higher diversity of endophytic bacteria in roots of soybean and common bean was observed as indicated by Shannon and Inverted Simpson (InvSimpson), whereas root nodules possessed less complex bacterial community structure. Co-inoculation of beneficial bacteria with *Rhizobium* reduced bacterial diversity in root and nodule tissues as indicated by Shannon and inverted Simpson indexes. The reduction in bacterial diversity in root and nodules was more prominent in common bean than in soybean (Bender et al. 2022).

Detection of 49 bacterial genera in root nodules of soybean and common bean indicated the importance of nodule endophytic bacteria in the specialized microecosystem of root nodules. Potential nitrogen-fixing bacteria such as *Methylobacterium*, *Paraburkholderia*, and *Cupriavidus* were found in the root nodules of soybean and revealed their contribution to the N-nutrition. Similarly, non-rhizobial endophytic bacteria of common bean root nodules include *Brevundimonas*, *Methylobacterium*, *Flavisolibacter*, *Lysobacter*, *Massilia*, and *Pseudarthrobacter*. Plant-associated microbiome is generally derived either from the rhizosphere (horizontal transfer) or transmitted vertically by seeds (Compant et al. 2019; Mendoza-Suarez et al. 2021). Vertical transmission of *Rhizobium*, *Pseudomonas*, *Klebsiella*, *Massilia*, *Acidovorax*, *Stenotrophomonas*, *Methylobacterium*, and *Serratia* from seeds to root tissues of common bean and soybean was established (Johnston-Monje et al. 2021).

5 Diversity of Root Nodule-Associated Microbiome

Bacterial community structures of soybean nodule samples from a single plant as well as from different locations were assessed (Mayhood and Mirza 2021). The significant difference was not observed in the clustering of bacterial sequences originating from the nodules of single plant or samples from different locations (permutational multivariate analysis of variance (PERMANOVA; P. 0.05). However, wider variation in the distribution of other inconsistent nodule endophytes was revealed by greater dispersion in nonmetric multidimensional scaling (NMDS) data points.

Plant-associated bacterial communities of soybean cultivars that contrast for nodulation potential were investigated across four developmental crop growth stages (Mayhood and Mirza 2021). The overall microbiome structures of super nodulating soybean were similar to those of normal nodulating soybean. Rhizosphere soil represents a heterogeneous environment that supports diverse microbial communities, while root nodules are considered as a truly distinct microhabitat (Vandenkoornhuyse et al. 2015) colonized predominantly by rhizobia, enriched with nitrogen and carbon, and depleted in oxygen to support effective nitrogen fixation (Yang et al. 2022; Fitzpatrick et al. 2020). Shannon diversity within soybean root nodules ranged from 2.03 to 2.83, which was significantly lower than rhizobacterial diversity (5.72 to 6.40). Taxonomically diverse bacterial communities of root and rhizosphere establish complex interactions of microbe-plant and microbe-microbe associations. The predominant nodule endophytic bacteria, i.e., *Rhizobium* forms highly specific binary microbe-plant interaction under strong selection pressure. However, microbe-microbe interactions within root nodules are unclear.

α-Diversity of the plant-associated bacterial community varied with the developmental stage of root nodule and plant, cultivar, and microhabitat. The bacterial CHAO 1 richness was higher, particularly in plant endosphere at seedling stage of soybean with normal nodulation potential in comparison with super-nodulating genotype. The variance partitioning analysis revealed that plant developmental stage influenced the nodule microhabitats in a greater extent than on root and rhizosphere. Similarly, rare taxa of nodule endophytic bacteria rather than abundant taxa were more influenced by nodule phenotype. PERMANOVA and Bray-Curtis dissimilarity index were used to compare ß-diversity among different soybean cultivar groups.

Plant-associated endophytic bacterial community in soybean genotypes with normal nodulation exhibited greater beta-diversity compared to those in super-nodulating soybean, and indicated greater dispersion. However, canonical analysis of principal coordinates (CAP) indicated that the beta-diversity of nodule endophytes at both ASV and genus resolution was not influenced by nodule number, while it had a significant effect on root endophytes as well as rhizobacterial community (Mayhood and Mirza 2021). The weighted beta-NTI revealed the significant role of deterministic processes in the assembly of the root-associated microbiome ($|\beta NTI| \geq 2$). The soybean-associated bacterial communities, particularly in root nodules, were strongly and consistently influenced by deterministic processes across cultivars and plant developmental stages. However, the relative importance of neutral processes was lower in the bacterial community inhabiting soybean roots with a super nodulated phenotype (Mayhood and Mirza 2021).

6 Factors Influencing Recruitment of Root Nodule Microbiome

Research efforts to unravel the selection pressure on the recruitment of nodule endophytic bacteria by legume plants are limited. It is necessary to know whether the nodule endophytic bacteria are recruited with selection pressure or colonize the roots passively. If the nodule endophytes are truly plant beneficial, they will be recruited and detected in all root nodules of a single plant. Otherwise, endophytic bacteria enter passively and may not be consistently present in all nodules. Root nodule endophytes of soybean such as *Nitrobacter* and *Tardiphaga* were consistently detected in almost all tested root nodules from individual plants (Mayhood and Mirza 2021). Since *Nitrobacter* and *Tardiphaga* are rarely isolated as plant endophytes (Bromfield et al. 2017), their detection in root nodules with relatively higher abundance indicated that endophytic bacteria are selectively enriched in root nodules.

The enriched genera in the phylum Actinobacteria are *Arthrobacter*, *Microbacterium*, *Agromyces*, *Knoellia*, and *Cellulomonas* of order *Micrococcales*,

Nocardioides, and *Aeromicrobium* of Propionibacteriales, *Kineosporia* of *Kineosporiales*, *Actinoplanes* and *Planosporangium* of *Micromonosporales*, *Streptomyces* of *Streptomycetales*, and *Pseudonocardia*, *Actinosynnema*, and *Mycobacterium* of order *Pseudonocardiales*. A few genera such as *Patulibacter*, *Terrimonas*, *Gaiella*, *Blastococcus*, *Rubellimicrobium*, and *Flavisolibacter* were observed with low abundance in alfalfa root nodules in comparison to root and indicated that they were not enriched in root nodules. Soybean root nodules are enriched with *Rubellimicrobium, Novosphingobium* of order Sphingomonadales, and *Rhizobium, Ensifer, Shinella, Devosia, Rhodoplanes, Microvirga, Methylobacterium, Bosea*, and *Bradyrhizobium* of Rhizobiales. *Massilia* and *Aquabacterium* of β-proteobacterium and *Pseudoxanthomonas* and *Pseudomonas* of γ-proteobacteria are highly enriched in soybean root nodules.

The variations in bacterial community structures are reported with root-associated microhabitats (Edwards et al. 2015). Soil type influenced the rhizosphere microbial community of *Medicago truncatula* and cowpea, while the host genotypes determine the structure of endophytic microbiota in plant tissues (Brown et al. 2020; Leite et al. 2017). Along with host genotypes, interactions between environment and microbiota also influence plant-associated microbial community structure (Trivedi et al. 2020). Plant development described the second source of variation in the bacterial communities sampled. The plant-associated bacterial community differ significantly at different growth stages of the plant, both in terms of α- and β-diversity. Plant host genotype provided a strong selection effect, enabling the colonization of distinct microbiota at different stages of plant growth (Xiong et al. 2021). Similarly, the nodule microbiome of soybean and alfalfa are primarily determined by host genotype, while the soil type determined their rhizobacterial community assembly (Xiao et al. 2017).

7 Interaction Between Endophytes and Host Plants

Colonization of legume root nodules by various non-rhizobial endophytic bacteria and their potential to improve plant growth were well established. Plant tissue-associated endophytes showed a wide range of relationships with hosts that include communalistic, symbiotic, mutuality, and trophobiotic. Plant growth promotional activities of endophytic bacteria can be host as well as cultivar-specific in nature. Endophytic bacteria can promote plant growth in one host while it do not show growth promotion or may inhibit the plant growth when they colonize other plant species. Nodule endophytes are reported to produce plant growth regulators, solubilize mineral phosphate, inhibit the growth of plant pathogens, and help the plants by enhancing stress regulation. Artificial inoculation of beneficial nodule endophytic bacteria can improve crop health, fitness, and agricultural yield. However, their role on improving biological nitrogen fixation is unclear. Benefits received by the plants from endophytic bacteria are comparatively higher than those from rhizobacteria (Chanway et al. 2000). By using active or passive mechanisms, endophytic bacteria translocate from rhizoplane to root cortex and then aerial parts with decreased population density as compared to root or rhizobacterial population. Endophytic bacteria can cross the endodermis of plants by producing cell-wall degrading enzymes.

8 Role of Nodule Endophytes on Symbiotic Efficiency

The role of nodule endophytic bacteria on improving nodulation and nitrogen fixation is unclear. It is also difficult to correlate the physiological functions of core endophytes of legume root nodules as beneficial or commensal. Beneficial reports of nodule endophytes are previously reported (Asaf et al. 2020; Segura et al. 2021). Bioinformatic tools have been developed to predict functional profiles of pathway genes of bacterial communities that can help to reveal the role of endophytes on symbiotic nitrogen fixation (Mongad et al. 2021). The sequence information on 16S rDNA, abundance of amplicon of sequence variants (ASV), and KEGG Orthology (KO) functional categories can be used to predict pathway genes related to SNF. Higher relative abundance of biological nitrogen fixation-related genes (*nif*H, *nif*D, and *nif*K), denitrification (*nor*B and *nor*C), and nitrate reduction-assimilatory pathway (*nir*A, *nas*A, and *nas*B) and dissimilatory pathway (*nir*B, *nap*A, and *nap*B) were reported in big nodules of peanut compared to small nodules (<1 mm). Small-sized nodules are uniquely enriched with glutamate synthase (ferredoxin; GLU, gltS), ferredoxin-nitrate reductase (*nar*B), nitrous-oxide reductase (*nos*Z), and hydroxylamine reductase (*hcp*). *Tardiphaga*, a common nodule endophyte of soybean and lathyrus was reported to possess *nod*M and *nod*T that are involved in signaling and root nodule initiation (Bromfield et al. 2017). It is also predicted to involve dissimilatory nitrate reduction under anaerobic conditions (Marcondes de Souza et al. 2014).

N-cycle-related genes with high relative abundance in small nodules of peanut are *nir*D (NADH) small subunit, nitrilase, *gdh*A (NADP+), *nas*B, GLUD1_2, *gdh*A (NAD(P)+), alpha subunit of nitrate reductase/nitrite oxidoreductase (*nar*G, *nar*Z, *nxr*A), beta subunit of nitrate reductase/nitrite oxidoreductase (*nar*H, *nar*Y, and *nxr*B), and nitrate reductase gamma subunit (*nar*I, *nar*V). Relative abundance of glutamate dehydrogenase (NADP+; gdhA) was higher in the small nodule microbiota, while glutamate synthase (NADPH+; gltB, and gltD) was higher in the big nodule microbiota of soybean. Microbial communities in amino acid-rich environments use NAD+ dependent GDH, while the microbiota utilizing inorganic nitrogen (nitrates or ammonia) use NADP+-specific GDH (Plaitakis et al. 2017).

The unique P-cycle-related genes enriched in small nodules were inositol oxygenase (MIOX), 1-phosphatidylinositol phosphodiesterase (*plc*), 2-keto-myoinositol isomerase (*iol*I), diphosphate-dependent phosphofructokinase (*pfp*), 6-phospho-3-hexuloisomerase (*hxl*B), 3-hexulose-6-phosphate synthase (*hxl*A), dehydrogluconokinase (*kgu*K), glucose-6-phosphate isomerase-archaeal (*pgi*1), and CDP-diacylglycerol-inositol 3-phosphatidyltransferase (CDIPT), and fructose-1,6-bisphosphatase III (*fbp*3). Higher relative abundance of 3-phytase, 4-phytase/acid phosphatase and phosphogluconate dehydrogenase, myoinositol 2-dehydrogenase, 6-phosphofructokinase, and glucose-6-phosphate isomerase were also reported in peanut nodule (Hossain et al. 2023). The presence of *Chitinophagales*, *Cytophagales*, *Sphingomonadales*, and *Xanthomonadales* as microbial indicators positively correlated with the availability of phosphorus in farming lands (Wu et al. 2022).

9 Effect of Non-rhizobial Endophytic Bacteria on Plant Health

The microbial community structure of root nodules varies across host plant species and even among the plants of the same species based upon the physical conditions of soil as pH, temperature, O_2 concentration, macro and micronutrients, etc. However, certain bacteria are more frequently associated with nodules such as *Arthrobacter*, *Bacillus*, *Chitinophaga*, *Chryseobacterium*, *Enterobacter*, *Flavobacterium*, *Niastella*, *Novosphingobium*, *Pseudomonas*, *Sphingobacterium*, *Sphingomonas*, and *Streptomyces*. These bacteria provide various plant growth promotion benefits to the host plant by alleviating abiotic stress, producing phytohormones, solubilizing minerals, and protecting plants from pathogens. Coniglio et al. (2022) found that the inoculation of *Azospirillum argentinense* strain Az39 increased the relative abundance of beneficial bacteria. The abundance of *Streptomyces* is positively correlated with drought tolerance of the plant, while the abundance of *Bacillus* and *Pseudomonas* increases phosphate solubilization efficiency.

There are certain bacterial species that are prevalent in various economically important plant hosts. Also, it can be noted that these are persistent throughout the plant's developmental stages. *Sphingomonas*, *Aeromicrobium*, *Pseudarthrobacter*, and *Rhodanobacter* are present in *Trifolium repens L.* (Wang et al. 2022) and *Arabidopsis thaliana* (Lundberg et al. 2012). Since they are widespread, it can be understood that they have sustained the selection pressure by host plants and adapted very well to establish themselves in the host tissue. Plant growth-promoting potential of many other NREs from soybean root nodules, such as *Pseudomonas* (Xiao et al. 2017; Sharaf et al. 2019; Tokgoz 2018; Zhao et al. 2018; Zhang et al. 2018), *Variovorax* (De-Almeida Lopes et al. 2016), *Novosphingobium* (Aserse et al. 2013), *Flavobacterium* (Sharaf et al. 2019), *Achromobacter* (Wedhastri et al. 2015), *Bacillus* (Sharaf et al. 2019; Zhao et al. 2018; Annapurna et al. 2013), etc., were well established. Members of *Novosphingobium*, *Sphingomonas*, and *Niastella* genera are known to produce beneficial phytohormones such as gibberellins and indole acetic acid, sphingan, and gellan gum and play a vital role in improving plant growth (Asaf et al. 2020; Segura et al. 2021). *Pseudomonas putida* and *Pseudomonas hibiscicola* are potential plant growth-promoting endophytes with the ability to produce IAA and are also involved in phosphate solubilization (Nhu and Diep 2014).

9.1 Enzymes and Vitamins Production

Endophytic bacterium *Niastella*, which comes under the order *Chitinophagales* of phylum *Bacteroidota* is known to degrade various carbohydrate-based biomass, making sugars available for themselves as well as other microbial communities (McKee et al. 2019). Similarly, *Variovorax* produced several hydrolytic enzymes including lipase, cellulase, and protease. It could solubilize phosphate, synthesize indole acetic acid (IAA), and be involved in heavy metal phytoremediation (Aserse et al. 2013; De-Almeida Lopes et al. 2016; Bessadok et al. 2020; Khan et al. 2023). Activation of D-glucarate and D-galactarate catabolism pathway of *Burkholderia-Caballeronia-Paraburkholderia* spp., leading to transport of carbon source across bacteroid and peribacteroid membrane and support nodule metabolism (Liu et al. 2018). *Rhizobium leguminosarum* and *Rhizobium etli* produced thiamine, vitamin B12, and biotin for plant growth promotion (Guillen-Navarro et al. 2005).

9.2 Anti-Microbial Activity

Bradyrhizobium, a nitrogen-fixing endophyte, inhibits the growth of phytopathogens such as *Macrophomina* and *Sclerotium* (Chan 2009). *Bacillus* sp. is one of the most abundant nodule endophytes that produced ß-1,3-glucanase to degrade fungal cell wall and showed biocontrol potential against *Rhizoctonia solani* and *Sclerotium rolfsii* (Compant et al. 2005). *Enterobacter kobei* is a root nodule endophyte of *Hedysarum*, inhibited anthracnose in banana and enhanced zinc and phosphate solubilization (Damasceno et al. 2019). *Pseudomonas* spp. and *Chryseobacterium indologenes* act as biocontrol agents against *Phytophthora* (Sang et al. 2018). *Clonostachys* spp. isolated from *Desmodium* acts as biocontrol agents against *Rosellinia* root rot of cocoa by *Clonostachys byssicola* and banana crown rot by *C. byssicola* (Krauss et al. 2013). It also effectively controls phytopathogens like *Phytophthora palmivora* (Krauss et al. 2013), *Moniliophthora roreri* (Harry et al. 2003), and *Fusarium graminearum* (Nygren et al. 2018). *Sarocladium kiliense* has been reported to confer resistance against pathogenic *Sclerotinia sclerotiorum* found in *Brachiaria* (Gama et al. 2020). *Bacillus amyloliquefaciens*, along with IAA and GA3 production, are known to produce volatile organic compounds and protect the plants against *Fusarium oxysporum* (Yuan et al. 2013). *Paenibacillus*, isolated from *Medicago sativa* showed chitinolytic activity for the control of *Rhizoctonia*, *Fusarium*, and *Macrophomina* (Lai et al. 2015).

10 Potential of Nodule Endophytes on Abiotic Stress Management

Certain endophytic microorganisms produce primary and secondary metabolites including ACC deaminase and mitigate abiotic stress effects to improve plant growth and yield. Under high saline conditions, the *Arthrobacter protophormiae* inhabiting the root nodules of *Pisum sativum* improved the plant growth by enhancing nodulation potential and mitigating the effect of salt stress (Barnawal et al. 2014). Plant beneficial traits of *Bacillus subtilis* BERA 71 on chickpea under high saline conditions was investigated and reported to enhance plant growth and reduce the oxidative damage caused by salinity. It also increased the availability of macronutrients such as N, P, K, Ca, and Mg alongside reducing the Na accumulation (Abd-Allah et al. 2018). When *Sinorhizobium meliloti* is exposed to salt stress, genes related to polysaccharide biosynthesis and transport of biomolecules such as amino acids, amines, peptides, anions, and alcohols were activated and conferred stress tolerance to plants (Ruberg et al. 2003). Most of these genes are located in chromosome and plasmid (pSmbB) of *Sinorhizobium meliloti* conferred adaptation to salt stress.

To reduce salt stress, endophytic bacteria play a critical role by adjusting cell osmolarity, detoxification, nutrient acquisition, and regulating plant hormones (Vaishnav et al. 2019). Endophytes with ACC deaminase activity, inhibit ethylene production and convert ethylene precursor ACC to α-ketobutyrate and ammonia which is later metabolized by bacteria during growth (Fig. 2) (Soni et al. 2018). Various genera of *Bacillus*, *Pseudomonas*, *Klebsiella*, *Serratia*, *Arthrobacter*, *Streptomyces*, *Isoptericola*, and *Microbacterium* show ACC deaminase activity under high salt conditions (Vaishnav et al. 2019). Such ACC deaminase producing bacterial isolate of *Pseudomonas* was used as a co-inoculant along with *Rhizobium* that enhanced mung bean growth under saline conditions (Ahmad et al. 2011). Actinobacteria adopted to thrive in extreme ecosystems including salinity, low and high pH, low water availability, extreme temperatures, radiation, and pressure. Hence, *Arthrobacter*, *Nocardioides*, *Microbacterium*, and *Streptomyces* of phylum

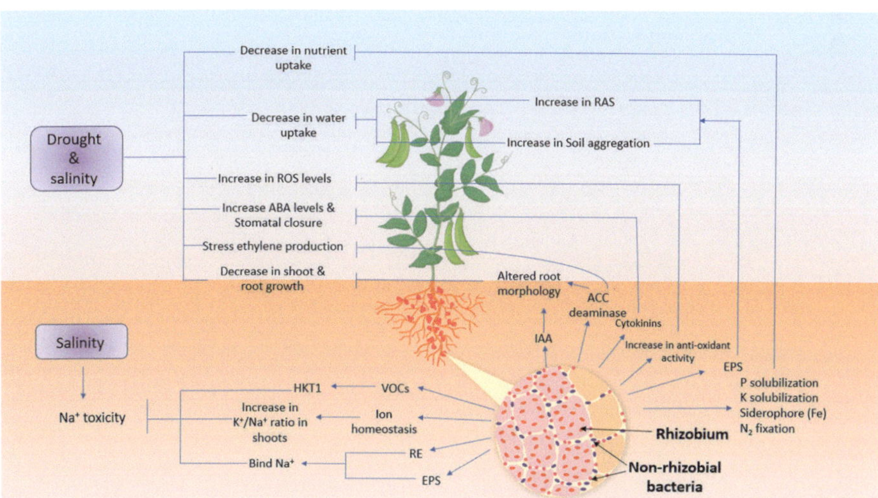

Fig. 2 Effect of root endophytic bacteria against drought and saline stress

Actinobacteria are identified as dominant bacterial genera associated with desert plants (Alsharif et al. 2020).

Root nodule-associated endophytes confer drought stress tolerance by improving root length and density to improve nutrient uptake and water use efficiency, regulate hormonal signaling in plants and soil-water-plant relationships, increase synthesis of proline and other organic and inorganic solutes and antioxidant enzymes, decrease the expression of stress-responsive genes, and produce hormones such as abscisic acid, indole-3-acetic acid, ACC deaminase, and volatile organic compounds (Fig. 2) (Ullah et al. 2019; Sofi et al. 2018; Sorty et al. 2018). Inoculation of root nodule endophytes from soybean such as *Bacillus* and *Pseudomonas* improved the growth of plants by maintaining membrane integrity, water status, and accumulation of compatible solutes and osmolytes (Dubey et al. 2021). *Bacillus subtilis* produced bacillibactin, a hexadentate siderophore that is similar to enterobactin of *Escherichia coli* (Ollinger et al. 2006). *Bacillus megaterium* had produced schizokinen, a derivative of citrates (N-hydroxyl-L-acetyl ornithine citrate) which act as siderophore in tomato plants. Utilizing siderophores of other microorganisms during iron-limiting conditions is of great advantage. *Bradyrhizobium japonicum* USDA 110 and 61A152 can also utilize ferrichrome, rhodotorulate (hydroxamatetate type siderophores), and pseudobactin ST3 (pyoverdine type siderophore).

Members of nodule endophytic bacteria have adapted to heavy metal pollution with tolerant mechanisms such as regulation of stress responsive genes, bioaccumulation of heavy metals, production of antioxidant enzymes, and regulating the level of stress related plant hormones (Fig. 3) (Tiwari and Lata 2018; Zhang et al. 2021). They also utilize strategies like adsorption of heavy metals, the expulsion of metal outside the cell, biotransformation, release of chelating agents, acidification of the surrounding environment, and ability to influence redox potential (Goel et al. 2018).

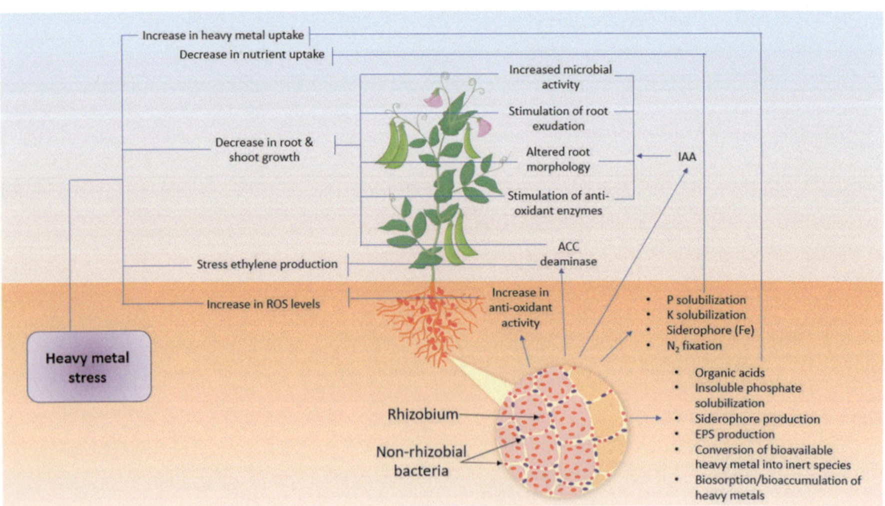

Fig. 3 Effect of root endophytic bacteria against heavy metal stress

Trifolium repens can tolerate Fe, Mn, Zn, and Cd, when the plant is associated with *Rhodococcus erythropolis*, *Achromobacter* sp., *Microbacterium* sp., and *Bacillus cereus* (Pereira et al. 2015). Similarly, *Lupinus luteus* when associated with *Bradyrhizobium* sp. 750, *Pseudomonas* sp., and *Ochrobactrum cytisi* can tolerate Fe, Mn, Zn, and Cd. *Ochrobactrum* increased the plant biomass when used in consortia with nodule-forming plant growth-promoting bacteria (Dary et al. 2020). *Vicia faba* in association with *Rhizobium* sp. CCNWSX0481, *Rhizobium leguminosarum* bv. *viciae*, *Enterobacter cloacae*, and *Pseudomonas* sp. were reported to have high seed yield and nodulation under moderate contamination with Cu (Fatnassi et al. 2015). *Cupriavidus* strains are resistant to heavy metals as well as they are known to fix nitrogen (Marques et al. 2013) and are considered as ideal for bioremediation.

11 Conclusion

Review of existing literature confirmed the presence of non-rhizobial endophytic bacteria in root nodules of legumes and their relative abundance through culture-independent metagenomic approaches. Plant inoculation experiments using non-rhizobial nodule endophytic bacteria along with rhizobia highlighted their significant role on plant growth promotion and fitness. Though the information on relative abundance of non-rhizobial nodule endophytes was available, non-availability-corresponding bacterial strains hinder their usage in agriculture. Physical availability of those agriculturally important bacteria will help to understand more about their specific role on plant health and sustainable agriculture. Since the non-rhizobial endophytic bacteria are selectively recruited by host genotypes and colonized a very specialized niche with low oxygen concentration, their interaction with *Rhizobium* and their influence on nitrogen fixation need to be explored. In-depth transcriptome and genome analysis of non-rhizobial endophytic bacteria may yield better insights related to biochemical pathways and molecular basis on conferring benefits to plants as well as symbiotic efficiency. Cultivation of legumes along with plant growth promoting bioinoculants including arbuscular mycorrhizae can restore the soil fertility and health. Hence, unique beneficial microbiome of root nodules and further developing and utilizing synthetic consortium of such microorganisms for improving soil and plant health need to be investigated.

Conflict of Interest Authors declare no conflict of interest.

References

Abd-Allah EF, Abdulaziz AA, Hashem A, Radhakrishnan R, Al-Huqail AS, Al-Otibi F, Malik J, Alharbi R, Egamberdieva D (2018) Endophytic bacterium *Bacillus subtilis* (BERA 71) improves salt tolerance in chickpea plants by regulating the plant defense mechanisms. J Plant Interact 13:37–44. https://doi.org/10.1080/17429145.2017.1414321

Adhikari D, Kaneto M, Itoh K, Suyama K, Pokharel BB, Gaihre YK (2012) Genetic diversity of soybean-nodulating rhizobia in Nepal in relation to climate and soil properties. Plant Soil 357:131–145. https://doi.org/10.1007/s11104-012-1134-6

Ahmad M, Zahir ZA, Asghar H, Arshad M (2011) The combined application of rhizobial strains and plant growth promoting rhizobacteria improves growth and productivity of mung bean (*Vigna radiata* L.) under salt-stressed conditions. Ann Microbiol 62:1321–1330. https://doi.org/10.1007/s13213-011-0380-9

Ali R, Chaluvadi SR, Wang X, Hazzouri KM, Sudalaimuthuasari N, Rafi M, Al-Nuaimi M, Sasi S, Antepenko E, Bennetzen JL, Amiri KMA (2024) Microbiome properties in the root nodules of *Prosopis cineraria*, a leguminous desert tree. Microbiol Spectr 12(6):e0361723. https://doi.org/10.1128/spectrum.03617-23

Alok D, Annapragada H, Singh S, Murugesan S, Singh NP (2020) Symbiotic nitrogen fixation and endophytic bacterial community structure in Bt-transgenic chickpea (*Cicer arietinum L*). Sci Rep 10(1):5453. https://doi.org/10.1038/s41598-020-62199-1

Alsharif W, Saad MM, Hirt H (2020) Desert microbes for boosting sustainable agriculture in extreme environments. Front Microbiol 11:1666. https://doi.org/10.3389/fmicb.2020.01666

Annapurna K, Ramadoss D, Bose P, Vithalkumar L (2013) *In situ* localization of *Paenibacillus polymyxa* HKA-15 in roots and root nodules of soybean (*Glycine max L.*). Plant Soil 373:641–648. https://doi.org/10.1007/s11104-013-1825-7

Asaf S, Numan M, Khan AL, Al-Harrasi A (2020) *Sphingomonas*: from diversity and genomics to functional role in environmental remediation and plant growth. Crit Rev Biotechnol 40(2):138–152. https://doi.org/10.1080/07388551.2019.1709793

Aserse AA, Rasanen LA, Aseffa F, Hailemariam A, Lindstrom K (2013) Diversity of sporadic symbionts and nonsymbiotic endophytic bacteria isolated from nodules of woody, shrub, and food legumes in Ethiopia. Appl Microbiol Biotechnol 97:10117–10134. https://doi.org/10.1007/s00253-013-5248-4

Barnawal D, Bharti N, Maji D, Chanotiya CS, Kalra A (2014) ACC deaminase-containing Arthrobacter protophormiae induces NaCl stress tolerance through reduced ACC oxidase activity and ethylene production resulting in improved nodulation and mycorrhization in *Pisum sativum*. J Plant Physiol 171(11):884–894. https://doi.org/10.1016/j.jplph.2014.03.007

Bender FR, Alves LC, da Silva JFM, Ribeiro RA, Pauli G, Nogueira MA, Hungria M (2022) Microbiome of nodules and roots of soybean and common bean: searching for differences associated with contrasting performances in symbiotic nitrogen fixation. Int J Mol Sci 23(19):12035. https://doi.org/10.3390/ijms231912035

Bessadok K, Navarro-Torre S, Pajuelo E, Mateos-Naranjo E, Redondo-Gomez S, Caviedes MA, Fterich A, Mars M, Rodriguez-Llorente ID (2020) The ACC-deaminase producing bacterium *Variovorax* sp. CT7.15 as a tool for improving *Calicotome villosa* nodulation and growth in arid regions of Tunisia. Microorganisms 8(4):541. https://doi.org/10.3390/microorganisms8040541

Bromfield ESP, Cloutier S, Tambong JT, Tran TT (2017) Soybeans inoculated with root zone soils of Canadian native legumes harbour diverse and novel *Bradyrhizobium* spp. that possess agricultural potential. Syst Appl Microbiol 40:440–447. https://doi.org/10.1016/j.syapm.2017.07.007

Brown SP, Grillo MA, Podowski JC, Heath KD (2020) Soil origin and plant genotype structure distinct microbiome compartments in the model legume *Medicago truncatula*. Microbiome 8(1):139. https://doi.org/10.1186/s40168-020-00915-9

Chan KG (2009) Exochelin production in *mycobacterium neoaurum*. Int J Mol Sci 10:345–353. https://doi.org/10.3390/ijms10010345

Chanway CP, Shishido M, Nairn J et al (2000) Endophytic colonization and field responses of hybrid spruce seedlings after inoculation with plant growth-promoting rhizobacteria. For Ecol Manag 133(1–2):81–88. https://doi.org/10.1016/S0378-1127(99)00300-X

Chizhevskaya EP, Naidenova EA, Onishchuk OP, Andronov EE, Simarov BV (2018) The melanin biosynthesis gene from the CA15-1 strain of alfalfa nodule bacteria: molecular analysis and phylogeny. Russ J Genet 54:925–932. https://doi.org/10.1134/S1022795418080045

Compant S, Duffy B, Nowak J, Clément C, Barka EA (2005) Use of plant growth-promoting bacteria for biocontrol of plant diseases: principles, mechanisms of action, and future prospects. Appl Environ Microbiol 71(9):4951–4959. https://doi.org/10.1128/AEM.71.9.4951-4959

Compant S, Samad A, Faist H, Sessitsch A (2019) A review on the plant microbiome: ecology, functions, and emerging trends in microbial application. J Adv Res 19:29–37

Coniglio LA, Molina G, Romina M, Torres MD, Avila A, de Carlan A, Erijman C, Figuerola L, Souza E, Cassan E, Fabricio (2022) Modulation of maize rhizosphere microbiota composition by inoculation with *Azospirillum argentinense* Az39 (Formerly *A. brasilense* Az39). J Soil Sci Plant Nutr 22. https://doi.org/10.1007/s42729-022-00909-6

Costa MR, Chibeba AM, Mercante FM, Hungria M (2018) Polyphasic characterization of rhizobia microsymbionts of common bean [*Phaseolus vulgaris* (L.)] isolated in Mato Grosso do Sul, a hotspot of Brazilian biodiversity. Symbiosis 76:163–176

Damasceno LC, Duarte EAA, dos Santos LBPR, de Oliveira TAS, de Jesus FN, de Oliveira LM, Goes-Neto A, Soares ACF (2019) Postharvest biocontrol of anthracnose in bananas by endophytic and soil rhizosphere bacteria associated with sisal (*Agave sisalana*) in Brazil. Biol Control 137:104016. https://doi.org/10.1016/j.biocontrol.2019.104016

Dary M, Chamber-Pérez MA, Palomares AJ, Pajuelo E (2020) *In situ* phyto-stabilisation of heavy metal polluted soils using *Lupinus luteus* inoculated with metal resistant plant-growth promoting rhizobacteria. J Hazard Mater 177(1–3):323–330. https://doi.org/10.1016/j.jhazmat.2009.12.035. Epub 6

De-Almeida Lopes KB, Carpentieri-Pipolo V, Oro TH, Stefani-Pagliosa E, Degrassi G (2016) Culturable endophytic bacterial communities associated with field-grown soybean. J Appl Microbiol 120:740–755. https://doi.org/10.1111/jam.13046

Dubey A, Saiyam D, Kumar A, Hashem A, Abd Allah EF, Khan ML (2021) Bacterial root endophytes: characterization of their competence and plant growth promotion in soybean (*Glycine max* (L.) Merr.) under drought stress. Int J Environ Res Public Health 18:931. https://doi.org/10.3390/ijerph18030931

Eaton WD, Hamilton DA, Chen W, Lemenze A, Soteropoulos P (2024) Use of high throughput DNA analysis to characterize the nodule-associated bacterial community from four ages of *Inga punctata* trees in a Costa Rican cloud forest. AIMS Microbiol 10(3):572–595. https://doi.org/10.3934/microbiol.2024027. PMID: 39219756; PMCID: PMC11362272

Edwards J, Johnson C, Santos-Medellín C, Lurie E, Podishetty NK, Bhatnagar S et al (2015) Structure, variation, and assembly of the root-associated microbiomes of rice. Proc Natl Acad Sci USA 112:E911–E920. https://doi.org/10.1073/pnas.1414592112

Ellman-Stortz LM (2021) Influence of cover crop selection on soil microbial activity during a transition to organic agriculture. Doctoral Dissertation. Texas A&M University

Fatnassi IC, Chiboub M, Saadani O, Jebara M, Jebara SH (2015) Phyto-stabilization of moderate copper contaminated soils using co-inoculation of *Vicia faba* with plant growth promoting bacteria. J Basic Microbiol 55(3):303–311. https://doi.org/10.1002/jobm.201300323. Epub 2013 Dec 12

Fitzpatrick CR, Salas-González I, Conway JM, Finkel OM, Gilbert S, Russ D et al (2020) The plant microbiome: from ecology to reductionism and beyond. Ann Rev Microbiol 74:81–100. https://doi.org/10.1146/annurev-micro-022620-014327

Gama DDS, Santos IAFM, De Abreu D, De Medeiros LM, Duarte FHVWF, Cardoso PG (2020) Endophytic fungi from Brachiaria grasses in Brazil and preliminary screening of *Sclerotinia sclerotiorum* antagonists. Sci Agric 77:210. https://doi.org/10.1590/1678-992x-2018-0210

Goel R, Suyal DC, Kumar V, Jain L, Soni R (2018) Stress-tolerant beneficial microbes for sustainable agricultural production. In: Panpatte D, Jhala Y, Shelat H, Vyas R (eds) Microorganisms for green revolution. Microorganisms for sustainability, vol 7. Springer, Singapore. https://doi.org/10.1007/978-981-10-7146-1_8

Guillen-Navarro K, Encarnacion S, Dunn MF (2005) Biotin biosynthesis, transport and utilization in rhizobia. FEMS Microbiol Lett 246(2):159–165. https://doi.org/10.1016/j.femsle.2005.04.020

Guo J, Ling N, Li Y, Li K, Ning H, Shen Q et al (2021) Seed-borne, endospheric and rhizospheric core microbiota as predictors of plant functional traits across rice cultivars are dominated by deterministic processes. New Phytol 230:2047–2060. https://doi.org/10.1111/nph.17297

Hamonts K, Trivedi P, Garg A, Janitz C, Grinyer J, Holford P et al (2018) Field study reveals core plant microbiota and relative importance of their drivers. Environ Microbiol 20:124–140. https://doi.org/10.1111/1462-2920.14031

Hansen BL, Pessotti RC, Fischer MS, Collins A, El-Hifnawi L, Liu MD, Traxler MF (2020) Cooperation, competition, and specialized metabolism in a simplified root nodule microbiome. MBio 11(4):e01917–e01920. https://doi.org/10.1128/mBio.01917-20. PMID: 32843548; PMCID: PMC7448283

Harry C, Keith A, Sarah E (2003) Endophytes and mycopara sites associated with an indigenous forest tree, *Theobroma gileri*, in Ecuador and a preliminary assessment of their potential as biocontrol agents of cocoa diseases. Mycol Prog 2:149–160. https://doi.org/10.1007/s11557-006-0053-4

Herridge DF, Giller KE, Jensen ES, Peoples MB (2022) Quantifying country-to-global scale nitrogen fixation for crop legumes. II. Coefficients, templates and estimates for soybean, groundnut and pulses. Plant Soil 474:1–15

Hossain MS, DeLaune PB, Gentry TJ (2023) Microbiome analysis revealed distinct microbial communities occupying different sized nodules in field-grown peanut. Front Microbiol 14:1075575. https://doi.org/10.3389/fmicb.2023.1075575

Ibanez F, Angelini J, Taurian T, Tonelli ML, Fabra A (2009) Endophytic occupation of peanut root nodules by opportunistic Gammaproteobacteria. Syst Appl Microbiol 32(1):49–55. https://doi.org/10.1016/j.syapm.2008.10.001. Epub 2008 Dec 2

Jesus JG, Maguas C, Dias R, Nunes M, Pascoal P, Pereira M, Trindade H (2023) What if root nodules are a guesthouse for a microbiome? The case study of *Acacia longifolia*. Biology (Basel) 12(9):1168. https://doi.org/10.3390/biology12091168. PMID: 37759568; PMCID: PMC10525506

Johnston-Monje D, Gutierrez JP, Lopez-Lavalle LAB (2021) Seed-transmitted bacteria and fungi dominate juvenile plant microbiomes. Front Microbiol 12:737616

Kapembwa R, Mweetwa AM, Ngulube M, Yengwe J, Yengwe J (2016) Morphological and biochemical characterization of soybean nodulating rhizobia indigenous to Zambia. Sustain Agric Res 5:84–92. https://doi.org/10.5539/sar.v5n3p84

Kemen AC, Agler MT, Kemen E (2015) Host–microbe and microbe–microbe interactions in the evolution of obligate plant parasitism. New Phytol 206:1207–1228. https://doi.org/10.1111/nph.13284

Khan N, Humm EA, Jayakarunakaran A, Hirsch AM (2023) Reviewing and renewing the use of beneficial root and soil bacteria for plant growth and sustainability in nutrient-poor, arid soils. Front Plant Sci 6(14):1147535. https://doi.org/10.3389/fpls.2023.1147535. PMID: 37089637; PMCID: PMC10117987

Krauss U, Rees R, Stirrup T, Argyle T, George A, Arroyo C et al (2013) Mycoparasitism by *Clonostachys byssicola* and *Clonostachys rosea* on *Trichoderma* spp. from cocoa (*Theobroma cacao*) and implication for the design of mixed biocontrol agents. Biol Control 67:317–327. https://doi.org/10.1016/j.biocontrol.2013.09.011

Lai WA, Hameed A, Lin SY, Hung MH, Hsu YH, Liu YC, Shahina M, Shen FT, Young CC (2015) *Paenibacillus medicaginis* sp. nov. a chitinolytic endophyte isolated from a root nodule of alfalfa (Medicago sativa L.). Int J Syst Evol Microbiol 65(11):3853–3860. https://doi.org/10.1099/ijsem.0.000505

Leite J, Fischer D, Rouws LF, Fernandes-Júnior PI, Hofmann A, Kublik S, Schloter M, Xavier GR, Radl V (2017) Cowpea nodules harbor non-rhizobial bacterial communities that are shaped by soil type rather than plant genotype. Front Plant Sci 20(7):2064. https://doi.org/10.3389/fpls.2016.02064. PMID: 28163711; PMCID: PMC5247471

Liu A, Contador CA, Fan K, Lam HM (2018) Interaction and regulation of carbon, nitrogen, and phosphorus metabolisms in root nodules of legumes. Front Plant Sci 9:1860. https://doi.org/10.3389/fpls.2018.01860. PMID: 30619423; PMCID: PMC6305480

Lundberg DS, Lebeis SL, Paredes SH, Yourstone S, Gehring J, Malfatti S, Tremblay J, Engelbrektson A, Kunin V, Del Rio TG, Edgar RC, Eickhorst T, Ley RE, Hugenholtz P, Tringe SG, Dangl JL (2012) Defining the core *Arabidopsis thaliana* root microbiome. Nature 488(7409):86–90. https://doi.org/10.1038/nature11237

Marcondes de Souza JA, Alves LMC, Varani AM, Lemos EGM (2014) The family *Bradyrhizobiaceae*. In: Rosenberg E, DeLong EF, Lory S, Stackebrandt E, Thompson F (eds) The prokaryotes: prokaryotic biology and symbiotic associations, vol 1, 4th edn. Springer, Berlin, pp 135–154. https://doi.org/10.1007/978-3-642-30197-1_253

Marques AP, Moreira H, Franco AR, Rangel AO, Castro PM (2013) Inoculating *Helianthus annuus* (sunflower) grown in zinc and cadmium-contaminated soils with plant growth-promoting bacteria—effects on phytoremediation strategies. Chemosphere 92(1):74–83

Martinez-Hidalgo P, Hirsch AM (2017) The nodule microbiome: N_2-fixing rhizobia do not live alone. Phytobiomes 1:70–82. https://doi.org/10.1094/PBIOMES-12-16-0019-RVW

Mayhood P, Mirza BS (2021) Soybean root nodule and rhizosphere microbiome: distribution of rhizobial and nonrhizobial endophytes. Appl Environ Microbiol 87(10):e02884–e02820. https://doi.org/10.1128/AEM.02884-20

McKee LS, Martínez-Abad A, Ruthes AC, Vilaplana F, Brumer H (2019) Focused metabolism of β-glucans by the soil *Bacteroidetes* species *Chitinophaga pinensis*. Appl Environ Microbiol 85:e02231–e02218. https://doi.org/10.1128/AEM.02231-18

Mendoza-Suarez M, Andersen SU, Poole PS, Sánchez-Cañizares C (2021) Competition, nodule occupancy, and persistence of inoculant strains: key factors in the *Rhizobium*-legume symbioses. Front Plant Sci 12:690567. https://doi.org/10.3389/fpls.2021.690567

Mongad DS, Chavan NS, Narwade NP, Dixit K, Shouche YS, Dhotre DP (2021) MicFunPred: a conserved approach to predict functional profiles from 16S rRNA gene sequence data. Genomics 113:3635–3643. https://doi.org/10.1016/j.ygeno.2021.08.016

Nhu V, Diep C (2014) Isolation, characterization, and phylogenetic analysis of endophytic bacteria in rice plants cultivated on soil of Phu Yen Province, Vietnam. Am J Life Sci 2(3):117–127

Nygren K, Dubey M, Zapparata A, Iqbal M, Tzelepis GD, Durling MB, Jensen DF, Karlsson M (2018) The mycoparasitic fungus *Clonostachys rosea* responds with both common and specific gene expression during interspecific interactions with fungal prey. Evol Appl 11(6):931–949. https://doi.org/10.1111/eva.12609

Ollinger J, Song KB, Antelmann H, Hecker M, Helmann JD (2006) Role of the *Fur*-regulon in iron transport in *Bacillus subtilis*. J Bacteriol 188(10):3664–3673. https://doi.org/10.1128/JB.188.10.3664-3673.2006

Pastorino G, Alcántara V, Malbrán I, Videira L, Sarinelli J, Balatti P (2015) *Ensifer (Sinorhizobium) fredii* interacted more efficiently than *Bradyrhizobium japonicum* with soybean. J Agric Ecol Res Int 2:10–19. https://doi.org/10.9734/JAERI/2015/13163

Pepe-Ranney C, Keyser C, Trimble JK, Bissinger B (2019) Surveying the sweet potato rhizosphere, endophyte, and surrounding soil microbiomes at two North Carolina farms reveals underpinnings of sweet potato microbiome community assembly. Phytobiomes J 4:75–89. https://doi.org/10.1094/PBIOMES-06-19-0029-R

Pereira SIA, Barbosa L, Castro PML (2015) Rhizobacteria isolated from a metal-polluted area enhance plant growth in zinc and cadmium-contaminated soil. Int J Environ Sci Technol 12:2127–2142. https://doi.org/10.1007/s13762-014-0614-z

Plaitakis A, Kalef-Ezra E, Kotzamani D, Zaganas I, Spanaki C (2017) The glutamate dehydrogenase pathway and its roles in cell and tissue biology in health and disease. Biol 6:11. https://doi.org/10.3390/biology6010011

Ruberg S, Tian ZX, Krol E, Linke B, Meyer F, Wang Y, Pühler A, Weidner S, Becker A (2003) Construction and validation of a *Sinorhizobium meliloti* whole-genome DNA microarray: genome-wide profiling of osmoadaptive gene expression. J Biotechnol 106(2–3):255–268. https://doi.org/10.1016/j.jbiotec.2003.08.005

Safronova VI, Kuznetsova IG, Sazanova AL, Kimeklis AK, Belimov AA, Andronov EE, Pinaev AG, Pukhaev AR, Popov KP, Akopian JA, Willems A, Tikhonovich IA (2015) Extra-slow-growing

Tardiphaga strains isolated from nodules of *Vavilovia formosa* (Stev.). Arch Microbiol 197:889–898. https://doi.org/10.1007/s00203-015-1122-3

Sang MK, Jeong JJ, Kim J, Kim KD (2018) Growth promotion and root colonization in pepper plants by phosphate-solubilizing *Chryseobacterium* sp. strain ISE14 that suppresses *Phytophthora* blight. Ann Appl Biol 172:208–223. https://doi.org/10.1111/aab.12413

Segura A, Udaondo Z, Molina L (2021) PahT regulates carbon fluxes in *Novosphingobium* sp. HR1a and influences its survival in soil and rhizospheres. Environ Microbiol 23:2969–2991. https://doi.org/10.1111/1462-2920.15509

Sharaf H, Rodrigues RR, Moon J, Zhang B, Mills K, Williams MA (2019) Unprecedented bacterial community richness in soybean nodules varies with cultivar and water status. Microbiome 7:63. https://doi.org/10.1186/s40168-019-0676-8

Sofi PA, Baba ZA, Hamid B, Meena RS (2018) Harnessing soil rhizobacteria for improving drought resilience in legumes. In: Meena R, Das A, Yadav G, Lal R (eds) Legumes for soil health and sustainable management. Springer, Singapore. https://doi.org/10.1007/978-981-13-0253-4_8

Soni R, Yadav SK, Rajput AS (2018) ACC-deaminase producing rhizobacteria: prospects and application as stress busters for stressed agriculture. In: Panpatte D, Jhala Y, Shelat H, Vyas R (eds) Microorganisms for green revolution. Microorganisms for sustainability, vol vol 7. Springer, Singapore. https://doi.org/10.1007/978-981-10-7146-1_9

Sorty AM, Bitla UM, Meena KK, Singh NP (2018) Role of microorganisms in alleviating abiotic stresses. In: Panpatte D, Jhala Y, Shelat H, Vyas R (eds) Microorganisms for green revolution. Microorganisms for sustainability, vol vol 7. Springer, Singapore. https://doi.org/10.1007/978-981-10-7146-1_6

Sriswasdi S, Yang CC, Iwasaki W (2017) Generalist species drive microbial dispersion and evolution. Nat Commun 8:1162. https://doi.org/10.1038/s41467-017-01265-1

Sturz AV, Christie BR, Matheson BG, Nowak J (1997) Biodiversity of endophytic bacteria which colonize red clover nodules, roots, stems, and foliage and their influence on host growth. Biol Fertil Soils 25(1):13–19. https://doi.org/10.1007/s003740050273

Tajima R, Lee ON, Abe J, Lux A, Morita S (2007) Nitrogen-fixing activity of root nodules in relation to their size in peanut (*Arachis hypogaea* L.). Plant Prod Sci 10:423–429. https://doi.org/10.1626/pps.10.423

Tiwari S, Lata C (2018) Heavy metal stress, signaling, and tolerance due to plant-associated microbes: an overview. Front Plant Sci 9:452. https://doi.org/10.3389/fpls.2018.00452

Tokgoz S (2018) Exploration of soybean nodule microbiome for plant health management. MS thesis, University of Nebraska, Lincoln

Trivedi P, Leach JE, Tringe SG, Sa T, Singh BK (2020) Plant-microbiome interactions: from community assembly to plant health. Nat Rev Microbiol 18:607–621. https://doi.org/10.1038/s41579-020-0412

Ullah A, Nisar M, Ali H, Hazrat A, Hayat K, Keerio AA, Ihsan M, Laiq M, Ullah S, Fahad S, Khan A, Khan AH, Akbar A, Yang X (2019) Drought tolerance improvement in plants: an endophytic bacterial approach. Appl Microbiol Biotechnol 103(18):7385–7397. https://doi.org/10.1007/s00253-019-10045-4

Vaishnav A, Shukla AK, Sharma A et al (2019) Endophytic bacteria in plant salt stress tolerance: current and future prospects. J Plant Growth Regul 38:650–668. https://doi.org/10.1007/s00344-018-9880-1

Vandenkoornhuyse P, Quaiser A, Duhamel M, Le Van A, Dufresne A (2015) The importance of the microbiome of the plant holobiont. New Phytol 206:1196–1206. https://doi.org/10.1111/nph.13312

Wang L, Gong L, Gan D, Li X, Yao J, Wang L et al (2022) Diversity, function and assembly of the *Trifolium repens* L. root-associated microbiome under lead stress. J Hazard Mater 438:129510. https://doi.org/10.1016/j.jhazmat.2022.129510

Wang L, Jiao Y, Bi Y, Hu Y, Jiang Y, Wang S, Wang S (2023) Nodulation number tempers the relative importance of stochastic processes in the assembly of soybean root-associated communities. ISME Commun 3(1):89. https://doi.org/10.1038/s43705-023-00296-8

Wedhastri S, Fardhani DM, Kabirun S, Widada J, Widianto D, Evizal R, Prijambada ID (2015) Legume nodulating bacterium, *Achromobacter xylosoxidans*, found in tropical shrub agroecosystem. Indones J Biotechnol 18:161–167. https://doi.org/10.22146/ijbiotech.7879

Wigley K, Moot D, Wakelin S, Laugraud A, Blond C, Seth K, Ridgway H (2017) Diverse bacterial taxa inhabit root nodules of lucerne (*Medicago sativa* L.) in New Zealand pastoral soils. Plant Soil 420:1–10. https://doi.org/10.1007/s11104-017-3395-6

Wu X, Rensing C, Han D, Xiao KQ, Dai Y, Tang Z et al (2022) Genome-resolved metagenomics reveals distinct phosphorus acquisition strategies between soil microbiomes. mSystems 7:e01107–e01121. https://doi.org/10.1128/msystems.01107-21

Xiao X, Chen W, Zong L, Yang J, Jiao S, Lin Y, Wang E, Wei G (2017) Two cultivated legume plants reveal the enrichment process of the microbiome in the rhizocompartments. Mol Ecol 26(6):1641–1651. https://doi.org/10.1111/mec.14027

Xiong C, Singh BK, He JZ, Han YL, Li PP, Wan LH et al (2021) Plant developmental stage drives the differentiation in ecological role of the maize microbiome. Microbiome 9:171. https://doi.org/10.1186/s40168-021-01118-6

Yan J, Han XZ, Ji ZH, Li Y, Wang ET, Xie ZH, Chen WF (2014) Abundance and diversity of soybean-nodulating rhizobia in black soil are impacted by land use and crop management. Appl Environ Microbiol 80:5394–5402. https://doi.org/10.1128/AEM.01135-14

Yang J, Lan L, Jin Y, Yu N, Wang D, Wang E (2022) Mechanisms underlying legume-*Rhizobium* symbioses. J Integr Plant Biol 64:244–267. https://doi.org/10.1111/jipb.13207

Yuan J, Ruan Y, Wang B, Zhang J, Waseem R, Huang Q, Shen Q (2013) Plant growth-promoting rhizobacteria strain *Bacillus amyloliquefaciens* NJN-6-enriched bio-organic fertilizer suppressed *Fusarium* wilt and promoted the growth of banana plants. J Agric Food Chem 61(16):3774–3780. https://doi.org/10.1021/jf400038z

Zhang YM, Li Y, Chen WF, Wang ET, Tian CF, Li QQ, Zhang YZ, Sui XH, Chen WX (2011) Biodiversity and biogeography of rhizobia associated with soybean plants grown in the North China Plain. Appl Environ Microbiol 77:6331–6342. https://doi.org/10.1128/AEM.00542-11

Zhang B, Du N, Li Y, Shi P, Wei G (2018) Distinct biogeographic patterns of rhizobia and non-rhizobial endophytes associated with soybean nodules across China. Sci Total Environ 643:569–578. https://doi.org/10.1016/j.scitotenv.2018.06.240

Zhang J, Cook J, Nearing JT, Zhang J, Raudonis R, Glick BR, Langille MGI, Cheng Z (2021) Harnessing the plant microbiome to promote the growth of agricultural crops. Microbiol Res 245:126690. https://doi.org/10.1016/j.micres.2020.126690

Zhao L, Xu Y, Lai X (2018) Antagonistic endophytic bacteria associated with nodules of soybean (*Glycine max L.*) and plant growth-promoting properties. Braz J Microbiol 49(1):269–278. https://doi.org/10.1016/j.bjm.2017.06.007

Multi-omics Approaches for Deciphering Plant-Microbe Dynamics in Sustainable Rice Cultivation

Nikita Bisht, Ritu Dixit, and Puneet Singh Chauhan

1 Introduction

Rice (*Oryza sativa* L.) is one of the most important cereal crops globally, feeding more than half of the world's population. Its cultivation sustains livelihoods, supports food security, and drives economies in many regions, particularly in Asia. However, the growing demands for rice production, combined with environmental challenges like climate change, nutrient depletion, salinization, and the spread of pests and diseases, have created an urgent need for sustainable agricultural practices (Kumar et al. 2021). Conventional high-input farming systems, which rely heavily on synthetic fertilizers and pesticides, have led to soil degradation, biodiversity loss, and increased greenhouse gas emissions, threatening long-term productivity and ecosystem health (Bisht and Chauhan 2020; Singh et al. 2024).

A promising alternative lies in harnessing the potential of plant-associated microbes, which naturally mediate numerous processes essential for plant growth and health (Bisht and Chauhan 2023; Ansari et al. 2024). These microorganisms inhabit the rhizosphere (root-associated soil), endosphere (plant tissues), and phyllosphere (aboveground plant parts). Through diverse mechanisms such as nitrogen fixation, phosphate solubilization, stress alleviation, and disease suppression, these microbes significantly contribute to plant performance under both optimal and adverse conditions (Bisht et al. 2024). In rice cultivation, these microbial communities play a significant role (Perera et al. 2021; Ansari et al. 2024). Rice fields are distinct ecosystems defined by periodic flooding, low oxygen levels, and fluctuating redox conditions, which significantly influence the diversity and activity of microbial communities. Understanding the intricate interactions between rice plants and

N. Bisht · R. Dixit · P. S. Chauhan (✉)
Microbial Technologies Division, CSIR-National Botanical Research Institute (CSIR-NBRI), Rana Pratap Marg, Lucknow, India
e-mail: puneet@nbri.res.in

Renu et al. (eds.), *Meta-omics in Crop Improvement*,
https://doi.org/10.1007/978-981-95-2294-1_12

their associated microbes is vital for devising innovative strategies to improve nutrient acquisition, enhance resilience to stress, and reduce the ecological footprint of rice farming.

Although significant advancements have been made in understanding plant-microbe interactions, the intricate and dynamic nature of these relationships presents considerable challenges. Conventional approaches, while valuable, often fall short in capturing the complete range of microbial diversity and functions, particularly under changing environmental conditions. In this context, multi-omics technologies have emerged as a revolutionary approach (Iqbal et al. 2021). Multi-omics refers to the use of high-throughput omics approaches including genomics, transcriptomics, proteomics, and metabolomics, to comprehensively analyse biological systems. By integrating these datasets, researchers can uncover the molecular and functional underpinnings of plant-microbe dynamics with unprecedented depth and precision.

The application of multi-omics technologies in rice cultivation is particularly promising (Iqbal et al. 2021). Recent studies have demonstrated their potential to identify microbial consortia with beneficial traits, such as nutrient cycling, pathogen suppression, and abiotic stress tolerance (Zargar et al. 2022; Sedeek et al. 2023; Bisht et al. 2023; Ansari et al. 2024). By decoding the molecular mechanism between rice plants and their associated microbiomes, these approaches can inform the development of targeted biofertilizers, biopesticides, and stress-mitigating strategies. Furthermore, multi-omics can guide the selection and engineering of rice varieties that optimally recruit and interact with beneficial microbes, thereby enhancing productivity and sustainability.

This chapter aims to explore the application of meta-omics technologies in deciphering the complexities of rice-microbe interactions. It provides an overview of the tools and methodologies involved, highlights recent advances in the field, and discusses the implications for sustainable rice cultivation. By bridging fundamental research with practical applications, the chapter seeks to advance our understanding of plant-microbe dynamics and their role in addressing the dual challenges of food security and environmental sustainability.

2 Multi-Omics for Deciphering Molecular Interconnections

Multi-omics involves multiple advanced omics technologies, including genomics, transcriptomics, proteomics, and metabolomics, to provide a comprehensive understanding of biological systems (Fig. 1). By integrating different layers of molecular data, this approach allows researchers to uncover the intricate interactions between genes, RNA, proteins, and metabolites, providing a comprehensive view of the functional architecture of living organisms (Satrio et al. 2024). Each of these approaches addresses specific aspects of biological systems, yet they are interconnected and complementary. Together, these approaches enable the exploration of genetic, transcriptional, proteomic, and metabolic networks that form the

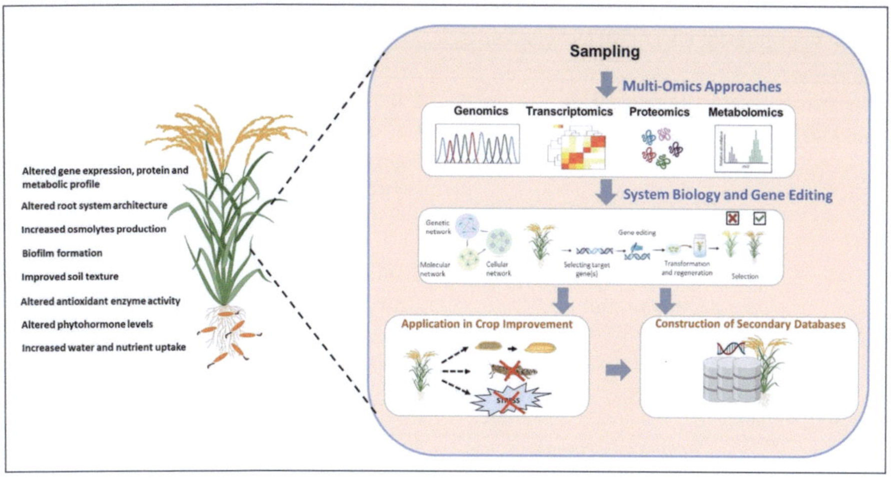

Fig. 1 Schematic representation elucidating the integration of multi-omics and system biology approaches, either as discrete entities or in a combinatorial framework, to unravel the intricate mechanisms driving microbe-induced enhancements in rice productivity

foundation of plant-microbe interactions and can be used to improve growth and productivity of rice plants (Table 1) (Kumar et al. 2023).

2.1 Genomics: The Foundation of Multi-omics

Genomics serves as the foundation of multi-omics studies, providing a complete genetic blueprint of an organism and insights into its functional capabilities. Advances in sequencing technologies, such as next-generation sequencing (NGS) and third-generation long-read sequencing, have made it possible to generate high-resolution genomics maps (Hu et al. 2021). These maps are instrumental in identifying genetic variants, regulatory elements, and structural variations that influence phenotypic traits.

In the context of plant-microbe interactions, genomics is applied to both plant genomes and microbial genomes to elucidate the genetic basis of symbiotic or associative relationships. In rice cultivation, microbial metagenomics, a genomics analysis of entire microbial communities, has been instrumental in identifying key microbial species and their functional genes (Krishnamoorthy et al. 2021). For example, studies have revealed the presence of plant growth-promoting rhizobacteria belonging to *Actinobacteria*, *Bacteroidetes*, *Firmicutes*, and proteobacteria and plant growth-promoting fungi (*Trichoderma*) in the rhizosphere and harbour genes that encode enzymes for nutrient cycling, phytohormone production, and stress tolerance (Mulani et al. 2021; Cheng et al. 2023).

Table 1 Molecular approaches employed to investigate the contribution of microbes to rice growth improvement under different conditions

Molecular approach	Microorganisms	Condition	Reference
Genomics	*Aquabacter pokkalii*	Nitrogen-poor brackish environments	Sunithakumari et al. (2024)
	Pseudomonas promysalinigenes	Salinity	Song et al. (2023)
	Klebsiella sp.	Salinity	Girma et al. (2022)
	Pseudomonas psychrotolerans	Heavy metal	Kang et al. (2020)
	Bacillus altitudinis	Drought	Narayanasamy et al. (2023)
Transcriptomics	*Bacillus velezensis*	Brown planthopper	Rashid et al. (2018)
	Stenotrophomonas rhizophila	Heavy metal	Gao et al. (2020)
	Herbaspirillum seropedicae Azospirillum brasilense	Growth promotion	Wiggins et al. (2022)
	Bacillus amyloliquefaciens	Salinity	Chauhan et al. (2019)
	Microbacterium ginsengiterrae	Salinity-alkalinity	Ji et al. (2024)
	Bacillus subtilis	Growth promotion	Xie et al. (2015)
Proteomics	*Bacillus subtilis*	Rhizoctonia solani	Durgadevi et al. (2021)
	Bacillus sp.	Salinity	Chompa et al. (2024)
	Bacillus megaterium	Growth promotion	Prabhukarthikeyan et al. (2023)
	Methylobacterium oryzae	Salinity	Roy et al. (2022)
	Pseudomonas alcaliphila	Salinity	Zeng et al. (2023)
Metabolomics	*Enterobacter ludwigii*	Drought	Zhang et al. (2024)
	Bacillus amyloliquefaciens	Drought, salinity	Joshi et al. (2023)
	Microbacterium ginsengiterrae	Salinity-slkalinity	Ji et al. (2024)
	Enterobacter	*Burkholderia glumae*	Peñaloza Atuesta et al. (2020)
	Bacillus amyloliquefaciens	Nutrient deficiency	Bisht et al. (2020)

2.2 Transcriptomics: Unveiling Gene Expression Dynamics

Transcriptomics focuses on the analysis of RNA molecules to determine gene expression patterns under specific conditions. In rice-microbe interactions, transcriptomics enables the identification of genes activated during symbiosis, nutrient exchange, or stress response (Chauhan et al. 2019). For instance, transcriptomics studies have shown that rice roots upregulate specific transporter genes in the presence of phosphate-solubilizing bacteria, facilitating enhanced phosphorus uptake (Mitra et al. 2023). Similarly, comparative transcriptomics studies of *Pseudomonas ogarae*

F113 in liquid medium versus the rhizosphere of alfalfa plants highlight the rhizosphere as a key driver of transcriptional changes in the bacterium. Transcriptomics analyses of F113 mutants for *amrZ* and *fleQ* genes revealed that *fleQ* mutations significantly alter the transcriptional profile compared to *amrZ* mutations. These changes involve genes linked to lifestyle transitions and adaptation to plant environments, affecting bacterial motility, biofilm formation, secretion systems, and metabolic pathways for nitrogen, sulfur, and amino acids. Notably, AmrZ and FleQ regulate these functions antagonistically, offering insights into the molecular mechanisms enabling PGPR to colonize and adapt to the rhizosphere effectively (Blanco-Romero et al. 2022).

Transcriptomics also offers a dynamic view of plant-microbe interactions under varying environmental conditions, such as drought, salinity, or pathogen attack (Pal et al. 2022). Further, by integrating transcriptomics data with genomics information, active regulatory networks and key transcription factors or non-coding RNAs driving cellular responses can be identified.

2.3 Proteomics: Connecting Genotype to Phenotype

Proteomics involves comprehensive study of proteins, which are the key functional components responsible for driving cellular activities. It investigates the structure, roles, and levels of proteins in cells or organisms (Mustafa et al. 2021). As proteins mediate biological processes, proteomics analysis is crucial for understanding how gene expression translates into observable traits. Comprehensive proteome analysis significantly enhances the annotation of plant genomes by enabling the validation and precise assignment of open reading frames (ORFs) to annotated or putative genes. Additionally, it facilitates the confirmation of alternative splicing patterns, thereby refining our understanding of gene structure and functional diversity within the genome (Diwan et al. 2022). Within the realm of plant-microbe interactions, proteomics serves as a pivotal tool for unravelling the intricate molecular dialogue, offering profound insights into the repertoire of proteins synthesized or secreted by both plants and microbial partners (Jain et al. 2021).

The advancements in mass spectrometry (MS) have revolutionized the identification and quantification of nearly the entire spectrum of expressed proteins within a biological system. This sophisticated analytical approach encompasses three fundamental steps: (1) the ionization of proteins and the generation of gas-phase ions, (2) the separation of these ions according to their mass-to-charge (m/z) ratio, and (3) the precise detection of the resulting ions (Diwan et al. 2022). These innovations have paved the way for high-resolution proteomics analyses, offering unparalleled insights into protein expression and functionality. For instance, Zhang et al. (2020) conducted a tandem mass tag (TMT)-based quantitative proteomics analysis to elucidate the proteomics alterations occurring during both compatible and incompatible interactions between *Xanthomonas oryzae* pv. oryzae and rice. This approach provided a detailed investigation into the dynamic protein expression profiles underpinning these distinct host-pathogen interaction scenarios. Thus, these tools enable the quantification and identification of proteins, providing a deeper understanding of the molecular networks underlying plant-microbe interactions.

2.4 Metabolomics: Capturing Biochemical Endpoints

Metabolites reflect the ultimate outcome of gene expression and enzymatic activities, serving as direct indicators of physiological states. Metabolomics focuses on the qualitative and quantitative analysis of metabolites in living systems, examining their dynamic responses to environmental changes (Gupta et al. 2022). The detection, analysis, and identification of metabolites form the foundation of any metabolomics study. These metabolites play critical roles in plant-microbe interactions, acting as signals, growth stimulators, or stress mitigators (Gupta et al. 2022; Dixit et al. 2023).

Rice roots, for instance, secrete a diverse array of metabolites, including sugars, amino acids, and secondary metabolites, that attract specific microbial taxa (Joshi et al. 2023). Conversely, microbes produce metabolites such as auxins, gibberellins, and siderophores, which enhance plant growth and nutrient acquisition. Recently, Ansari et al. (2024) identified metabolites like β-D-galactopyranoside and β-D-glucopyranose as potential contributors to energy metabolism, signaling, and plant defense against *Rhizoctonia solani* infection in rice. Furthermore, metabolomics has been instrumental in clarifying the distinctions and connections between disease and symbiosis. Studies on interactions between *Arabidopsis*, plant growth-promoting bacteria, and other microbes have demonstrated the effectiveness of integrating various omics datasets to explore these complex communications (Gupta et al. 2022). Advanced platforms such as gas chromatography-mass spectrometry (GC-MS) and liquid chromatography-mass spectrometry (LC-MS) allow for detailed profiling of these metabolites. By mapping metabolic pathways, key biochemical exchanges that underpin beneficial plant-microbe interactions can be identified.

Although metabolomics is an expanding field in plant sciences, its application in studying plant-microbe interactions remains less developed compared to other omics approaches. This limitation presents significant opportunities to advance our understanding of these interactions. However, leveraging metabolomics in plant-microbe studies comes with challenges, such as identifying the source of analyzed metabolites, deciphering the metabolic complexity of interacting organisms, and integrating metabolomics data with other omics layers, including transcriptomics, proteomics, and phenomics.

3 Rice Microbiome

The microbiome associated with rice represents a diverse and dynamic community of microorganisms that inhabit various ecological niches, including the rhizosphere, endosphere, and phyllosphere (Kim and Lee 2020). These microbes play an integral role in plant growth, development, and resilience by mediating nutrient acquisition, suppressing pathogens, and improving stress tolerance. Characterizing the rice microbiome and its functional contributions is essential for leveraging its potential to support sustainable rice cultivation.

3.1 Diversity and Composition of the Rice Microbiome

Rice plants host a vast array of microorganisms, including bacteria, fungi, archaea, and viruses. The rhizosphere, or the soil region immediately surrounding the roots, harbors the highest microbial diversity due to the availability of root exudates that serve as a nutrient source. Key bacterial groups include nitrogen-fixing diazotrophs such as *Rhizobium*, *Azospirillum*, and *Herbaspirillum*, as well as phosphate-solubilizing bacteria like *Pseudomonas* and *Bacillus*. Similarly, arbuscular mycorrhizal fungi (AMF) and endophytic fungi such as *Trichoderma* are prominent in rice ecosystems (Kim and Lee 2020). The endosphere, comprising the internal tissues of rice plants, hosts endophytes that establish symbiotic relationships, enhancing nutrient uptake and stress resistance. The phyllosphere, which includes aboveground plant parts, is primarily inhabited by bacteria and fungi that play roles in pathogen defense and abiotic stress tolerance. The composition of these microbial communities varies depending on soil type, environmental conditions, and the developmental stage of the rice plant (Santos-Medellín et al. 2017).

3.2 Functional Roles of the Rice Microbiome

Microbes associated with rice plants provide numerous ecosystem services that are vital for plant health and productivity. Plant-associated microbes produce phytohormones such as auxins, cytokinins, and gibberellins, which influence plant growth and development. Some microbes can also modulate ethylene levels in plants by producing enzymes such as ACC deaminase, thereby alleviating stress-induced ethylene accumulation (Kim and Lee 2020). Microbes contribute to nutrient acquisition by mobilizing essential elements such as nitrogen, phosphorus, and potassium. Nitrogen-fixing bacteria convert atmospheric nitrogen into forms that plants can assimilate, while phosphate-solubilizing bacteria release inorganic phosphorus from soil-bound complexes. Potassium-solubilizing microorganisms also enhance the bioavailability of this vital nutrient (Mahala et al. 2020). These beneficial microorganisms enhance resilience of rice plants to abiotic stresses such as drought, salinity, and heavy metal toxicity (Table 2) (Zhao et al. 2023).

3.3 Factors Shaping the Rice Microbiome

Several factors influence the structure and functionality of the rice microbiome, including (1) Soil type, moisture, temperature, and pH significantly affect microbial diversity and activity. For example, acidic soils may favor the growth of specific bacterial genera, while salinity impacts the composition of both bacteria and fungi. (2) Different rice cultivars exhibit distinct microbiome profiles due to variations in root exudate composition and other physiological traits. Selective breeding and genetic engineering of rice varieties can enhance their ability to recruit beneficial microbes. (3) Irrigation methods, fertilizer use, and crop rotation schemes influence

Table 2 Beneficial microbial associations with rice and their mechanism of action in mitigating diverse biotic and abiotic stresses

Stress	PGPR	Mechanism	Reference
Rhizoctonia solani (sheath blight)	*Bacillus subtilis*, *Trichoderma* sp.	Secretes chitinases and glucanases, produces antimicrobial peptides, and enhances plant resistance	Datta et al. (2021)
Fusarium oxysporum (wilt disease)	*Pseudomonas fluorescens*, *Bacillus amyloliquefaciens*	Induces systemic resistance and produces antifungal metabolites	Damodaran et al. (2020)
Magnaporthe oryzae (rice blast)	*Pseudomonas* sp.	Induces plant defense responses, produces reactive oxygen species	Oliveira-Garcia et al. (2024)
Xanthomonas oryzae (bacterial blight)	*Pseudomonas*	Produces antibiotics, induces systemic resistance, and promotes plant growth	Ngalimat et al. (2021)
Salinity	*Pseudomonas promysalinigenes*	Enhances salt tolerance by producing osmolytes and improves nutrient uptake	Ren et al. (2024)
Drought	*Bacillus amyloliquefaciens*, *Azospirillum brasilense*	Produces abscisic acid to regulate water loss and protects plants from oxidative stress	Chieb and Gachomo (2023)
Heavy metal toxicity	*Pseudomonas putida*, *Bacillus thuringiensis*,	Binds to heavy metals and reduces their uptake, promoting plant growth	Zhu et al. (2024)
Osmotic stress	*Azotobacter chroococcum*, *Rhizobium* sp.	Produces osmolytes like trehalose and enhances water retention in plants	AbuQamar et al. (2024)
Nutrient deficiency (nitrogen)	*Rhizobium* sp.	Fixes atmospheric nitrogen and increases nitrogen uptake by plants	Akhtar et al. (2024)
Nutrient deficiency (phosphorus)	*Bacillus subtilis*, *Pseudomonas fluorescens*	Solubilizes bound phosphorus and makes it available for plant uptake	Pan and Cai (2023)

microbial communities in rice fields. For instance, organic amendments such as compost or biochar can enhance the abundance of beneficial microbes (Wu et al. 2018).

The exploration of the rice microbiome offers immense potential for advancing sustainable agriculture. By leveraging meta-omics technologies to decode the complexities of microbial communities, innovative strategies to enhance nutrient use efficiency, improve stress resilience, and mitigate greenhouse gas emissions in rice cultivation can be developed.

4 Microbial Consortia and Functional Traits in Rice Cultivation

Microbial consortia have gained significant attention for their potential to enhance sustainable agricultural practices. These consortia consist of a combination of multiple fungal and bacterial species or strains, offering distinct advantages over the use of single-species or single-strain applications (Santoyo et al. 2021). Microorganisms collaborate to create synergistic enzyme systems, within microbial community and different enzymes together to catalyze and facilitate the substrates breakdown (Dixit et al. 2023). Moreover, microbial consortia affect the texture of soil and are involved in its health and function, by contributing to the decomposition of organic matter, toxin removal, nutrient cycling, and disease suppression and C, P, and N cycling. Recent meta-omics advancements have identified key microbial functional traits critical for rice growth (Nunes et al. 2024).

5 Data Integration and Computational Challenges

Microbes are important for ecosystem dynamics and sustainable agriculture and are influenced by several factors like host characteristics, environmental conditions, and human activities. The rapid advancements in high-throughput omics technologies have revolutionized the study of biological systems. However, the true potential of these technologies lies in their integration, which enables a holistic understanding of cellular and organismal functions (Satrio et al. 2024). Integrative data analysis combines information from various omics layers (genomics, transcriptomics, proteomics, metabolomics, and others) bridging the gaps between molecular insights and system-level processes (Fig. 1). Simultaneously, systems biology leverages integrated omics datasets to build computational models that simulate biological systems. These models capture the dynamic, multifaceted interactions that govern cellular behavior, providing predictive insights into responses to environmental or genetic perturbations. Moreover, translating findings from meta-omics studies into practical applications requires robust computational tools to analyze interactions and predict microbial functionality within specific agricultural contexts (Jain et al. 2024; Kimotho and Maina 2024). Thus, advancements in computational power, machine learning algorithms, and user-friendly bioinformatics platforms are progressively being used and have significant potential for enhancing agricultural practices by enhancing plant-microbe interactions.

6 Challenges and Future Perspectives

The application of multi-omics technologies in deciphering plant-microbe interactions has significantly advanced our understanding of these intricate relationships. However, several challenges must be addressed to maximize their potential in sustainable rice cultivation. One of the most significant challenges is integrating diverse omics datasets, as each dataset delivers distinct yet complementary insights, but the lack of standardized analytical pipelines and tools often hampers their combined interpretation. Furthermore, assigning functional roles to genes, proteins, and metabolites in complex microbial communities remains difficult due to incomplete reference databases. Multi-omics analyses involve sophisticated equipment and expertise, which can be prohibitively expensive, particularly for resource-limited regions where rice cultivation is most prevalent. Moreover, the resolution of these technologies may not yet capture rare microbial taxa or low-abundance metabolites that could be critical in plant-microbe interactions.

While multi-omics identifies a wealth of microbial taxa and functions, many of these microorganisms remain unculturable under laboratory conditions. This limitation hinders efforts to validate their roles experimentally and develop them into microbial inoculants. The transition from controlled laboratory studies to field applications poses another challenge. Microbial interactions observed in the lab may not replicate under varying environmental conditions, such as fluctuating soil moisture, temperature, and nutrient levels. The deployment of microbial consortia or genetically engineered microbes faces regulatory scrutiny and public acceptance. Addressing biosafety concerns and demonstrating the long-term benefits of such interventions are critical for widespread adoption.

To overcome these challenges and fully harness the potential of multi-omics to investigate microbe-mediated enhancements in rice cultivation necessitate a multidisciplinary approach combining advances in technology, ecology, and agronomy. Developing robust computational frameworks for integrating multi-omics datasets will be crucial for gaining holistic insights into rice-microbe dynamics. Machine learning and artificial intelligence (AI) tools offer promising avenues for identifying patterns and predicting functional interactions within large datasets. Innovations in sequencing technologies and analytical tools are expected to reduce the costs associated with meta-omics studies. Portable sequencing devices and simplified workflows could make these technologies accessible in developing countries.

Moreover, advances in synthetic biology can enable the design of synthetic microbial consortia with optimized functional traits. These engineered communities could be tailored to specific rice cultivars and environmental conditions, enhancing their efficacy as biofertilizers and biocontrol agents. The identification of plant genetic traits that enhance beneficial plant-microbe interactions offers an exciting opportunity for breeding microbiome-friendly rice varieties. Marker-assisted selection and gene-editing technologies, such as CRISPR-Cas9, could accelerate the development of these varieties. Comprehensive field trials across diverse agroecological zones are essential for evaluating the performance and stability of microbial inoculants and microbiome-optimized rice varieties. Efforts must be made to address potential environmental risks associated with the introduction of novel microbial strains or genetically engineered crops.

The integration of multi-omics technologies into rice research represents a paradigm shift in sustainable agriculture. By decoding the molecular networks governing plant-microbe interactions, these tools have the potential to enhance nutrient use efficiency, reduce dependency on chemical inputs, and mitigate the impacts of climate change. However, achieving these goals will require a concerted effort from researchers, policymakers, and industry stakeholders. Interdisciplinary collaborations and investments in capacity building will be pivotal in translating scientific discoveries into measurable outcomes for farmers and ecosystem. Furthermore, transparent communication of research findings and the benefits of these technologies can help build public trust and acceptance. By addressing the challenges and seizing emerging opportunities, meta-omics-driven strategies can play a central role in ensuring global food security and environmental sustainability, particularly for rice, a staple crop that sustains more than half of the global population.

7 Conclusion

The application of multi-omics technologies to study plant-microbe interactions has revolutionized our understanding of the complex relationships between rice plants and their associated microbial communities. By providing a comprehensive and integrative view of the genetic, functional, and biochemical processes that govern these interactions, these approaches have uncovered novel insights into microbial roles in nutrient cycling, stress tolerance, and disease suppression. These technologies are poised to drive the development of innovative strategies for enhancing sustainable rice cultivation. Microbial consortia tailored to improve nutrient acquisition, resilience to environmental stresses, and pest management offer promising alternatives to conventional chemical inputs, thereby supporting the principles of agroecological intensification. Additionally, the discovery of microbial traits that can be harnessed through genetic and genomics interventions opens avenues for breeding microbiome-friendly rice varieties that naturally recruit beneficial microbes, enhancing productivity while maintaining ecological balance. However, several challenges remain in translating meta-omics discoveries into practical, field-level applications. Data integration across multiple omics layers, high costs, the inability to culture many microbes, and variable performance in diverse environmental conditions all present hurdles to large-scale adoption. Nevertheless, ongoing advancements in multi-omics integration, cost-reduction technologies, and field-based validation will likely overcome these challenges in the coming years.

Conflict of Interest Authors declare no conflict of interest.

References

AbuQamar SF, El-Saadony MT, Saad AM, Desoky ESM, Elrys AS, Abd El-Mageed TA, El-Tarabily KA (2024) Halotolerant plant growth-promoting rhizobacteria improve soil fertility and plant salinity tolerance for sustainable agriculture—A review. Plant Stress 12: 100482.

Akhtar K, Ain N, Vara Prasad PV, Naz M, Aslam MM, Djalovic I, Riaz M, Ahmad S, Varshney RK, He B, Wen R (2024) Physiological, molecular, and environmental insights into plant nitrogen uptake, and metabolism under abiotic stresses. The Plant Genome:e20461

Ansari MM, Bisht N, Singh T, Mishra SK, Anshu A, Singh PC, Chauhan PS (2024) *Bacillus amyloliquefaciens* modulate autophagy pathways to control *Rhizoctonia solani* infection in rice. Plant Physiol Biochem 20:109317

Bisht N, Chauhan PS (2020) Excessive and disproportionate use of chemicals cause soil contamination and nutritional stress. Soil contamination-threats and sustainable solutions:1–10

Bisht N, Chauhan PS (2023) Harvesting tomorrow: embracing new innovations in microbial biotechnology for smart farming and sustainable agriculture. Int J Con Microbio 9(2)

Bisht N, Mishra SK, Chauhan PS (2020) *Bacillus amyloliquefaciens* inoculation alters physiology of rice (*Oryza sativa* L. var. IR-36) through modulating carbohydrate metabolism to mitigate stress induced by nutrient starvation. Int J Biol Macromol 143:937–951

Bisht N, Anshu A, Singh PC, Chauhan PS (2023) Comprehensive analysis of *OsJAZ* gene family deciphers rhizobacteria-mediated nutrient stress modulation in rice. Int J Biol Macromol 253:126832

Bisht N, Singh T, Ansari MM, Chauhan PS (2024) Tailored microbial inoculants in nutrient recycling and soil health maintenance. In the potential of microbes for a circular economy. Academic Press, pp 201–218

Blanco-Romero E, Durán D, Garrido-Sanz D, Rivilla R, Martín M, Redondo-Nieto M (2022) Transcriptomic analysis of *Pseudomonas ogarae* F113 reveals the antagonistic roles of AmrZ and FleQ during rhizosphere adaption. Microb Genom 8(1):000750

Chauhan PS, Lata C, Tiwari S, Chauhan AS, Mishra SK, Agrawal L, Chakrabarty D, Nautiyal CS (2019) Transcriptional alterations reveal *Bacillus amyloliquefaciens*-rice cooperation under salt stress. Sci Rep 9(1):11912

Cheng Z, Zheng Q, Shi J, He Y, Yang X, Huang X, Wu L, Xu J (2023) Metagenomic and machine learning-aided identification of biomarkers driving distinctive Cd accumulation features in the root-associated microbiome of two rice cultivars. ISME Commun 3(1):14

Chieb M, Gachomo EW (2023) The role of plant growth promoting rhizobacteria in plant drought stress responses. BMC Plant Biol 23:407

Chompa SS, Zuan AT, Amin AM, Hun TG, Ghazali AH, Sadeq BM, Akter A, Rahman ME, Rashid HO (2024) Growth and protein response of rice plant with plant growth-promoting rhizobacteria inoculations under salt stress conditions. Dermatol Int Microbiol:1–8

Damodaran T, Rajan S, Muthukumar M, Gopal R, Yadav K, Kumar S et al (2020) Biological management of banana Fusarium wilt caused by *Fusarium oxysporum* f. sp. cubense tropical race 4 using antagonistic fungal isolate CSR-T-3 (*Trichoderma reesei*). Front Microbiol 11:595845

Datta S, Sarkar M, Chowdhury A, Rakwal R, Agrawal GK, Sarkar AA (2021) Comprehensive insight into the biology of *Rhizoctonia solani* AG1-IA Kühn, the causal organism of the sheath blight disease of rice. J Plant Pathol 104:79–98

Diwan D, Rashid MM, Vaishnav A (2022) Current understanding of plant-microbe interaction through the lenses of multi-omics approaches and their benefits in sustainable agriculture. Microbiol Res 265:127180

Dixit R, Bisht N, Misra S, Gupta SC, Chauhan PS (2023) Bacillus consortia modulate transcriptional and metabolic machinery of Arabidopsis plants for salt tolerance. Curr Microbiol 80(2):77

Durgadevi D, Harish S, Manikandan R, Prabhukarthikeyan SR, Alice D, Raguchander T (2021) Proteomic profiling of defense/resistant genes induced during the tripartite interaction of *Oryza sativa*, *Rhizoctonia solani* AG1-1A, and *Bacillus subtilis* against rice sheath blight. Physiol Mol Plant Path 115:101669

Gao J, Wu S, Liu Y, Wu S, Jiang C, Li X, Wang R, Bai Z, Zhuang G, Zhuang X (2020) Characterization and transcriptomic analysis of a highly Cr (VI)-resistant and-reductive plant-growth-promoting rhizobacterium *Stenotrophomonas rhizophila* DSM14405T. Environ Pollut 263:114622

Girma B, Panda AN, Roy PC, Ray L, Mohanty S, Chowdhary G (2022) Molecular, biochemical, and comparative genome analysis of a rhizobacterial strain *Klebsiella* Sp. KBG6. 2 imparting salt stress tolerance to *Oryza sativa* L. Environ Exp Bot 203:105066

Gupta S, Schillaci M, Roessner U (2022) Metabolomics as an emerging tool to study plant–microbe interactions. Emerg Topic Life Sci 6(2):175–183
Hu T, Chitnis N, Monos D, Dinh A (2021) Next-generation sequencing technologies: an overview. Hum Immunol 82(11):801–811
Iqbal Z, Iqbal MS, Khan MI, Ansari MI (2021) Toward integrated multi-omics intervention: rice trait improvement and stress management. Front Plant Sci 12:741419
Jain A, Singh HB, Das S (2021) Deciphering plant-microbe crosstalk through proteomics studies. Microbiol Res 242:126590
Jain A, Sarsaiya S, Singh R, Gong Q, Wu Q, Shi J (2024) Omics approaches in understanding the benefits of plant-microbe interactions. Front Microbiol 15:1391059
Ji H, Qi Y, Zhang X, Yang G (2024) Transcriptomics and metabolomics analysis revealed the ability of *Microbacterium ginsengiterrae* S4 to enhance the saline-alkali tolerance of Rice (*Oryza sativa* L.) seedlings. Agronomy 14(4):649
Joshi H, Bisht N, Mishra SK, Prasad V, Chauhan PS (2023) *Bacillus amyloliquefaciens* modulate carbohydrate metabolism in rice-PGPR cross-talk under abiotic stress and phytohormone treatments. J Plant Growth Regul 42(7):4466–4483
Kang SM, Asaf S, Khan AL, Lubna, Khan A, Mun BG, Khan MA, Gul H, Lee IJ (2020) Complete genome sequence of *Pseudomonas psychrotolerans* CS51, a plant growth-promoting bacterium, under heavy metal stress conditions. Microorganisms 8(3):382
Kim H, Lee YH (2020) The rice microbiome: a model platform for crop holobiome. Phytobio J 4(1):5–18
Kimotho RN, Maina S (2024) Unraveling plant-microbe interactions: can integrated omics approaches offer concrete answers? J Exp Bot 75:1289–1313
Krishnamoorthy A, Gupta A, Sar P, Maiti MK (2021) Metagenomics of two gnotobiotically grown aromatic rice cultivars reveals genotype-dependent and tissue-specific colonization of endophytic bacterial communities attributing multiple plant growth promoting traits. World J Microbiol Biotechnol 37(4):59
Kumar N, Chhokar RS, Meena RP, Kharub AS, Gill SC, Tripathi SC, Gupta OP, Mangrauthia SK, Sundaram RM, Sawant CP, Gupta A (2021) Challenges and opportunities in productivity and sustainability of rice cultivation system: a critical review in Indian perspective. Cereal Res Commun 28:1–29
Kumar U, Raj S, Sreenikethanam A, Maddheshiya R, Kumari S, Han S, Kapoor KK, Bhaskar R, Bajhaiya AK, Gahlot DK (2023) Multi-omics approaches in plant–microbe interactions hold enormous promise for sustainable agriculture. Agronomy 13(7):1804
Mahala DM, Maheshwari HS, Yadav RK, Prabina BJ, Bharti A, Reddy KK, Kumawat C, Ramesh A (2020) Microbial transformation of nutrients in soil: an overview. Soil and Plant Functions, Rhizosphere Microbes, pp 175–211
Mitra D, Nayeri FD, Sansinenea E, Ortiz A, Bhatta BB, Adeyemi NO, Janeeshma E, Tawfeeq Al-Ani LK, Sharma SB, Boutaj H, Priyadarshini A (2023) Unraveling arbuscular mycorrhizal fungi interaction in rice for plant growth development and enhancing phosphorus use efficiency through recent development of regulatory genes. J Plant Nutr 46(13):3184–3220
Mulani R, Mehta K, Saraf M, Goswami D (2021) Decoding the mojo of plant-growth-promoting microbiomes. Physiol Mol Plant Path 115:101687
Mustafa G, Komatsu S (2021) Plant proteomic research for improvement of food crops under stresses: a review. Mol Omics 17(6):860–880
Narayanasamy S, Thankappan S, Kumaravel S, Ragupathi S, Uthandi S (2023) Complete genome sequence analysis of a plant growth-promoting phylloplane *Bacillus altitudinis* FD48 offers mechanistic insights into priming drought stress tolerance in rice. Genomics 115(1):110550
Ngalimat MS, Mohd Hata E, Zulperi D, Ismail SI, Ismail MR, Mohd Zainudin NAI et al (2021) Plant growth-promoting bacteria as an emerging tool to manage bacterial rice pathogens. Microorganisms 9:682
Nunes PS, Junior GVL, Mascarin GM, Guimarães RA, Medeiros FH, Arthurs S et al (2024) Microbial consortium of biological products: do they have a future? Biol Control 188:105439

Oliveira-Garcia E, Yan X, Oses-Ruiz M, de Paula S, Talbot NJ (2024) Effector-triggered susceptibility by the rice blast fungus *Magnaporthe oryzae*. New Phytol 241:1007–1020

Pal G, Bakade R, Deshpande S, Sureshkumar V, Patil SS, Dawane A, Agarwal S, Niranjan V, Prasanna Kumar MK, Vemanna RS (2022) Transcriptomic responses under combined bacterial blight and drought stress in rice reveal potential genes to improve multi-stress tolerance. BMC Plant Biol 22(1):349

Pan L, Cai B (2023) Phosphate-solubilizing bacteria: advances in their physiology, molecular mechanisms and microbial community effects. Microorganisms 11(12):2904

Peñaloza Atuesta GC, Murillo Arango W, Eras J, Oliveros DF, Méndez Arteaga JJ (2020) Rice-associated rhizobacteria as a source of secondary metabolites against *Burkholderia glumae*. Molecules 25(11):2567

Perera TA, Tirimanne S (2021) Role of microbial communities in sustainable rice cultivation. Role of microbial communities for sustainability:189–223

Prabhukarthikeyan SR, Parameswaran C, Sawant SB, Keerthana U, Yadav MK, Raghu S, Baite MS, Mahanty A, Panneerselvam P, Anandan A, Rath PC (2023) Unraveling the molecular basis of *Bacillus megaterium* interactions in rice for plant growth promotion through proteomics and gene expression. J Plant Growth Regul 42(5):2827–2839

Rashid MHO, Kim HJ, Yeom SI, Yu HA, Manir MM, Moon SS, Kang YJ, Chung YR (2018) *Bacillus velezensis* YC7010 enhances plant defenses against brown planthopper through transcriptomic and metabolic changes in rice. Front Plant Sci 9:1904

Ren L, Zhang Y, Zhou J, Wang G, Mo Y, Ling Y et al (2024) RL-WG26 mediated salt stress tolerance in rice seedlings: a new insight into molecular mechanisms. Plant Stress 11:100306

Roy Choudhury A, Roy SK, Trivedi P, Choi J, Cho K, Yun SH, Walitang DI, Park JH, Kim K, Sa T (2022) Label-free proteomics approach reveals candidate proteins in rice (*Oryza sativa* L.) important for ACC deaminase producing bacteria-mediated tolerance against salt stress. Environ Microbiol 24(8):3612–3624

Santos-Medellín C, Edwards J, Liechty Z, Nguyen B, Sundaresan V (2017) Drought stress results in a compartment-specific restructuring of the rice root-associated microbiomes. mBio 8:e00764–e00717

Santoyo G, Guzmán-Guzmán P, Parra-Cota FI, Santos-Villalobos SD, Orozco-Mosqueda MD, Glick BR (2021) Plant growth stimulation by microbial consortia. Agronomy 11(2):219

Satrio RD, Fendiyanto MH, Miftahudin M (2024) Tools and techniques used at global scale through genomics, transcriptomics, proteomics, and metabolomics to investigate plant stress responses at the molecular level. In: Molecular dynamics of plant stress and its management. Springer, Singapore, pp 555–607

Sedeek K, Zuccolo A, Fornasiero A, Weber AM, Sanikommu K, Sampathkumar S, Rivera LF, Butt H, Mussurova S, Alhabsi A, Nurmansyah N (2023) Multi-omics resources for targeted agronomic improvement of pigmented rice. Nature Food 4(5):366–371

Singh T, Bisht N, Bhowmick S, Chauhan PS (2024) Biofertilizer and its application for sustainable bioeconomy in agriculture. In: The potential of microbes for a circular economy. Academic Press, pp 185–200

Song J, Wang Y, Weng L, Ling Y, Zhang Y, Xue Y, Hu H, Ren L (2023) Complete genome sequence of *Pseudomonas promysalinigenes* strain RL-WG26, a PGPR isolated from the rice rhizosphere for studying the promotion of plant growth under salt stress. Phyto Frontiers 3(4):906–910

Sunithakumari VS, Menon RR, Suresh GG, Krishnan R, Rameshkumar N (2024) Characterization of a novel root-associated diazotrophic rare PGPR taxa, *Aquabacter pokkalii* sp. nov., isolated from pokkali rice: new insights into the plant-associated lifestyle and brackish adaptation. BMC Genomics 25(1):424

Wiggins G, Thomas J, Rahmatallah Y, Deen C, Haynes A, Degon Z, Glazko G, Mukherjee A (2022) Common gene expression patterns are observed in rice roots during associations with plant growth-promoting bacteria, *Herbaspirillum seropedicae* and *Azospirillum brasilense*. Sci Rep 12(1):8827

Wu Z, Liu Q, Li Z, Cheng W, Sun J, Guo Z, Li Y, Zhou J, Meng D, Li H, Lei P (2018) Environmental factors shaping the diversity of bacterial communities that promote rice production. BMC Microbiol 18:1–1

Xie S, Wu H, Chen L, Zang H, Xie Y, Gao X (2015) Transcriptome profiling of *Bacillus subtilis* OKB105 in response to rice seedlings. BMC Microbiol 15:1–4

Zargar SM, Mir RA, Ebinezer LB, Masi A, Hami A, Manzoor M, Salgotra RK, Sofi NR, Mushtaq R, Rohila JS, Rakwal R (2022) Physiological and multi-omics approaches for explaining drought stress tolerance and supporting sustainable production of rice. Front Plant Sci 12:803603

Zeng Q, Man X, Huang Z, Zhuang L, Yang H, Sha Y (2023) Effects of rice blast biocontrol strain *Pseudomonas alcaliphila* Ej2 on the endophytic microbiome and proteome of rice under salt stress. Front Microbiol 14:1129614

Zhang F, Zhang F, Huang L, Zeng D, Cruz CV, Li Z, Zhou Y (2020) Comparative proteomic analysis reveals novel insights into the interaction between rice and *Xanthomonas oryzae* pv. oryzae. BMC Plant Biol 20:1–4

Zhang J, Jia F, Song K, Wang F, Li J, Huang L, Qu T (2024) *Enterobacter ludwigii* b3 in the rhizosphere of wild rice assists cultivated rice in mitigating drought stress by direct and indirect methods. Biochem Biophys Res Commun 735:150489

Zhao J, Yu X, Zhang C, Hou L, Wu N, Zhang W, Wang Y, Yao B, Delaplace P, Tian J (2023) Harnessing microbial interactions with rice: strategies for abiotic stress alleviation in the face of environmental challenges and climate change. Sci Total Environ 28:168847

Zhu X, Beiyuan J, Ju W, Qiu T, Cui Q, Li C, Chao H, Shen Y, Fang L (2024) Inoculation with *Bacillus thuringiensis* reduces uptake and translocation of Pb/Cd in soil-wheat system: a life cycle study. Sci Total Environ 945:174032

Meta-omics in Tobacco: Spearheading for Trait Tailoring

K. Prabhakara Rao, P. Venkata Diwakar, Bishnupriya Gouda, K. Sarala, and M. Sheshu Madhav

1 Introduction

Tobacco stands out as a crucial commercial crop cultivated on a global scale, celebrated for its economic value and its significance in plant research, fostering economic growth for millions of farmers. The crop covers approximately 4.2 million hectares worldwide, producing around 5.78 million metric tons annually both in semiarid and rainfed regions (FAOSTAT 2022). In India, tobacco is grown across 4.33 lakh hectares in 15 states, generating approximately 760 million kilograms (ICAR-CTRI 2024). Despite occupying less than 1% of the world's agricultural land, tobacco sustains the livelihoods of 45.7 million people globally and contributes more than Rs. 12,000 crores in terms of foreign exchange. A distinct feature of tobacco production in India is that myriad styles of tobacco are cultivated under widely diverse agroecological situations. This diversity has resulted in a strong international demand for Indian tobacco, recognized for its different styles, qualities, and price ranges. India is the third largest exporter of unmanufactured tobacco in the world. Among the existing species, *Nicotiana tabacum* and *Nicotiana rustica* are two cultivated species in the country. The Indian tobacco is found to be highly polymorphic with wide range of morphological types and diversified utilities. Tobacco types are classified primarily based on the curing methods applied to the green leaf, which includes flue-cured Virginia (FCV) and non-flue cured Virginia (non-FCV) categories. The non-FCV types are further subdivided into air/sun-cured, smoke-cured, fire-cured, and pit-cured. Burley, oriental, and Maryland types are identified on the basis of leaf type. Based on the ultimate use of leaf, tobacco types are termed cigarette, cigar filler, cigar wrapper, binder, cheroot, bidi, chewing, snuff, hookah, etc. Several tobacco varieties, including burley and Turkish, are

K. Prabhakara Rao (✉) · P. Venkata Diwakar · B. Gouda · K. Sarala · M. Sheshu Madhav
ICAR-National Institute for Research on Commercial Agriculture (formerly ICAR-CTRI), Rajahmundry, India

Renu et al. (eds.), *Meta-omics in Crop Improvement*,
https://doi.org/10.1007/978-981-95-2294-1_13

cultivated to meet diverse consumer preferences, bolstering its economic significance in both local and global markets.

Meta-omics encompasses the integration of various omics technologies such as meta genomics, metatranscriptomics, metaproteomics, and metametabolomics to analyze the collective biological data from microbial communities associated with tobacco. This approach enables researchers to characterize not only the diversity of organisms present but also their functional roles and interactions within the tobacco ecosystem (Segata et al. 2013). The metagenomics studies utilizing 16S rRNA sequencing have revealed the presence of both culturable and non-culturable microorganisms in tobacco products, providing insights into how these microbes may impact carcinogenic compounds found in tobacco (Dollive et al. 2012). Moreover, understanding the microbial composition of tobacco can apprise the breeding strategies aimed at developing varieties that are not only more resilient to diseases but also possess reduced levels of harmful compounds. By identifying beneficial microbes that enhance plant health and resistance to pathogens, meta-omics can facilitate trait tailoring in tobacco breeding programs. This integrative approach holds significant potential for improving the sustainability and quality of tobacco crop while addressing public health concerns.

2 The Microbiome of Tobacco

The tobacco microbiome, which encompasses the diverse microbial communities present in both the phyllosphere (leaf surface) and the rhizosphere (root zone), plays a pivotal role in the growth, development, and overall health of tobacco plants. Studies have demonstrated that these microbial communities are not merely passive entities; rather, they play an active role in numerous processes that affect plant performance. These interactions are vital for the plant's nutrient acquisition, disease resistance, and resilience to environmental stressors. An extensive insight into the tobacco microbiome may result in the development of more sustainable farming practices, improved plant traits, and a decreased dependence on chemical fertilizers and pesticides (Li et al. 2024b; Zhou et al. 2020).

2.1 Major Constituents of Tobacco Microbiome

Microbial communities in the tobacco phyllosphere and rhizosphere differ markedly across various growth stages. Studies using 16S rRNA sequencing have revealed substantial changes in the bacterial populations during the plant's lifecycle. Early growth stages, such as seedling and squaring, are dominated by *Proteobacteria* and *Actinobacteria*, which are involved in initial plant establishment and nutrient cycling. As tobacco plants mature, there is a marked shift in the microbiome composition, with *Firmicutes* and *Bacteroidetes* becoming more prevalent (Gao et al. 2023; Wang and Cernava 2023). These bacteria are linked to later-stage nutrient uptake and the breakdown of organic matter, influencing plant health and growth.

Similarly, the fungal community in the phyllosphere also evolves as the plant matures. Studies have identified two major fungal phyla *Ascomycota* and *Basidiomycota* which exhibit changes in their relative abundance across different growth stages. Notably, *Basidiomycota* becomes more dominant during the squaring stage, indicating that fungi may play an important role in specific stages of tobacco development. The rhizosphere, where tobacco plants interact with soil microbes, is equally important, as soil microbes facilitate essential processes like nitrogen fixation and phosphorus solubilization, directly influencing plant growth and overall health (Ahmed et al. 2022; Zhu et al. 2022).

2.2 Functional Characteristics of the Tobacco Microbiome

The ecological functions of the tobacco microbiome are central to its role in improving plant health and productivity. In the phyllosphere, microbial communities influence the availability of key nutrients, such as nitrogen, potassium, and leaf sugars. This is vital for optimal plant growth and quality, as these nutrients are crucial for various physiological processes. Some bacteria, such as those from the genus *Rhizobium*, fix atmospheric nitrogen, making it available to the plant. Additionally, other microbes enhance the solubility of phosphorus, a nutrient essential for tobacco's development (Gao et al. 2023; Wang and Cernava 2023). These microbial communities are also linked to enhancing the plant's resilience to stress by promoting efficient nutrient cycling and improving the overall nutritional profile of the plant.

Furthermore, the microbiome plays a significant role in disease resistance. Beneficial microbes protect tobacco plants from pathogens such as *Ralstonia solanacearum*, the causative agent of bacterial wilt. These microbes achieve this by degrading harmful compounds in the soil, as well as triggering the plant's immune response. Research has shown that microbial inoculations can reduce disease incidence, up to 60% in the case of bacterial wilt, while simultaneously boosting plant growth (Zhu et al. 2022). These interactions between plants and microbes are crucial for enhancing the plant's natural defenses and improving the resilience of tobacco crops.

One of the key implications of understanding the tobacco microbiome is its potential for improving sustainable agriculture. By harnessing the power of beneficial microbes, growers can reduce their reliance on synthetic fertilizers and pesticides. Microbial inoculations or microbial fertilizers can help optimize nutrient uptake, improve plant growth, and increase disease resistance. Specifically, bacterial strains like *Bacillus* and *Pseudomonas* have been shown to enhance nutrient cycling and protect plants from soil-borne diseases (Zhou et al. 2024). The use of composite microbial fertilizers, which contain a mix of beneficial microbes, has proven to be more effective than using single strains, as they promote synergistic interactions that improve plant health and growth.

3 Meta-omics Tools for Decoding the Microbiome of Tobacco

Meta-omics comprising metagenomics, metatranscriptomics, metaproteomics, and metametabolomics plays an essential role in advancing our understanding of microbial communities associated with tobacco. These tools allow researchers to explore microbial diversity, functionality, and their impact on tobacco plants and products. Integration of these meta-omics techniques facilitates the enhancement of the traits, such as resistance to diseases, efficient nutrient uptake, and the reduction of detrimental substances like carcinogens, which ultimately contributes to the improvement of tobacco farming and the safety of its products (Aguiar-Pulido et al. 2016).

3.1 Metagenomics

Metagenomics, the study of genetic material recovered directly from environmental samples, has emerged as a powerful tool for understanding the genetic variability of endophytic microbes within tobacco microbiomes (Handelsman 2004). These studies offer significant potential for exploring the structural diversity and capabilities of microbes that have not yet been cultured from environmental samples (Krishnamoorthy et al. 2021). Traditional culture-based techniques for microbial isolation have several limitations, as they can only recover a small fraction of microbes (<1%), restricting their application in agricultural studies. In recent times, information on the bacteria, fungi, and archaea dynamics and functions in the root endosphere has been revealed from metagenomics studies (Fadiji and Babalola 2020; Samuel et al. 2022; Yuan et al. 2022). In this context, shotgun metagenomics sequencing to explore the phyllosphere microbiome of cigar tobacco revealed a considerable diversity of microbial species that are essential for plant health and resilience. This approach has identified key genes associated with abiotic stress resistance, such as those related to drought and heavy metal tolerance, highlighting the functional potential of microbial communities in promoting plant growth (Enebe and Babalola 2019; Xue et al. 2023). The extensive genetic catalogue obtained from these analyses includes over 444,000 protein-coding genes, showcasing the vast microbial species richness and functional capabilities present in the tobacco phyllosphere. Such insights are crucial for developing strategies that leverage beneficial microbes to enhance tobacco cultivation practices and improve stress resilience. In addition to that, metagenomics analysis has also uncovered the functional genes related to nitrogen metabolism, phosphate solubilization, ACC deaminase, IAA, and tryptophan biosynthesis, in various crops (Nikolic et al. 2011; Hong et al. 2019). These studies are also applicable in discovering variable genes involved in bacterial attachment, stress responses, and metabolic processes for tobacco cultivation also.

One notable application is the characterization of the tobacco microbiome through metagenomics sequencing, which has enabled researchers to identify both cultivable and non-cultivable microorganisms associated with tobacco products. This approach has revealed the presence of various microbial communities that can influence the production of carcinogenic compounds in tobacco (Li et al. 2024a). These findings aid in understanding the role of microbes in the formation of carcinogenic compounds, potentially guiding strategies to mitigate health risks associated with tobacco use. By understanding these microbial dynamics, strategies can be developed to reduce harmful substances in tobacco products, ultimately aiming to improve public health outcomes.

3.2 Metatranscriptomics

Metatranscriptomics focuses on analyzing the RNA transcripts present in microbial communities, providing insights into gene expression patterns under specific conditions. This technique has been used to investigate the dynamic interactions between tobacco plants and their associated microbiomes, revealing differences in taxonomic abundances between metagenomics and metatranscriptomics datasets. By employing high-throughput sequencing methods, researchers can capture the active microbial populations and their metabolic activities during various stages of tobacco processing. A study highlighted the use of combined metagenomics and metatranscriptomics analysis to identify differences in microbial taxonomic abundances and to detect RNA viruses that were not revealed in DNA-only studies (Tyx et al. 2020). This approach enables a more comprehensive understanding of how microbial constituents contribute to the formation of harmful compounds, such as tobacco-specific nitrosamines (TSNAs), which are potent carcinogens. Metatranscriptomics is particularly effective in distinguishing between viable and nonviable organisms within tobacco products. By analyzing RNA, researchers can determine which microbes are actively metabolizing and contributing to biochemical processes (Tyx et al. 2020). This distinction is crucial for evaluating the potential health risks associated with smokeless tobacco products, as viable bacteria may produce harmful substances during storage.

3.3 Metaproteomics

Alongside metagenomics and metatranscriptomics, metaproteomics plays a vital role in genomics research by focusing on the analysis of proteins and gene expression within microbial communities in complex biological samples (Valenzuela et al. 2006). Unlike metagenomics, which focuses on genetic data, metaproteomics directly measures protein abundance and activity, offering a clearer understanding of active biological processes. This approach has been utilized to classify and quantify proteins from environmental microbes (Kaul et al. 2016), offering insights into gene functions, DNA transcription to mRNA, and protein synthesis. Additionally, it aids in understanding microbial community interactions, particularly within host organisms such as humans or plants, and helps clarify how microbes impact health, disease, and environmental processes (Zhang et al. 2022). By examining protein structures and functions through metaproteomics, researchers can better understand the molecular profiles and gene expression of endophytic microbiomes under various experimental conditions (Kaul et al. 2016). By analyzing protein profiles in various ecosystems, metaproteomics has proven to be a powerful tool for uncovering the molecular mechanisms that drive microbial contributions to ecosystem functioning. This technique is particularly influential in studying plant-microbe interactions, such as those between microorganisms and tobacco plants, where it aids in identifying the proteins and pathways that influence essential traits like nicotine synthesis, stress tolerance, and overall plant quality. Interestingly, metaproteomics features, which include identification of proteins, enzymes involved in metabolic pathways, and osmotic balancing can further reveal microbial functions under drought stress.

In tobacco research, metaproteomics provides direct, functional insights into the complex interactions between microorganisms and tobacco plants. By examining protein expression in various cultivars, environmental conditions, and microbial communities, metaproteomics helps identify proteins responsible for critical traits like nicotine content and stress responses. This approach has revealed how factors such as shading and microbial communities contribute to tobacco growth, quality, and resilience (Yan et al. 2019). Beyond tobacco, metaproteomics is essential in microbiome research, where it explores how microbial proteins affect human health and disease and in environmental monitoring to assess microbial activity in ecosystems. It also plays a pivotal role in clinical research, particularly in developing microbiota-directed therapies. Despite challenges like sample complexity and the need for advanced bioinformatics tools, metaproteomics continues to show promise across diverse scientific fields. As technology advances, its applications—especially in tobacco research—are expected to expand, providing deeper insights into plant-microbe interactions and improving agricultural practices and product quality.

Tobacco consumption, particularly smoking, significantly impacts the human oral microbiome, promoting a pro-inflammatory environment. A review of 36 studies found that smoking leads to shifts in microbial communities, such as an increase in bacterial taxa like *Fusobacterium*, *Actinobacteria*, *Streptococcus*, *Prevotella*, and *Veillonella*. These changes are accompanied by alterations in metabolic pathways related to amino acid metabolism and xenobiotic biodegradation, which have broader implications for human health. Metaproteomics enables a deeper understanding of these microbial shifts by identifying proteins expressed by microbial communities in response to tobacco exposure. By quantifying these proteins, researchers gain insights into how tobacco affects microbial functions at a molecular level. These findings also hold relevance for agriculture, allowing optimization of tobacco cultivation to address the health risks posed by tobacco use.

3.4 Metametabolomics

Metametabolomics, an emerging branch of metabolomics, focuses on the interactions between metabolites across multiple organisms within complex biological systems. Unlike traditional metabolomics, which examines metabolites in a single organism, metametabolomics explores the dynamic interactions of metabolites within ecosystems, such as host-microbiota relationships and multi-organism communities, offering insights into biological processes like disease mechanisms, ecological dynamics, and environmental responses (Chakraborty 2024). This field investigates the collective metabolic activities arising from interactions between a host organism and its microbiota, as well as other organisms within the same ecosystem. Through the study of how metabolites are exchanged, altered, or synthesized in these communities, metametabolomics provides a more comprehensive view of metabolic networks. By integrating various omics technologies, including genomics, transcriptomics, and proteomics, researchers gain a better understanding of how metabolic pathways function in complex environments such as the human gut microbiome or soil ecosystems (Muller et al. 2021; Jendoubi 2021). Central to this approach is the concept of "meta-organisms," which refers to the host and its microbiota working together to support survival, adapt to environmental changes, and influence disease processes (Chakraborty 2024). The integration of data from metagenomics, metatranscriptomics, and metaproteomics with metabolomics enhances the understanding of metabolic regulation and the functionality of these pathways across different organisms within a biological system (Fig. 1).

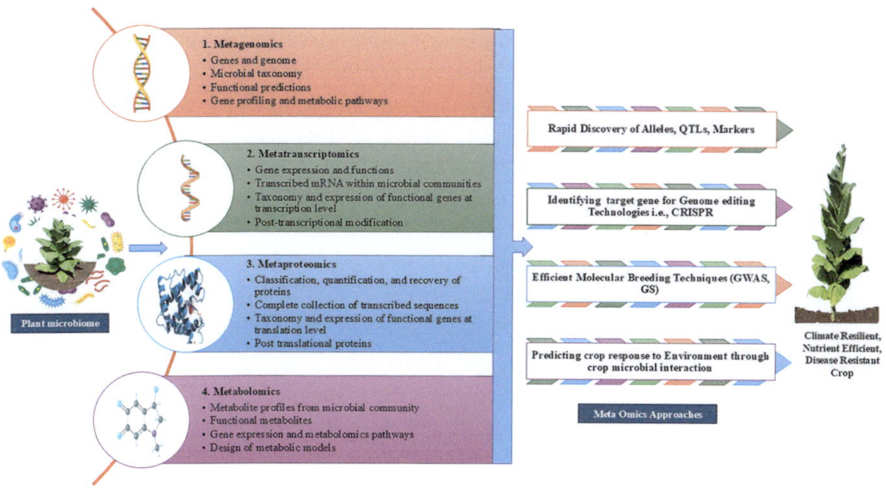

Fig. 1 Flowchart illustrates the integrated application of metagenomics, metatranscriptomics, metaproteomics, and metabolomics in plant microbiome research. Each layer serves a distinct function: metagenomics focuses on gene identification and genome profiling; metatranscriptomics captures gene expression patterns and mRNA transcription; metaproteomics enables protein classification and functional analysis; and metabolomics provides insights into metabolite profiling and metabolic pathway modeling. These interconnected processes collectively inform key applications such as allele discovery, CRISPR-based genome editing, advanced molecular breeding techniques, and the prediction of plant responses to environmental conditions. The overall framework emphasizes the role of integrated meta-omics in developing climate-resilient, nutrient-efficient, and disease-resistant crop varieties, thereby advancing sustainable agricultural practices

Recent advancements in metametabolomics have significantly enhanced our understanding of tobacco's impact on both health and agriculture, paving the way for improved trait tailoring in tobacco cultivation. A 2016 study employed untargeted metabolomics to analyze smokers' blood, identifying 31 metabolites influenced by smoking, including the novel detection of menthol-glucuronide. This discovery, along with insights into gender and racial differences in metabolomics profiles, provides new avenues for tailoring tobacco products to meet specific consumer preferences and health considerations, especially in menthol-containing varieties (Nair et al. 2020). Moreover, past studies on the role of microbial interactions within tobacco plants highlight how these communities influence plant health and metabolite production, which could be harnessed to improve crop resilience and overall quality (Kumar et al. 2022).

4 Meta-omics Data Resources of Tobacco

Databases are essential tools in advancing meta-omics research related to tobacco, particularly in elucidating the intricate interactions between tobacco metabolites, the microbiome, and genetic determinants that influence specific traits like nicotine dependence and smoking behaviors. A key resource in this area is the *Genetic Resources of Nicotine and Smoking (GRONS)*, which aggregates genetic data associated with nicotine exposure and smoking-related diseases, thereby enabling comprehensive analyses of the genes implicated in these behaviors (Fang et al. 2017). The application of multi-omics approaches is exemplified by databases such as *Plant-Reactome*, *Gramene*, and *KaPPA-View4 KEGG* which integrate functional genomics, transcriptomics, and metabolic pathway data, facilitating a deeper understanding of the genetic and metabolic networks in tobacco and its associated microbiota.

The integration of multiple omics platforms is crucial for identifying genetic loci and metabolic pathways that govern unique traits like nicotine metabolism and consumption behaviors. Genome-wide association studies (GWAS) have elucidated the complex interplay between genetic predispositions and environmental exposures that influence tobacco use, a relationship that is further explored through metabolomics databases like *KOMICS* and *KNApSAcK*, which provide extensive data on plant metabolites including tobacco (Buchwald et al. 2020; Palmer et al. 2021). Recent advances in meta-analysis methodologies allow for the synthesis of findings from diverse omics studies, thereby enhancing our understanding of the metabolic networks involved in tobacco.

These diverse and comprehensive databases, spanning functional genomics, transcriptomics, proteomics, and metabolomics, offer critical insights into the biological mechanisms underlying tobacco use and its associated health risks. By providing access to integrated omics data, they enable the development of more targeted interventions and strategies for improvement of specific traits and resilience in tobacco (Table 1).

Table 1 Multi-omics data resources for tobacco

S. no.	Database	Features and functionality	URL
1	Solgenomics	Multi-omics: functional genomics, transcriptomics, proteomics, and metabolic pathways	https://solgenomics.net/
2	Gene Expression Omnibus	Multi-omics: comparative functional genomics, transcriptomics, and metabolic pathways	https://www.ncbi.nlm.nih.gov/geoprofiles/
3	Gramene	Multi-omics: comparative functional genomics, transcriptomics, and metabolic pathways	http://gramene.org/
4	Faprotax	Metatranscriptomics	FAPROTAX 1.2.10
5	iTRAQ	Meta-omics approach	https://www.creative-proteomics.com/services/itraq-based-proteomics-analysis.htm
6	KBase	Metagenomics	https://www.kbase.us/
7	GabiPD	Multi-omics	http://www.gabipd.org/
8	KOMICS	Metabolomics	http://www.kazusa.or.jp/kpv4/kegg/
9	KNApSAcK	Metabolomics	http://kanaya.naist.jp/KNApSAcK/Family

5 Integration of Meta-omics for Trait Tailoring

The integration of multi-omics data is crucial for developing a systems biology approach to crop improvement. By combining data from various meta-omics layers such as genomics, transcriptomics, proteomics, and metabolomics researchers can uncover insights that would not be apparent from any single dataset alone. By linking genomics data with transcriptomics and proteomics information, researchers can construct models that predict how changes at the molecular level affect phenotypic outcomes

Phenotype to Genotype: This approach helps identify genetic markers linked to desirable traits.

Genotype to Phenotype: Understanding how specific genetic variations manifest as observable traits assists in selecting appropriate breeding strategies.

The integration of omics data is a powerful approach that enhances our understanding of complex biological systems in tobacco research. By leveraging high-throughput phenotyping techniques and advanced computational methods, researchers can uncover insights into the genetic and molecular mechanisms underlying important traits. Despite the challenges associated with multi-omics integration, ongoing advancements in technology and methodology hold great promise for improving tobacco crop resilience and quality.

6 Metadata Derivatives for Trait Tailoring in Tobacco

6.1 Enhancing Growth and Resilience

The primary application of metagenomics in tobacco is the identification and utilization of plant growth-promoting bacteria (PGPB) found in the rhizosphere. These beneficial microbes improve tobacco plant growth by enhancing nutrient availability, promoting root development, and boosting resilience to abiotic stresses like drought, salinity, and temperature extremes (Jha and Saraf 2015; Zhang et al. 2022). Investigations revealed that beneficial microbes can also help mitigate the effects of environmental stress by activating genes responsible for stress tolerance. Genes such as *proA*, involved in proline biosynthesis, and *phoD*, which plays a role in phosphorus solubilization, are commonly found in beneficial microbes. By enhancing microbial communities in the rhizosphere, metagenomics allows farmers to optimize plant health, resulting in higher yields and improved quality of tobacco products.

6.2 Disease Resistance

Metagenomics offers valuable insights into how microbial communities contribute to disease resistance in tobacco. The microbes, such as *Bacillus* spp., produce antibiotics that inhibit the growth of pathogenic fungi and bacteria. Others, like *Pseudomonas* spp., help in controlling soil-borne pathogens and nematodes, such as *Meloidogyne incognita*, which causes root-knot disease (Zhang et al. 2023). Furthermore, metagenomics analysis has revealed the presence of genes involved in plant immune responses, such as *PR-1* (pathogenesis-related protein), which can be activated by beneficial microbes to enhance the plant's defense mechanisms (Alattas et al. 2024).

6.3 Flavor Enhancement

The fermentation process is crucial in developing the characteristic flavors of tobacco products. Microbial communities involved in fermentation contribute significantly to the breakdown of complex compounds, releasing flavor-enhancing chemicals. Metagenomics studies have identified various microbes responsible for producing compounds that shape the flavor profile of tobacco. The species of *Lactobacillus* and *Aspergillus* play a key role in breaking down sugars and amino acids, thereby contributing to the flavor complexity during tobacco curing and fermentation (Tao et al. 2024; Jiang et al. 2024).

6.4 Reduction of Harmful Constituents

The presence of tobacco-specific nitrosamines (TSNAs) in cured and processed tobacco is a significant health concern. Metagenomics offers promising strategies for reducing carcinogenic compounds. It identified specific bacteria that facilitate this process, allowing for potential interventions to reduce TSNA formation by targeting these nitrate-reducing bacteria with antimicrobial treatments or microbial management techniques (Jiang et al. 2021; Wang et al. 2017). Similarly, in smokeless tobacco products (STPs), metagenomics profiling of the microbial communities involved in fermentation has revealed that some microbes promote TSNA formation, and others can inhibit or reduce their production. By manipulating these microbial populations, safer STPs can be developed by reducing harmful microbes or promoting beneficial ones that help mitigate TSNA levels (Tyx et al. 2020; Sajid et al. 2023; Yang et al. 2023).

7 Designing Synthetic Microbial Communities for Enhanced Tobacco Traits

Meta-omics holds transformative potential for tailoring plant traits through a deeper understanding of plant-microbe interactions and the development of synthetic microbial communities. This exciting frontier of metagenomics aid in creation of synthetic microbial communities (synbiotics) tailored to enhance specific traits in tobacco plants. These communities can be engineered to improve disease resistance, enhance stress tolerance, or optimize fermentation for desired flavor profiles. By assembling a consortium of beneficial microbes, researchers can introduce these communities into the tobacco plant's environment, enhancing plant health, resilience, and overall productivity.

The development of synthetic microbial communities also holds promise for reducing the environmental impact of tobacco farming. By decreasing the need for chemical fertilizers and pesticides, this approach promotes more sustainable and environmentally friendly agricultural practices. Moreover, the use of tailored microbial communities could help regulate the production of TSNAs and optimize the fermentation process, resulting in safer, more consistent tobacco products (Sajid et al. 2023).

8 Conclusion

The future of meta-omics in tobacco research is promising, with several advancements anticipated to enhance tobacco cultivation and product safety. As metagenomics and sequencing technologies evolve, detailed microbial characterization will become possible, enabling the identification of microbial taxa that influence both beneficial traits and harmful compounds like tobacco-specific nitrosamines (TSNAs). By integration of meta-omics data, a comprehensive understanding of plant-microbe interactions will emerge, supporting precision breeding programs with trait tailoring aimed at improving tobacco quality and safety. Moreover, meta-omics analysis will facilitate the development of targeted microbial inoculants to optimize nutrient uptake, enhance disease resistance, and reduce harmful compound formation. In addition, metabolomics will play a critical role in improving flavor profiles and product quality, ensuring that consumer preferences for better-tasting, healthier tobacco are met. Future research will include longitudinal studies and real-time monitoring of microbial dynamics, which will provide insights into optimizing plant health and product safety over time. Ultimately, the continued development of meta-omics holds the potential to revolutionize tobacco production by making it more sustainable, healthier, and aligned with evolving consumer and regulatory demands.

Conflict of Interest Authors declare no conflict of interest.

References

Aguiar-Pulido V, Huang W, Suarez-Ulloa V, Cickovski T, Mathee K, Narasimhan G (2016) Metagenomics, metatranscriptomics, and metabolomics approaches for microbiome analysis. Evol Bioinformatics Online 12(Suppl 1):5–16. https://doi.org/10.4137/EBO.S36436

Ahmed W, Dai Z, Liu Q, Munir S, Yang J, Karunarathna SC et al (2022) Microbial cross-talk: dissecting the core microbiota associated with flue-cured tobacco (Nicotiana tabacum) plants under healthy and diseased state. Front Microbiol 13:845310

Alattas H, Glick BR, Murphy DV, Scott C (2024) Harnessing Pseudomonas spp. for sustainable plant crop protection. Front Microbiol 15

Buchwald J, Chenoweth MJ, Palviainen T, Zhu G, Benner C, Gordon SD, Korhonen T, Ripatti S, Madden PA, Lehtimäki T, Raitakari OT, Salomaa VV, Rose RJ, George TP, Lerman C, Pirinen M, Martin NG, Kaprio JA, Loukola A, Tyndale RF (2020) Genome-wide association meta-analysis of nicotine metabolism and cigarette consumption measures in smokers of European descent. Mol Psychiatry 26:2212–2223

Chakraborty N (2024) Metabolites: a converging node of host and microbe to explain meta-organism. Front Microbiol 15

Dollive S, Peterfreund GL, Sherrill-Mix S, Bittinger K, Sinha R, Hoffmann C, Nabel CS, Hill DA, Artis D, Bachman MA, Custers-Allen R, Grunberg S, Wu GD, Lewis JD, Bushman FD (2012) A tool kit for quantifying eukaryotic rRNA gene sequences from human microbiome samples. Genome Biol 13(7):R60. https://doi.org/10.1186/gb-2012-13-7-r60

Enebe MC, Babalola OO (2019) The impact of microbes in the orchestration of plants' resistance to biotic stress: a disease management approach. Appl Microbiol Biotechnol 103:9–25

Fadiji AE, Babalola O (2020) Metagenomics methods for the study of plant-associated microbial communities: a review. J Microbiol Methods:105860

Fang Z, Yang Y, Hu Y, Li MD, Wang J (2017) GRONS: a comprehensive genetic resource of nicotine and smoking. J Biol Databases Curation 2017

FAOSTAT (2022) https://www.fao.org/statistics/en

Gao J, Uwiringiyimana E, Zhang D (2023) Microbial composition and diversity of the tobacco leaf phyllosphere during plant development. Front Microbiol 14:1199241

Handelsman J (2004) Metagenomics: application of genomics to uncultured microorganisms. Microbiol Mol Biol Rev 68:669–685. https://doi.org/10.1128/mmbr.68.4.669-685.2004

Hong CE, Kim JU, Lee JW, Bang KH, Jo IH (2019) Metagenomic analysis of bacterial endophyte community structure and functions in Panax ginseng at different ages. 3 Biotech 9. https://doi.org/10.1007/s13205-019-1838-x, 200

https://openebench.bsc.es/tool/grons

ICAR-CTRI (2024) https://ctri.icar.gov.in/digital_library-Panoramic View of Indian Tobacco

Jendoubi T (2021) Approaches to integrating metabolomics and multi-omics data: A primer. Meta 11

Jha CK, Saraf M (2015) Plant growth promoting rhizobacteria (PGPR): a review. J Agri Res Dev 5(2):108–119

Jiang Y, Gong J, Chen Y, Hu B, Sun J, Zhu Y, Xia Z, Zou C (2021) Biodegradation of nicotine and TSNAs by Bacterium sp. strain J54. Iran J Biotechnol 19:e2812

Jiang C, Xie Y, Tian W, Gao W, Wu F, Li Z, Han G, Pan W (2024) Investigation on rapid fermentation technology of seasonally effective high-end tobacco-specific organic fertilizers. J Ecol Eng 25:221

Kaul S, Sharma T, Dhar MK (2016) "Omics" tools for better understanding the plant–endophyte interactions. Front Plant Sci 7:955. https://doi.org/10.3389/2Ffpls.2016.00955

Krishnamoorthy A, Gupta A, Sar P, Maiti MK (2021) Metagenomics of two gnotobiotically grown aromatic rice cultivars reveals genotype-dependent and tissue-specific colonization of endophytic bacterial communities attributing multiple plant growth promoting traits. World J Microbiol Biotechnol 37:59. https://doi.org/10.1007/s11274-021-03022-5

Kumar RS, Mishra N, Kumar A (2022) Characterization of tobacco microbiome by metagenomics approach. Methods Mol Biol (Clifton, N.J.) 2413:229–244. https://doi.org/10.1007/978-1-0716-1896-7_22

Li A, Jin K, Zhang Y, Deng X, Chen Y, Wei X et al (2024a) Root exudates and rhizosphere microbiota in responding to long-term continuous cropping of tobacco. Sci Rep 14(1):11274

Li X, Zheng X, Yadav N, Saha S, Salama E-S, Li X, Wang L, Jeon B-H (2024b) Rational management of the plant microbiome for the second green revolution. Plant Commun 5

Muller E, Algavi YM, Borenstein E (2021) A meta-analysis study of the robustness and universality of gut microbiome-metabolome associations. Microbiome 9

Nair V, Tran M, Behar RZ, Zhai S, Cui X, Phandthong R, Wang Y, Pan S, Luo W, Pankow JF, Volz DC, Talbot P (2020) Menthol in electronic cigarettes: a contributor to respiratory disease? bioRxiv

Nikolic B, Schwab H, Sessitsch A (2011) Metagenomic analysis of the 1-aminocycloprop ane-1-carboxylate deaminase gene (acdS) operon of an uncultured bacterial endophyte colonizing Solanum tuberosum L. Arch Microbiol 193:665–676. https://doi.org/10.1007/s00203-011-0703-z

Palmer RH, Benca-Bachman CE, Huggett SB, Bubier JA, McGeary JE, Ramgiri N, Srijeyanthan J, Yang J, Visscher PM, Yang J, Knopik VS, Chesler EJ (2021) Multi-omic and multi-species meta-analyses of nicotine consumption. Transl Psychiatry:11

Sajid M, Srivastava S, Yadav RK, Joshi L, Bharadwaj M (2023) Fungal community composition and function associated with loose smokeless tobacco products. Curr Microbiol 80(4):131

Samuel SO, Suzuki K, Asiloglu R, Harada N (2022) Soil-root interface influences the assembly of the endophytic bacterial community in rice plants. Biol Fertil Soils 58(1):35–48. https://doi.org/10.1007/s00374-021-01611-y

Segata N, Boernigen D, Tickle TL, Morgan XC, Garrett WS, Huttenhower C (2013) Computational meta'omics for microbial community studies. Mol Syst Biol 9:666. https://doi.org/10.1038/msb.2013.22

Tao J, Chen S, Jiang Z, Wang C, Zhang E, Liang H, Xu Y, Cao P, Ding N, Zhang M, He W, Chen Q (2024) Fermentation process of tobacco leaves drives the specific changes of microbial community. BMC Microbiol 24

Tyx RE, Rivera AJ, Keong LM, Stanfill SB (2020) An exploration of smokeless tobacco product nucleic acids: a combined metagenome and metatranscriptome analysis. Appl Microbiol Biotechnol 104(2):751–763. https://doi.org/10.1007/s00253-019-10232-3

Valenzuela L, Chi A, Beard S, Orell A, Guiliani N, Shabanowitz J, Hunt DF, Jerez CA (2006) Genomics, metagenomics and proteomics in biomining microorganisms. Biotechnol Adv 24:197–211. https://doi.org/10.1016/j.biotechadv.2005.09.004

Wang M, Cernava T (2023) The phyllosphere microbiome. Front Plant Sci 14:1234843

Wang J, Yang H, Shi H, Zhou J, Bai R, Zhang M, Jin T (2017) Nitrate and nitrite promote formation of tobacco-specific nitrosamines via nitrogen oxides intermediates during postcured storage under warm temperature. J Chem 2017(1):6135215

Xue F, Yang J, Luo C, Li D, Shi G, Song G, Li Y (2023) Metagenomic insight into the biodegradation of biomass and alkaloids in the aging process of cigar. Bioresour Bioprocess 10(1):45. https://doi.org/10.1186/s40643-023-00667-y

Yan L, Zhu J, Zhao X, Shi J, Jiang C, Shao D (2019) Beneficial effects of endophytic fungi colonization on plants. Appl Microbiol Biotechnol 103:3327–3340. https://doi.org/10.1007/s00253-019-09713-2

Yang Y, Xu R, Yang M, Xu Q, Miao C, Guo J, Mou W, Du H, Wei G, Hu L, Hu Z (2023) Characterization of bacterial community in tobacco leaves at flue-curing and redrying processing stages. Sci Rep 13(1):13333. https://doi.org/10.1038/s41598-023-40502-0

Yuan Y, Zuo J, Zhang H, Zu M, Liu S (2022) The Chinese medicinal plants rhizosphere: metabolites, microorganisms, and interaction. Rhizosphere 22:100540. https://doi.org/10.1016/j.rhisph.2022.100540

Zhang M, Wang W, Bai SH, Xu Z, Yun Z, Zhang W (2022) Linking Phyllostachys edulis (moso bamboo) growth with soil nitrogen cycling and microbial community of plant-soil system: effects of plant age and niche differentiation. Ind Crop Prod 177:114520. https://doi.org/10.1016/j.indcrop.2022.114520

Zhang J, Rong S, Qin Y, Wu J, Wang J, Li J et al (2023) Effects of different harvesting methods on TSNAs and precursors content of tobacco lamina and stem after modulation. J Henan Agri Sci 52(2):57

Zhou J, Yu L, Zhang J, Zhang X, Xue Y, Liu J, Zou X (2020) Characterization of the core microbiome in tobacco leaves during aging. Microbiology 9(3):e984. https://doi.org/10.1002/mbo3.984

Zhou Z, Wang C, Cha X, Zhou T, Pang X, Zhao F, Han X, Yang G, Wei G, Ren C (2024) The biogeography of soil microbiome potential growth rates. Nat Commun 15:19472

Zhu YG, Xiong C, Wei Z, Chen QL, Ma B, Zhou SYD et al (2022) Impacts of global change on the phyllosphere microbiome. New Phytol 234(6):1977–1986

Bioinformatics Approaches for Meta-omics in Crop Improvement

Sharanbasappa D. Madival, Nitesh Kumar Sharma, Dwijesh Chandra Mishra, and Girish Kumar Jha

1 Introduction

Robust collection of technologies for comprehending microbial populations and their functions in plant health and production is meta-omics, an integrated approach that combines metagenomics, metatranscriptomics, metaproteomics, and metabolomics (Kelaher et al. 2018; Sharuddin et al. 2022). Compared to conventional single-omics techniques, meta-omics offers a more comprehensive understanding of microbial dynamics by simultaneously studying the genomic, transcriptomic, proteomic, and metabolomic profiles of microbial communities. Whole-genome shotgun sequencing (WGS) and amplicon sequencing, respectively, are two examples of next-generation sequencing (NGS) technologies that are essential to meta-omics research because they allow for the in-depth investigation of microbial diversity and function in diverse environmental samples (Staley et al. 2014; Zuñiga et al. 2017).

Researchers can investigate the genetic diversity and functional potentials of microbial communities by using metagenomics, the study of genetic material extracted directly from environmental samples (Handelsman 2004). This method is essential to comprehending how microbes contribute to disease prevention, nutrient cycling, and environmental stress reduction. Nevertheless, metagenomics does not identify which genes are actively expressed in a given environment, even while it offers insights into the makeup of microorganisms and their possible roles.

Contrarily, metatranscriptomics examines the RNA transcripts found in microbial communities, offering real-time information on patterns of gene expression and assisting in determining how functionally active microorganisms are in response to

S. D. Madival · D. C. Mishra · G. K. Jha (✉)
ICAR-Indian Agricultural Statistics Research Institute, New Delhi, India

N. K. Sharma
ICAR-Indian Agricultural Statistics Research Institute, New Delhi, India

Dr Rajendra Prasad Central Agricultural University, Pusa, Samastipur, Bihar, India

Renu et al. (eds.), *Meta-omics in Crop Improvement*,
https://doi.org/10.1007/978-981-95-2294-1_14

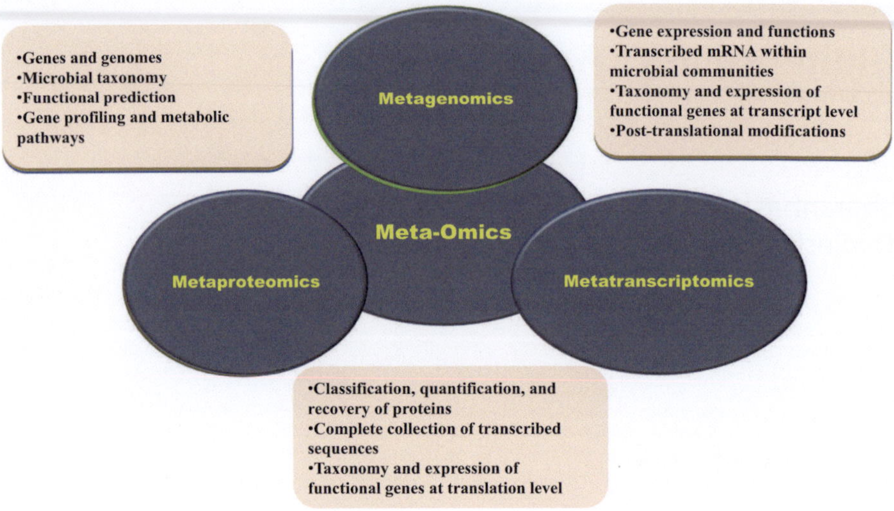

Fig. 1 Meta-omics

different stresses (Peimbert and Alcaraz 2016). Understanding how microbes react to abiotic stresses like drought, salt, and temperature extremes—all of which are major issues in agriculture as a result of climate change—is essential.

Finally, metaproteomics further extends these insights by examining the proteins produced by microbial communities. Proteomic analysis helps link gene expression with functional outcomes, such as identifying enzymes involved in nutrient cycling, pathogen resistance, or stress tolerance (Wilmes and Bond 2006). Understanding microbial protein production allows for deeper insights into the biochemical pathways that influence plant growth and resilience (Fig. 1).

The integration of these meta-omics approaches provides a comprehensive understanding of microbial communities and their contributions to crop productivity and sustainability. For crop improvement, this integrated approach is transformative. Meta-omics enables the identification of beneficial microbes that enhance plant growth, resilience to stress, and nutrient uptake, which can be harnessed for developing microbial-based solutions, such as biofertilizers and biopesticides. By utilizing bioinformatics tools to analyze the vast amounts of data generated by these technologies, researchers can develop targeted strategies for optimizing microbial communities tailored to specific crops and environmental conditions (Segata et al. 2013) (Fig. 2).

Furthermore, the SDGs are directly aligned with the use of meta-omics for agricultural development. These bio-based solutions complement SDG 2 (Zero Hunger) and SDG 12 (Responsible Consumption and Production) by lowering the demand for chemical pesticides and fertilizers, thereby promoting more sustainable agricultural practices. Additionally, by helping to design crops that are more tolerant to environmental pressures, meta-omics insights can promote SDG 13 (Climate Action) and contribute to food security in the face of climate change.

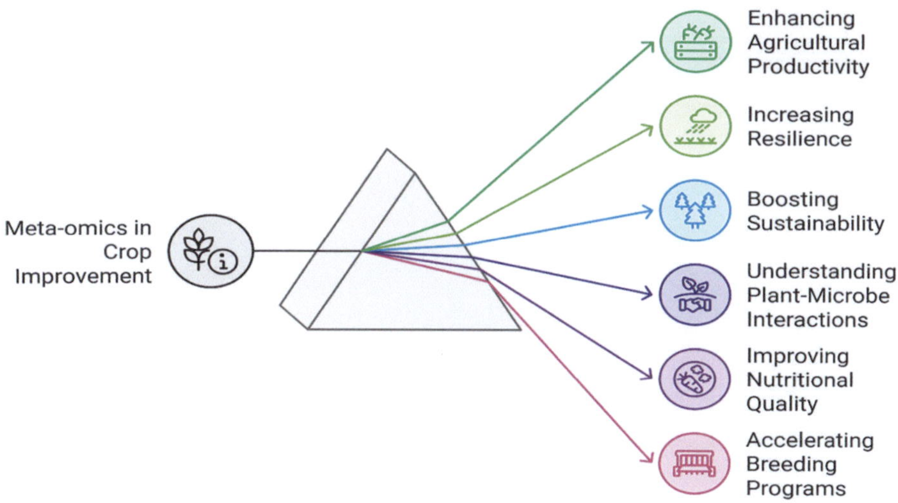

Fig. 2 Application of meta-omics in crop improvement

2 Core Bioinformatics Tools and Workflows

In the analysis of meta-omics data, bioinformatics plays a pivotal role, transforming raw sequence data into meaningful biological insights (Fig. 3). This section details the core bioinformatics tools and workflows commonly used in meta-omics studies, each contributing to the accurate interpretation of complex datasets.

2.1 Metagenomics Tools

The study of genetic material extracted directly from environmental samples is known as metagenomics, and it provides a comprehensive approach of examining microbial populations in their native environments. Unlike culture-dependent methods for microbial isolation have significant limitations, recovering less than 1% of microbes, which restricts their application in agricultural research, metagenomics enables the simultaneous investigation of multiple organisms, including those that are unculturable.

This approach has transformed our knowledge of the ecology, diversity, and function of microorganisms by shedding light on their roles in environmental processes, illness, and health. DNA extraction, sequencing, data preprocessing, assembly, annotation, taxonomic classification, functional analysis, and statistical evaluation are some of the crucial processes involved in metagenomic research. Researchers are now able to produce enormous volumes of data thanks to high-throughput sequencing technologies like PacBio and Illumina, which make it necessary to process, analyze, and interpret the results using sophisticated bioinformatics tools.

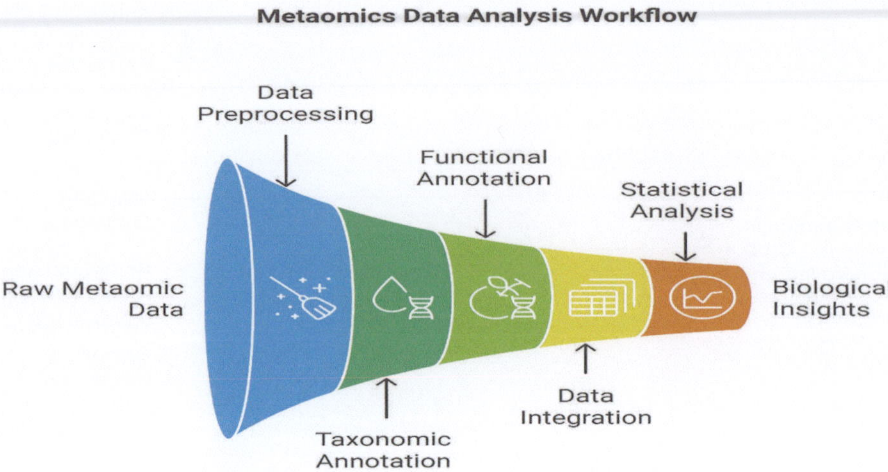

Fig. 3 General bioinformatics workflow of meta-omics data

Bioinformatics tools in metagenomics address various aspects of the workflow, including sequence trimming, assembly, taxonomic profiling, annotation, and statistical analysis. Some of the widely used tools and their respective roles in metagenomics studies are given in Table 1.

These tools enable the exploration of metagenomic datasets to uncover microbial diversity, ecosystem dynamics, and their functional potential. Together, they form the backbone of computational approaches in metagenomics, providing the precision and scalability required to handle large and complex datasets.

2.2 Metatranscriptomics Tools

Metatranscriptomics is the study of the complete set of RNA transcripts obtained from a microbial community within a specific environment. This field provides a snapshot of the actively expressed genes, allowing researchers to understand functional activity, microbial interactions, and responses to environmental changes. Unlike metagenomics, which focuses on DNA, metatranscriptomics delves into gene expression profiles, offering insights into dynamic microbial functions rather than their mere presence.

The advent of high-throughput sequencing platforms such as Illumina and Oxford Nanopore, coupled with sophisticated bioinformatics tools, has significantly advanced the field by enabling the analysis of vast transcriptomic datasets. Key steps in metatranscriptomics workflows include RNA extraction, sequencing, quality control, assembly, annotation, taxonomic profiling, functional analysis, and statistical interpretation. The key bioinformatics tools and their roles in metatranscriptomics analysis are given in Table 2.

Table 1 Bioinformatics tools for analysis of metagenomics data

Functional category	Name of tool	References	Application	Link
Preprocessing	FASTQC	Andrews (2010)	Quality control of high-throughput sequencing data, checks for issues like adapter contamination	https://www.bioinformatics.babraham.ac.uk/projects/fastqc/
	Trimmomatic	Bolger et al. (2014)	Trimming of Illumina reads to remove low-quality bases and adapters	http://www.usadellab.org/cms/?page=trimmomatic
	PRINSEQ	Cantu et al. (2019)	Quality filtering and preprocessing of high-throughput sequence data, supports various platforms	http://prinseq.sourceforge.net/
	SolexaQA	Cox et al. (2010)	Quality control for Illumina, Ion Torrent, and 454 data, detects and visualizes errors	http://solexaqa.sourceforge.net/
	FASTX-Toolkit	Hao et al. (2022)	Set of command-line tools for preprocessing, including trimming, filtering, and quality control	http://hannonlab.cshl.edu/fastx_toolkit/

(continued)

Table 1 (continued)

Functional category	Name of tool	References	Application	Link
De novo assembly	RAY	Boisvert et al. (2012)	Assembling short-read data from high-throughput sequencing for genome and transcriptome analysis	http://denovoassembler.sf.net/
	MetaVelvet	Namiki et al. (2011)	de novo assembly of metagenomic data using velvet with enhanced features for handling large datasets	http://metavelvet.dna.bio.keio.ac.jp/
	Genovo	Laserson et al. (2011)	Assembler for high-throughput sequencing, focused on de novo assembly of large genomes	https://pubmed.ncbi.nlm.nih.gov/21385045/
	CLC Genomics Workbench	Matvienko (2015)	Comprehensive toolset for DNA sequence analysis, including de novo assembly and RNA-seq analysis	https://digitalinsights.qiagen.com/clc-genomics-workbench-features/
	Meta-IDBA	Peng et al. (2011)	Metagenomic assembly tool for handling large and complex datasets	https://i.cs.hku.hk/~alse/hkubrg/projects/metaidba/

	SOAPdenovo[	Xie et al. (2014)	Efficient assembly of short reads from Illumina sequencing data	https://github.com/aquaskyline/SOAPdenovo2
	Newbler	Miller et al. (2010)	Assembly software for data from 454 GS-series pyrosequencing platforms	https://help.rc.ufl.edu/doc/Newbler
	ABySS	Simpson et al. (2009)	Scalable de novo assembler for high-throughput sequence data	http://www.bcgsc.ca/downloads/abyss/
	ALLPATHS-LG	Gnerre et al. (2011)	Assembling large genomes using Illumina data	https://github.com/danforthcenter/bioinformatics/blob/master/docs/allpaths.md
Reference-based alignment	BWA	Jung and Han (2022)	Mapping of short DNA reads to a reference genome, particularly for Illumina sequencing	https://sourceforge.net/projects/bio-bwa/
	Bowtie	Langmead and Salzberg (2012)	Fast alignment of short DNA sequences to large reference genomes, widely used for Illumina data	https://sourceforge.net/projects/bowtie-bio/

(continued)

Table 1 (continued)

Functional category	Name of tool	References	Application	Link
	MUMer	Delcher et al. (1999)	Aligning large sequences to a reference genome, used for genomic comparison and mapping	https://sourceforge.net/projects/mummer/
	BFAST	Homer et al. (2009)	Sequence alignment tool focused on high-throughput sequencing reads	http://bfast.sourceforge.net/
	MrFAST	Hach et al. (2010)	Illumina read alignment tool designed for high-speed processing	http://mrfast.sourceforge.net/pubs.html
	CloudBurst	Sreekanti et al. (2020)	Cloud-based DNA read alignment for high-throughput sequencing	https://sourceforge.net/p/cloudburst-bio/wiki/CloudBurst/
	SOAP	Gudgin et al. (2003)	A read mapping tool for Illumina-Solexa data, known for speed and memory efficiency	https://academic.oup.com/bioinformatics/article/24/5/713/203564
	BLAST	Madden (2013)	Sequence alignment tool widely used for comparing biological sequences	https://blast.ncbi.nlm.nih.gov/Blast.cgi
	Novoalign	Ruffalo et al. (2012)	A short read aligner optimized for Illumina sequencing data	http://www.novocraft.com/products/novoalign/

	MOSAIK	Lee et al. (2014)	A flexible tool for aligning short DNA sequences, supporting multiple platforms	https://code.google.com/archive/p/mosaik-aligner/
Annotation and comparison	MG-RAST	Keegan et al. (2016)	A web-based server for functional annotation and comparative analysis of metagenomic datasets.	https://www.mg-rast.org/
	METAREP	Goll et al. (2010)	Taxonomic, functional, and comparative analysis for metagenomic datasets	https://www.jcvi.org/research/metarep
	DIYA	Stewart et al. (2009)	Functional annotation of microbial communities and environmental datasets	https://sourceforge.net/projects/diyg/
	CloVR	Agrawal et al. (2017)	Cloud-based functional and taxonomic analysis for metagenomics	http://clovr.org/
	RATT	Otto et al. (2011)	Functional analysis tool designed for microbial annotation and comparison	http://ratt.sourceforge.net/

(continued)

Table 1 (continued)

Functional category	Name of tool	References	Application	Link
	CAMERA	Kuhl et al. (2012)	Taxonomic, phylogenetic, functional, and comparative analysis tool for metagenomic data	https://journals.plos.org/plosbiology/article?id=10.1371/journal.pbio.0050075#s4
	Eragatis	Orvis et al. (2010)	Provides tools for functional and comparative analysis of sequencing data	http://ergatis.sourceforge.net/
	IMG-M	Markowitz et al. (2007)	Taxonomic and comparative analysis tool for genomic data	http://img.jgi.doe.gov/m
	Blast against COG database	Tatusov et al. (2000)	Functional annotation using the Clusters of Orthologous Groups (COG) database	http://www.ncbi.nlm.nih.gov/COG/
	PICRUSt	Douglas et al. (2020)	Phylogenetic investigation of communities by reconstructing functional profiles	http://picrust.github.io/picrust/
	MEGAN	Huson et al. (2007)	Taxonomic, functional, and comparative analysis tool for metagenomic and sequencing data	https://uni-tuebingen.de/fakultaeten/mathematisch-naturwissenschaftliche-fakultaet/fachbereiche/informatik/lehrstuehle/algorithms-in-bioinformatics/software/megan6/

	LEfSe	Segata et al. (2011)	Tool for taxonomic analysis based on the linear discriminant analysis	https://huttenhower.sph.harvard.edu/lefse/
	DADA2	Callahan et al. (2016)	Tool for analyzing high-resolution amplicon sequence data for taxonomic and phylogenetic analysis	http://www.metagenomics.wiki/tools/16s/dada2
	QIIME	Bolyen et al. (2019)	Comprehensive toolset for taxonomic, phylogenetic, and functional analysis of microbial communities	http://qiime.org/
Statistical analysis	STAMP	Parks et al. (2014)	Statistical analysis of taxonomic and functional profiles in ecological and metagenomic datasets	http://kiwi.cs.dal.ca/Software/STAMP
	PAST	Hammer and Harper (2001)	Provides tools for statistical analysis, including time-series and multivariate analysis	https://past.en.lo4d.com/windows

(continued)

Table 1 (continued)

Functional category	Name of tool	References	Application	Link
	FAST UniFrac	Hamady et al. (2010)	Analyzes diversity in microbial communities by comparing phylogenetic distances	https://www.nature.com/articles/ismej200997
	R software Package		Comprehensive statistical computing and graphics package widely used in bioinformatics	https://www.r-project.org/
	XLSTAT	Vidal et al. (2020)	Advanced statistical analysis tool integrated with Excel, covering a variety of methods	https://www.xlstat.com/
	CANOCO	ter Braak and Šmilauer (2012)	Multivariate data analysis and visualization tool, often used in ecology and environmental studies	http://www.canoco5.com/

Table 2 Bioinformatics tools for analysis of metatranscriptomics data

Functional category	Name of tool	Application	Link
Quality control and preprocessing	Trimmomatic	Used for removing low-quality reads, adapters, and contaminants from sequencing data	http://www.usadellab.org/cms/?page=trimmomatic
	FASTQC	Provides a visual summary of sequencing quality metrics to identify potential issues in raw data	https://www.bioinformatics.babraham.ac.uk/projects/fastqc/
	Cutadapt (Martin 2011)	Trims adapters and other unwanted sequences from high-throughput sequencing reads	https://cutadapt.readthedocs.io/en/stable/
	SolexaQA	Performs quality control and visualization for sequencing data, offering easy identification of low-quality regions	http://solexaqa.sourceforge.net/
	PRINSEQ	Filters and formats raw reads for quality improvement by removing duplicates, low-complexity, or low-quality sequences	http://prinseq.sourceforge.net/
Assembly	Trinity	De novo RNA-seq assembly tool designed to reconstruct transcriptomes from short-read sequences	https://github.com/trinityrnaseq/trinityrnaseq
	MetaSPAdes	Optimized for assembling metagenomic and RNA-seq data with high precision and scalability	https://github.com/ablab/spades
	IDBA-Tran	Focuses on transcriptome assembly, specifically designed for RNA-seq data	https://github.com/loneknightpy/idba
	rnaSPAdes (Bushmanova et al. 2019)	An RNA-specific assembler within SPAdes, suitable for single-cell and metatranscriptome data	https://github.com/1dayac/cloudrnaSPAdes-release

(continued)

Table 2 (continued)

Functional category	Name of tool	Application	Link
Taxonomic assignment	Kraken2 (Lu et al. 2022)	Classifies reads to taxonomic categories with high speed and accuracy using a k-mer-based approach	https://ccb.jhu.edu/software/kraken2/
	MetaPhlAn (Blanco-Míguez et al. 2023)	Characterizes microbial communities using marker gene analysis, providing accurate taxonomic profiling	https://huttenhower.sph.harvard.edu/metaphlan/
	Kaiju (Menzel et al. 2016)	A fast taxonomic classifier for metagenomic reads that assigns sequences at the protein level	http://kaiju.binf.ku.dk/
	Centrifuge (kim et al. 2016)	Rapidly classifies metagenomic sequences by mapping them to a reference database	https://ccb.jhu.edu/software/centrifuge/
Functional annotation	HUMAnN3 (Beghini et al. 2021)	Performs functional profiling of microbial communities to identify and quantify their pathways	https://huttenhower.sph.harvard.edu/humann/
	PROKKA (Seemann 2014)	Annotates prokaryotic genomes quickly, including gene identification and functional prediction	https://github.com/tseemann/prokka
	eggNOG-mapper (Cantalapiedra et al. 2021)	Annotates functional genes and protein families based on orthology assignments	http://eggnog-mapper.embl.de/
	MetaErg (Dong and Strous 2019)	Supports automated functional annotation of metagenomic datasets with user-friendly workflows	https://github.com/xiaoli-dong/metaerg

Differential expression analysis	DESeq2 (Love et al. 2014)	Performs robust differential expression analysis for RNA-seq data	https://bioconductor.org/packages/release/bioc/html/DESeq2.html
	edgeR (Robinson et al. 2010)	Detects differentially expressed genes using count-based statistical methods	https://bioconductor.org/packages/release/bioc/html/edgeR.html
	Limma (Smyth 2005)	Conducts statistical analysis of gene expression changes for various high-throughput experiments	https://bioconductor.org/packages/release/bioc/html/limma.html
	NOISeq (Tarazona et al. 2011)	Provides differential expression analysis based on noise modeling for non-replicated data	https://www.bioconductor.org/packages/release/bioc/html/NOISeq.html
Pathway analysis	KEGG Mapper (Kanehisa and Sato 2020)	Maps genes and proteins to pathways using the KEGG database	https://www.genome.jp/kegg/mapper.html
	GSEA (Shi and Walker 2007)	Identifies gene sets associated with biological processes or conditions through enrichment analysis	https://www.gsea-msigdb.org/
	DAVID(Dennis et al. 2003)	Provides functional annotation tools to analyze gene lists for biological meaning	https://david.ncifcrf.gov/
	Reactome (Fabregat et al. 2018)	Offers pathway-based insights into molecular interactions and cellular processes	https://reactome.org/
Statistical analysis	R	A versatile programming language for statistical computing, data analysis, and visualization	https://www.r-project.org/
	STAMP	Statistical analysis tool for metagenomic profiles, focusing on comparative studies	http://kiwi.cs.dal.ca/Software/STAMP
	PAST	A statistical package for scientific data analysis with diverse visualization options	https://past.en.lo4d.com/windows

(continued)

Table 2 (continued)

Functional category	Name of tool	Application	Link
Visualization	Krona (Edalati et al. 2022)	Provides interactive and hierarchical visualizations of complex data sets, especially taxonomic data	https://github.com/marbl/Krona
	MEGAN	Allows visualization and analysis of taxonomic and functional data derived from metagenomics	https://uni-tuebingen.de/fakultaeten/mathematisch-naturwissenschaftliche-fakultaet/fachbereiche/informatik/lehrstuehle/algorithms-in-bioinformatics/software/megan6/
	ggplot2 (R) (Valero-Mora 2010)	A widely used data visualization package in R, enabling the creation of complex, multilayered graphics	https://ggplot2.tidyverse.org/

Metatranscriptomics has emerged as a critical approach for studying microbial functionality and their interactions with the environment. By leveraging these tools, researchers can unravel the active metabolic pathways and dynamics of microbial ecosystems under various conditions.

2.3 Metaproteomics Tools

Metaproteomics is the large-scale study of all proteins expressed by microbial communities within a specific environment. By analyzing the protein complements of microbial populations, metaproteomics provides insights into their functional activity, metabolic pathways, and interactions. Unlike metagenomics or metatranscriptomics, which focus on genetic material and RNA, respectively, metaproteomics directly examines the proteins, offering a closer understanding of the actual molecular machinery at work.

This field plays a vital role in studying microbial ecology, host-microbe interactions, and environmental responses. Its applications include identifying biomarkers, understanding nutrient cycling, and monitoring community dynamics in diverse environments (Armengaud 2023). Metaproteomics workflows typically include protein extraction, digestion, mass spectrometry-based identification, and bioinformatics analyses for functional annotation, taxonomic profiling, and statistical evaluation summarizing. The key bioinformatics tools and their roles in metaproteomics analysis are given in Table 3.

Metaproteomics provides a direct and functional perspective of microbial communities by analyzing their protein expression profiles. Using these bioinformatics tools, researchers can investigate the active roles of microbes, enabling the discovery of novel functions and mechanisms in diverse ecosystems.

3 Case Studies

3.1 Case Studies in Metagenomics

Metagenomics has become an essential tool for uncovering microbial diversity and functional roles within rhizosphere microbiomes, significantly advancing agricultural research. For instance, a study in the Upper Indo-Gangetic Plains analyzed the microbial composition of wheat rhizosphere soils using 16S rRNA gene sequencing. Prominent bacterial taxa, including *Proteobacteria*, *Chloroflexi*, *Actinobacteria*, and *Firmicutes*, were identified, with soil properties like pH, organic carbon, and nutrient levels strongly influencing microbial diversity and abundance. This research provides a foundation for developing microbial solutions to enhance crop yields (Kumar et al. 2023). Similarly, in the Ghazipur region of Eastern Uttar Pradesh, metagenomic sequencing of wheat rhizosphere samples highlighted *Proteobacteria* as the dominant phylum, followed by *Firmicutes* and *Actinobacteria*. The study demonstrated the functional potential of microbial communities in promoting plant

Table 3 Bioinformatics tools used in metaproteomics analysis

Category	Name of tool	Application	Link
Protein identification	Max Quant (Tyanova et al. 2016a)	Protein identification and quantification through mass spectrometry (MS)	https://www.maxquant.org/
	Mascot	Protein identification using mass spectrometry (MS) data	http://www.matrixscience.com/
	Sequest	Protein identification by analyzing mass spectrometry (MS) data	http://fields.scripps.edu/sequest/
	X!Tandem (Bjornson et al. 2008)	Protein identification from mass spectrometry (MS) data	https://www.thegpm.org/tandem/
	MS-GF+ (Kim and Pevzner 2014)	Protein identification utilizing mass spectrometry (MS) data	https://omics.pnl.gov/software/ms-gf
Protein quantification	Progenesis QI (Sanches et al. 2023)	Quantitative proteomics for identifying and measuring proteins through MS	https://www.nonlinear.com/progenesis/qi/
	Skyline (MacLean et al. 2010)	Targeted proteomics and quantification using mass spectrometry (MS) data	https://skyline.ms/
Taxonomic profiling	MetaPro-IQ (Zhang et al. 2016)	Taxonomic profiling of metaproteomics data from MS analysis	https://github.com/compomics/metaProIQ
	UniPept (Gurdeep Singh et al. 2018)	Taxonomic profiling by analyzing peptide sequences from mass spectrometry (MS)	https://unipept.ugent.be/
Functional annotation	Blast2GO (Conesa et al. 2005)	Functional annotation of protein sequences through sequence alignments	https://www.blast2go.com/
	InterProScan (Jones et al. 2014)	Functional and domain annotation of protein sequences using sequence comparison	https://www.ebi.ac.uk/interpro/
	eggNOG-mapper	Functional annotation of proteins based on orthology using sequence data	http://eggnog-mapper.embl.de/
	Pfam (Mistry et al. 2021)	Domain annotation of protein sequences based on conserved domains	http://pfam.xfam.org/
	MEROPS (Rawlings et al. 2010)	Protease annotation and classification based on protein sequence data	https://www.ebi.ac.uk/merops/

Metaproteomic pipelines	MetaProteomeAnalyzer (Muth et al. 2015)	Comprehensive metaproteomic analysis for large-scale studies	http://mpa.ovgu.de/
	ProteoStorm (Beyter et al. 2018)	Automation of metaproteomics workflows for efficient analysis	https://github.com/ProtMet/
	Galaxy-P (Sheynkman et al. 2014)	Open-source proteomics pipeline for managing and analyzing MS data	https://galaxyproject.org/use/galaxy-p/
	MetaLab (Cheng et al. 2017)	Comprehensive metaproteomic analysis for functional and taxonomic profiling	https://metaproteomics.pnnl.gov/
Pathway analysis	KEGG Mapper	Pathway mapping of proteins to known biological pathways	https://www.genome.jp/kegg/mapper.html
	Reactome	Pathway mapping for biological processes and molecular pathways	https://reactome.org/
	iPath (Letunic et al. 2008)	Visualization of metabolic and signaling pathways based on protein data	https://pathways.embl.de/
Statistical analysis	Perseus (Tyanova et al. 2016b)	Statistical analysis and visualization of quantitative proteomics data	https://www.maxquant.org/perseus/
	R Software	Statistical computing and visualization across various omics data	https://www.r-project.org/
	MetaboAnalyst (Xia and Wishart 2016)	Statistical and pathway analysis for quantitative data, especially in metabolomics	https://www.metaboanalyst.ca/

health and soil functionality (Srivastava et al. 2020). In maize, shotgun metagenomics revealed key genes related to carbon and nitrogen fixation, phosphorus solubilization, and stress tolerance, emphasizing the rhizosphere microbiome's role in enhancing plant growth and resilience (Omotayo et al. 2022). A study of lentil cultivars under rice-fallow ecology further highlighted cultivar-specific microbiome compositions. While "Farmer-2" showed microbial adaptations favoring phosphorus solubilization, "Moitree" exhibited a higher abundance of genes associated with nitrogen cycling and stress adaptation. These findings underscore the significance of microbial interactions in improving nutrient management and crop productivity (Pramanik et al. 2020).

Collectively, these studies demonstrate metagenomics' pivotal role in understanding rhizosphere dynamics and its application in sustainable agriculture to improve soil health and crop performance.

3.2 Case Studies in Metatranscriptomics

In a study agricultural soils exposed to long-term chemical fertilizers and pesticides were arranged to generate ~20 Gb of data via Illumina HiSeq 2000 sequencing. Dominant bacterial genera, including *Achromobacter*, *Pseudomonas*, *Bacillus*, and *Sphingobium*, were linked to pesticide degradation, while *Ralstonia* and *Thiobacillus* played roles in heavy metal detoxification. Functional analysis revealed transcripts encoding enzymes such as catechol 1,2-dioxygenase and mercuric ion reductase, along with phosphate transport-related genes, highlighting microbial adaptability to chemically stressed environments and potential for bioremediation strategies (Singh et al. 2018).

A subsequent study built on these findings, analyzing soils for pesticide and heavy metal contamination. Sequencing yielded ~11.49 and ~8.43 Gb of data, identifying genera like *Streptomyces* and *Desulfobacterium* involved in chemical degradation. Transcripts for enzymes such as benzoate 1,2-dioxygenase and thioredoxin reductase suggested microbial reliance on organosulfonated compounds and phosphate stress adaptation, further showcasing the resilience of microbial communities (Singh et al. 2018).

Beyond soil ecosystems, metatranscriptomics has advanced the understanding of traditional fermentation. In Dajiang, a fermented soybean paste, metatranscriptomics and metabolite profiling linked core microbiota (e.g., *Lactobacillus* and *Tetragenococcus*) to the synthesis of key volatile compounds contributing to flavor and chromaticity. These findings reveal the microbiota's role in fermentation and opportunities for process optimization (Mukherjee and Reddy 2020). In agricultural rhizospheres, a metatranscriptomic study of wheat uncovered 21 pathways for aromatic compound degradation and six for xenobiotics, driven by microbial taxa like *Proteobacteria*, *Firmicutes*, and *Actinobacteria*. The abundance of transcripts linked to amine aromatics and benzoates highlights the bioremediation potential of rhizosphere microbial communities in chemically stressed soils (An et al. 2021). In polluted environments, metatranscriptomics identified genes encoding laccases, cellulases, and peroxidases for organic matter degradation and metallothionein and phytochelatin synthase for heavy metal detoxification, demonstrating microbial potential for environmental remediation (Sharma and Sharma 2018).

3.3 Case Studies in Metaproteomics

In a study on rice, rhizospheric soil was optimized protein extraction for various soil types and identified nearly 1000 protein spots using 2-DE and MALDI-TOF/TOF-MS. Among these, 122 proteins involved in metabolism, signal transduction, and resistance were characterized, while many remained unidentified, hinting at novel proteins. This study highlighted the intricate plant-microbe interactions in the rhizosphere and the potential of metaproteomics to enhance soil health and crop productivity (Shrestha et al. 2021; Wang et al. 2011). Another study examined interactions within a synthetic microbial community of ten bacterial strains from *Populus deltoides* rhizosphere, grown in complex and defined media. Proteomic analysis revealed that despite compositional changes, a few strains dominated under both conditions, showcasing functional stability through adaptive mechanisms like motility, metabolic shifts, and antibiotic production. Posttranslational modifications offered insights into regulatory mechanisms supporting microbial resilience (Renu et al. 2023). In maize rhizospheric soil, metaproteomics optimized protein extraction and identified 696 proteins representing 244 genera and 393 species. Conducted at the ICAR-IISS farm, this study provided the first maize rhizosphere proteome reference dataset, revealing biological processes crucial for plant growth. The dataset, accessible via ProteomeXchange (PXD014519), serves as a foundation for future research (Gupta et al. 2019). Comparative metaproteomics of wheat rhizospheric soils under saline and nonsaline conditions identified 1538 and 891 proteins, respectively. Salinity induced unique proteins like DNA-directed RNA polymerase and enzymes for proline and spermidine biosynthesis, while osmotic regulators like inositol transporter and HSP90A were specific to saline conditions. Key pathways involved sphingolipid and phenazine metabolism, highlighting microbial strategies for stress adaptation. Data is available on ProteomeXchange (PXD015387), offering insights into microbial contributions to salinity tolerance (Adeleke and Babalola 2022).

4 Conclusion

Meta-omics, encompassing metagenomics, metatranscriptomics, metaproteomics, and metabolomics, has revolutionized the study of microbial communities and their interactions with crops. By enabling a multidimensional analysis of microbial diversity, functional dynamics, and biochemical processes, meta-omics provides critical insights into plant-microbe interactions and environmental influences. These technologies, when integrated with advanced bioinformatics tools, have demonstrated their potential to address agricultural challenges such as nutrient cycling, stress tolerance, disease resistance, and crop productivity.

The transformative impact of meta-omics in unraveling the complexities of rhizosphere microbiomes, stress responses, and biochemical pathways, bridging the gap between fundamental research and applied agricultural solutions. Furthermore, integrating meta-omics data with genomic approaches like GWAS and gene editing technologies has laid the foundation for precision breeding and the development of sustainable farming practices. Despite significant advancements, challenges remain in data integration, analysis, and interpretation, emphasizing the need for interdisciplinary collaboration.

5 Future Scope

To fully harness the potential of meta-omics, future research should prioritize several key areas. Developing standardized protocols and comprehensive frameworks for multi-omics data integration and analysis is essential to enable holistic insights into microbial and plant ecosystems. Incorporating emerging technologies such as long-read sequencing, single-cell omics, and spatial transcriptomics can significantly enhance our understanding of functional and structural interactions at unprecedented resolutions. Additionally, artificial intelligence and machine learning offer immense potential for analyzing complex datasets, predicting functional outcomes, and identifying biomarkers, necessitating their integration into bioinformatics pipelines for more efficient and accurate interpretations.

Experimental validation remains critical for confirming the roles of identified genes, proteins, and metabolites in crop improvement. Future studies should emphasize functional assays and in situ experiments to bridge this gap. Leveraging meta-omics data for microbial community engineering also holds promise, enabling the design of synthetic microbiomes tailored to address agricultural challenges like disease suppression and stress mitigation. As climate change increasingly impacts agriculture, research should explore the adaptive mechanisms of microbial communities to environmental shifts, providing strategies to maintain crop resilience and productivity. Finally, translating meta-omics discoveries into practical applications such as biofertilizers, biopesticides, and genetically enhanced crops is vital to bridge the gap between laboratory findings and real-world solutions accessible to farmers. By addressing these areas, meta-omics can drive innovations that enhance agricultural sustainability, mitigate environmental impacts, and ensure global food security.

Conflict of Interest Authors declare no conflict of interest.

References

Adeleke BS, Babalola OO (2022) Meta-omics of endophytic microbes in agricultural biotechnology. Biocatal Agric Biotechnol 42:102332

Agrawal S, Arze C, Adkins RS, Crabtree J, Riley D, Vangala M et al (2017) CloVR-comparative: automated, cloud-enabled comparative microbial genome sequence analysis pipeline. BMC Genomics 18:1–11

An F, Li M, Zhao Y, Zhang Y, Mu D, Hu X et al (2021) Metatranscriptome-based investigation of flavor-producing core microbiota in different fermentation stages of dajiang, a traditional fermented soybean paste of Northeast China. Food Chem 343:128509

Andrews S (2010) FastQC: a quality control tool for high throughput sequence data

Armengaud J (2023) Metaproteomics to understand how microbiota function: the crystal ball predicts a promising future. Environ Microbiol 25(1):115–125

Beghini F, McIver LJ, Blanco-Míguez A, Dubois L, Asnicar F, Maharjan S et al (2021) Integrating taxonomic, functional, and strain-level profiling of diverse microbial communities with bioBakery. 3 elife 10:e65088

Beyter D, Lin MS, Yu Y, Pieper R, Bafna V (2018) Proteostorm: an ultrafast metaproteomics database search framework. Cell Syst 7(4):463–467

Bjornson RD, Carriero NJ, Colangelo C, Shifman M, Cheung KH, Miller PL, Williams K (2008) X!! Tandem, an improved method for running X! Tandem in parallel on collections of commodity computers. J Proteome Res 7(1):293–299

Blanco-Míguez A, Beghini F, Cumbo F, McIver LJ, Thompson KN, Zolfo M et al (2023) Extending and improving metagenomic taxonomic profiling with uncharacterized species using MetaPhlAn 4. Nat Biotechnol 41(11):1633–1644

Boisvert S, Raymond F, Godzaridis É, Laviolette F, Corbeil J (2012) Ray meta: scalable de novo metagenome assembly and profiling. Genome Biol 13:1–13

Bolger AM, Lohse M, Usadel B (2014) Trimmomatic: a flexible trimmer for Illumina sequence data. Bioinformatics 30(15):2114–2120

Bolyen E, Rideout JR, Dillon MR, Bokulich NA, Abnet CC, Al-Ghalith GA et al (2019) Reproducible, interactive, scalable and extensible microbiome data science using QIIME 2. Nat Biotechnol 37(8):852–857

Bushmanova E, Antipov D, Lapidus A, Prjibelski AD (2019) rnaSPAdes: a de novo transcriptome assembler and its application to RNA-Seq data. GigaScience 8(9):giz100

Callahan BJ, McMurdie PJ, Rosen MJ, Han AW, Johnson AJA, Holmes SP (2016) DADA2: high-resolution sample inference from Illumina amplicon data. Nat Methods 13(7):581–583

Cantalapiedra CP, Hernández-Plaza A, Letunic I, Bork P, Huerta-Cepas J (2021) eggNOG-mapper v2: functional annotation, orthology assignments, and domain prediction at the metagenomic scale. Mol Biol Evol 38(12):5825–5829

Cantu VA, Sadural J, Edwards R (2019) PRINSEQ++, a multi-threaded tool for fast and efficient quality control and preprocessing of sequencing datasets

Cheng K, Ning Z, Zhang X, Li L, Liao B, Mayne J et al (2017) MetaLab: an automated pipeline for metaproteomic data analysis. Microbiome 5:1–10

Conesa A, Götz S, García-Gómez JM, Terol J, Talón M, Robles M (2005) Blast2GO: a universal tool for annotation, visualization and analysis in functional genomics research. Bioinformatics 21(18):3674–3676

Cox MP, Peterson DA, Biggs PJ (2010) SolexaQA: at-a-glance quality assessment of Illumina second-generation sequencing data. BMC Bioinfor 11:1–6

Delcher AL, Kasif S, Fleischmann RD, Peterson J, White O, Salzberg SL (1999) Alignment of whole genomes. Nucleic Acids Res 27(11):2369–2376

Dennis G, Sherman BT, Hosack DA, Yang J, Gao W, Lane HC, Lempicki RA (2003) DAVID: database for annotation, visualization, and integrated discovery. Genome Biol 4:1–11

Dong X, Strous M (2019) An integrated pipeline for annotation and visualization of metagenomic contigs. Front Genet 10:999

Douglas GM, Maffei VJ, Zaneveld JR, Yurgel SN, Brown JR, Taylor CM et al (2020) PICRUSt2 for prediction of metagenome functions. Nat Biotechnol 38(6):685–688

Edalati A, Tahaei M, Kobyzev I, Nia VP, Clark JJ, Rezagholizadeh M (2022) Krona: parameter efficient tuning with kronecker adapter. arXiv preprint arXiv:2212.10650

Fabregat A, Jupe S, Matthews L, Sidiropoulos K, Gillespie M, Garapati P et al (2018) The reactome pathway knowledgebase. Nucleic Acids Res 46(D1):D649–D655

Gnerre S, MacCallum I, Przybylski D, Ribeiro FJ, Burton JN, Walker BJ et al (2011) High-quality draft assemblies of mammalian genomes from massively parallel sequence data. Proc Natl Acad Sci 108(4):1513–1518

Goll J, Rusch DB, Tanenbaum DM, Thiagarajan M, Li K, Methé BA, Yooseph S (2010) METAREP: JCVI metagenomics reports—an open source tool for high-performance comparative metagenomics. Bioinformatics 26(20):2631–2632

Gudgin M, Hadley M, Mendelsohn N, Moreau JJ, Nielsen HF, Karmarkar A, Lafon Y (2003) SOAP Version 1.2. W3C Recommendation 24:12

Gupta SK, Rai AK, Sarim KM, Sharma A, Budhlakoti N, Arora D et al (2019) Metaproteomic data of maize rhizosphere for deciphering functional diversity. Data Brief 27:104574

Gurdeep Singh R, Tanca A, Palomba A, Van der Jeugt F, Verschaffelt P, Uzzau S et al (2018) Unipept 4.0: functional analysis of metaproteome data. J Proteome Res 18(2):606–615

Hach F, Hormozdiari F, Alkan C, Hormozdiari F, Birol I, Eichler EE, Sahinalp SC (2010) mrsFAST: a cache-oblivious algorithm for short-read mapping. Nat Methods 7(8):576–577

Hamady M, Lozupone C, Knight R (2010) Fast UniFrac: facilitating high-throughput phylogenetic analyses of microbial communities including analysis of pyrosequencing and PhyloChip data. ISME J 4(1):17–27

Hammer Ø, Harper DA (2001) Past: paleontological statistics software package for education and data analysis. Palaeontol Electron 4(1):1

Handelsman J (2004) Metagenomics: application of genomics to uncultured microorganisms. Microbiol Mol Biol Rev 68(4):669–685

Hao Z, Liang X, Li G (2022) Quality control and preprocessing of sequencing reads. Bio-Protocol 12:13

Homer N, Merriman B, Nelson SF (2009) BFAST: an alignment tool for large scale genome resequencing. PLoS One 4(11):e7767

Huson DH, Auch AF, Qi J, Schuster SC (2007) MEGAN analysis of metagenomic data. Genome Res 17(3):377–386

Jones P, Binns D, Chang HY, Fraser M, Li W, McAnulla C et al (2014) InterProScan 5: genome-scale protein function classification. Bioinformatics 30(9):1236–1240

Jung Y, Han D (2022) BWA-MEME: BWA-MEM emulated with a machine learning approach. Bioinformatics 38(9):2404–2413

Kanehisa M, Sato Y (2020) KEGG mapper for inferring cellular functions from protein sequences. Protein Sci 29(1):28–35

Keegan KP, Glass EM, Meyer F (2016) MG-RAST, a metagenomics service for analysis of microbial community structure and function. Microb Environ Genomics (MEG):207–233

Kelaher BP, Birrer SC, Dafforn KA, Sun MY, Potts J, Scanes P et al (2018) Using meta-omics of contaminated sediments to monitor changes in pathways relevant to climate regulation. Environ Microbiol 21:389

Kim S, Pevzner PA (2014) MS-GF+ makes progress towards a universal database search tool for proteomics. Nat Commun 5(1):5277

Kim D, Song L, Breitwieser FP, Salzberg SL (2016) Centrifuge: rapid and sensitive classification of metagenomic sequences. Genome Res 26(12):1721–1729

Kuhl C, Tautenhahn R, Bottcher C, Larson TR, Neumann S (2012) CAMERA: an integrated strategy for compound spectra extraction and annotation of liquid chromatography/mass spectrometry data sets. Anal Chem 84(1):283–289

Kumar M, Ansari WA, Zeyad MT, Singh A, Chakdar H, Kumar A, Farooqi MS, Sharma A, Srivastava S, Srivastava AK (2023) Core microbiota of wheat rhizosphere under Upper Indo-Gangetic plains and their response to soil physicochemical properties. Front Plant Sci 14:1186162. https://doi.org/10.3389/fpls.2023.1186162

Langmead B, Salzberg SL (2012) Fast gapped-read alignment with Bowtie 2. Nat Methods 9(4):357–359

Laserson J, Jojic V, Koller D (2011) Genovo: de novo assembly for metagenomes. J Comput Biol 18(3):429–443
Lee WP, Stromberg MP, Ward A, Stewart C, Garrison EP, Marth GT (2014) MOSAIK: a hash-based algorithm for accurate next-generation sequencing short-read mapping. PLoS One 9(3):e90581
Letunic I, Yamada T, Kanehisa M, Bork P (2008) iPath: interactive exploration of biochemical pathways and networks. Trends Biochem Sci 33(3):101–103
Love M, Anders S, Huber W (2014) Differential analysis of count data–the DESeq2 package. Genome Biol 15(550):10–1186
Lu J, Rincon N, Wood DE, Breitwieser FP, Pockrandt C, Langmead B et al (2022) Metagenome analysis using the Kraken software suite. Nat Protoc 17(12):2815–2839
MacLean B, Tomazela DM, Shulman N, Chambers M, Finney GL, Frewen B et al (2010) Skyline: an open source document editor for creating and analyzing targeted proteomics experiments. Bioinformatics 26(7):966–968
Madden T (2013) The BLAST sequence analysis tool. NCBI Handbook 2(5):425–436
Markowitz VM, Ivanova NN, Szeto E, Palaniappan K, Chu K, Dalevi D et al (2007) IMG/M: a data management and analysis system for metagenomes. Nucleic Acids Res 36(suppl_1):D534–D538
Martin M (2011) Cutadapt removes adapter sequences from high-throughput sequencing reads. EMBnet J 17(1):10–12
Matvienko M (2015) CLC genomics workbench. Plant and Animal Genome. Sr. Field Application Scientist, CLC Bio, Aarhus, DE
Menzel P, Ng KL, Krogh A (2016) Kaiju: Fast and sensitive taxonomic classification for metagenomics. Nat Commun 7:11257
Miller JR, Koren S, Sutton G (2010) Assembly algorithms for next-generation sequencing data. Genomics 95(6):315–327
Mistry J, Chuguransky S, Williams L, Qureshi M, Salazar GA, Sonnhammer EL et al (2021) Pfam: the protein families database in 2021. Nucleic Acids Res 49(D1):D412–D419
Mukherjee A, Reddy MS (2020) Metatranscriptomics: an approach for retrieving novel eukaryotic genes from polluted and related environments. 3 Biotech 10(2):71
Muth T, Behne A, Heyer R, Kohrs F, Benndorf D, Hoffmann M et al (2015) The MetaProteomeAnalyzer: a powerful open-source software suite for metaproteomics data analysis and interpretation. J Proteome Res 14(3):1557–1565
Namiki T, Hachiya T, Tanaka H, Sakakibara Y (2011) MetaVelvet: an extension of velvet assembler to de novo metagenome assembly from short sequence reads. In: Proceedings of the 2nd ACM conference on bioinformatics, computational biology and biomedicine, pp 116–124
Omotayo OP, Igiehon ON, Babalola OO (2022) Microbial genes of agricultural importance in maize rhizosphere unveiled through shotgun metagenomics. Span J Soil Sci 12:10427
Orvis J, Crabtree J, Galens K, Gussman A, Inman JM, Lee E et al (2010) Ergatis: a web interface and scalable software system for bioinformatics workflows. Bioinformatics 26(12):1488–1492
Otto TD, Dillon GP, Degrave WS, Berriman M (2011) RATT: rapid annotation transfer tool. Nucleic Acids Res 39(9):e57–e57
Parks DH, Tyson GW, Hugenholtz P, Beiko RG (2014) STAMP: statistical analysis of taxonomic and functional profiles. Bioinformatics 30(21):3123–3124
Peimbert M, Alcaraz LD (2016) A hitchhiker's guide to metatranscriptomics. Field guidelines for genetic experimental designs in high-throughput sequencing. Springer, Cham, pp 313–342
Peng Y, Leung HC, Yiu SM, Chin FY (2011) Meta-IDBA: a de Novo assembler for metagenomic data. Bioinformatics 27(13):i94–i101
Pramanik K, Das A, Banerjee J, Das A, Chatterjee S, Sharma R et al (2020) Metagenomic insights into rhizospheric microbiome profiling in lentil cultivars unveils differential microbial nitrogen and phosphorus metabolism under rice-fallow ecology. Int J Mol Sci 21(23):8895
Rawlings ND, Barrett AJ, Bateman A (2010) MEROPS: the peptidase database. Nucleic Acids Res 38(suppl_1):D227–D233
Renu, Sarim KM, Gupta SK et al (2023) A comparative analysis of rhizospheric metaproteome of wheat grown in saline and non-saline soils identifies proteins linked with characteristic functions. Plant Growth Regul 101:415–426. https://doi.org/10.1007/s10725-023-01027-8

Robinson MD, McCarthy DJ, Smyth GK (2010) edgeR: a Bioconductor package for differential expression analysis of digital gene expression data. Bioinformatics 26(1):139–140

Ruffalo M, Koyutürk M, Ray S, LaFramboise T (2012) Accurate estimation of short read mapping quality for next-generation genome sequencing. Bioinformatics 28(18):i349–i355

Sanches PH, Oliveira DCD, Reis IGD, Fernandes AM, Silva AA, Eberlin MN et al (2023) Fitting structure-data files (. SDF) libraries to progenesis QI identification searches. J Braz Chem Soc 34(7):1013–1019

Seemann T (2014) Prokka: rapid prokaryotic genome annotation. Bioinformatics 30(14):2068–2069

Segata N, Izard J, Waldron L, Gevers D, Miropolsky L, Garrett WS, Huttenhower C (2011) Metagenomic biomarker discovery and explanation. Genome Biol 12:1–18

Segata N, Boernigen D, Tickle TL, Morgan XC, Garrett WS, Huttenhower C (2013) Computational meta'omics for microbial community studies. Mol Syst Biol 9(1):666

Sharma R, Sharma PK (2018) Metatranscriptome sequencing and analysis of agriculture soil provided significant insights about the microbial community structure and function. Ecol Genetics Genomics 6:9–15

Sharuddin SS, Ramli N, Yusoff MZM, Muhammad NAN, Ho LS, Maeda T (2022) Advancement of metatranscriptomics towards productive agriculture and sustainable environment: a review. Int J Mol Sci 23(7):3737

Sheynkman GM, Johnson JE, Jagtap PD, Shortreed MR, Onsongo G, Frey BL et al (2014) Using Galaxy-P to leverage RNA-Seq for the discovery of novel protein variations. BMC Genomics 15:1–9

Shi J, Walker MG (2007) Gene set enrichment analysis (GSEA) for interpreting gene expression profiles. Curr Bioinforma 2(2):133–137

Shrestha HK, Appidi MR, Villalobos Solis MI, Wang J, Carper DL, Burdick L et al (2021) Metaproteomics reveals insights into microbial structure, interactions, and dynamic regulation in defined communities as they respond to environmental disturbance. BMC Microbiol 21:1–17

Simpson JT, Wong K, Jackman SD, Schein JE, Jones SJ, Birol I (2009) ABySS: a parallel assembler for short read sequence data. Genome Res 19(6):1117–1123

Singh DP, Prabha R, Gupta VK, Verma MK (2018) Metatranscriptome analysis deciphers multifunctional genes and enzymes linked with the degradation of aromatic compounds and pesticides in the wheat rhizosphere. Front Microbiol 9:1331

Smyth GK (2005) Limma: linear models for microarray data. In: Bioinformatics and computational biology solutions using R and Bioconductor. Springer, New York, pp 397–420

Sreekanti V, Wu C, Lin XC, Schleier-Smith J, Faleiro JM, Gonzalez JE et al (2020) Cloudburst: Stateful functions-as-a-service. arXiv preprint arXiv:2001.04592

Srivastava R, Srivastava AK, Ramteke PW, Gupta VK, Srivastava AK (2020) Metagenome dataset of wheat rhizosphere from Ghazipur region of Eastern Uttar Pradesh. Data Brief 28:105094. https://doi.org/10.1016/j.dib.2019.105094

Staley C, Gould TJ, Wang P, Phillips J, Cotner JB, Sadowsky MJ (2014) Core functional traits of bacterial communities in the Upper Mississippi River show limited variation in response to land cover. Front Microbiol 5:414

Stewart AC, Osborne B, Read TD (2009) DIYA: a bacterial annotation pipeline for any genomics lab. Bioinformatics 25(7):962–963

Tarazona S, García F, Ferrer A, Dopazo J, Conesa A (2011) NOIseq: a RNA-seq differential expression method robust for sequencing depth biases. EMBnet J 17(B):18–19

Tatusov RL, Galperin MY, Natale DA, Koonin EV (2000) The COG database: a tool for genome-scale analysis of protein functions and evolution. Nucleic Acids Res 28(1):33–36

ter Braak CJ, Šmilauer P (2012) Canoco reference manual and user's guide: software for ordination, version 5.0

Tyanova S, Temu T, Cox J (2016a) The MaxQuant computational platform for mass spectrometry-based shotgun proteomics. Nat Protoc 11(12):2301–2319

Tyanova S, Temu T, Sinitcyn P, Carlson A, Hein MY, Geiger T et al (2016b) The Perseus computational platform for comprehensive analysis of (prote) omics data. Nat Methods 13(9):731–740

Valero-Mora PM (2010) ggplot2: elegant graphics for data analysis. J Stat Softw 35:1–3

Vidal NP, Manful CF, Pham TH, Stewart P, Keough D, Thomas R (2020) The use of XLSTAT in conducting principal component analysis (PCA) when evaluating the relationships between sensory and quality attributes in grilled foods. MethodsX 7:100835

Wang HB, Zhang ZX, Li H, He HB, Fang CX, Zhang AJ et al (2011) Characterization of metaproteomics in crop rhizospheric soil. J Proteome Res 10(3):932–940

Wilmes P, Bond PL (2006) Metaproteomics: studying functional gene expression in microbial ecosystems. Trends Microbiol 14(2):92–97

Xia J, Wishart DS (2016) Using MetaboAnalyst 3.0 for comprehensive metabolomics data analysis. Curr Protoc Bioinformatics 55(1):14–10

Xie Y, Wu G, Tang J, Luo R, Patterson J, Liu S et al (2014) SOAPdenovo-trans: de novo transcriptome assembly with short RNA-Seq reads. Bioinformatics 30(12):1660–1666

Zhang X, Ning Z, Mayne J, Moore JI, Li J, Butcher J et al (2016) MetaPro-IQ: a universal metaproteomic approach to studying human and mouse gut microbiota. Microbiome 4:1–12

Zuñiga C, Zaramela L, Zengler K (2017) Elucidation of complexity and prediction of interactions in microbial communities. Microb Biotechnol 10(6):1500–1522

Meta-omics Approaches in Crop and Agriculture Research: A Scientometric Analysis of the Paradigm Shift

Ratna Prabha, Renu, Amrender Kumar, and Dhananjaya Pratap Singh

1 Omics Approaches in Crop Improvement

Agricultural research is being revolutionized by different "omics" technologies that enable rapid analysis of plant characteristics and accelerate crop enhancement while promoting environmental sustainability (Castillejo et al. 2023; Cortés et al. 2023). With the advent of diverse omics technologies in recent decades, the landscape of biological research has been revolutionized which enable researchers to broadly study plants' responses to environmental stresses and physiological challenges across different molecular levels. From DNA alterations to fluctuations in transcript abundance, protein expression, metabolite profiles, and mineral content, these approaches provide extraordinary insights into crop development and adaptation (Muthamilarasan et al. 2019; Yang et al. 2021). The approaches such as genomics, transcriptomics, proteomics, metabolomics, and phenomics are assisting to gather extensive datasets through various large-scale analytical approaches, spanning from genetic material to observable traits (Cortés et al. 2023). The integration of various omics disciplines—including genomics, metagenomics, transcriptomics, proteomics, metabolomics, phenomics, and ionomics—has created a powerful approach for understanding the complicated molecular networks that govern plant systems (Salt et al. 2008; Houle et al. 2010; Talukdar and Sinjushin 2015; Wu et al. 2017; Muthamilarasan et al. 2019; Yang et al. 2021). Advanced computational methods

R. Prabha · A. Kumar
Agricultural Knowledge Management Unit (AKMU), ICAR-Indian Agricultural Research Institute, New Delhi, Delhi, India

Renu
Crop Science Division, Indian Council of Agricultural Research, New Delhi, India

D. P. Singh (✉)
ICAR-Indian Institute of Vegetable Research, Varanasi, Uttar Pradesh, India
e-mail: dhananjaya.singh@icar.org.in

Renu et al. (eds.), *Meta-omics in Crop Improvement*,
https://doi.org/10.1007/978-981-95-2294-1_15

and bioinformatics approaches, including machine learning algorithms, are proving invaluable assistance in decoding these complex, multidimensional datasets, offering insights into the intricate relationships between genes, traits, and broader ecological interactions, investigating how crops interact with their microbial communities (Zhang et al. 2021) and respond to varying environmental conditions (Resende et al. 2021; Costa-Neto and Fritsche-Neto 2021). This comprehensive approach is enhancing our understanding of crop behavior across multiple scales (Cortés et al. 2023).

Genomics is a rapidly evolving field that has significantly transformed biological research and its applications across various domains. Genomics has progressed from its inception in the late 1980s to the present day, driven by innovative technologies such as high-throughput sequencing and microarray analysis (Descombes and Oakeley 2016; Carlson and Choudhuri 2008). These advancements have enabled comprehensive genome-wide studies, shifting research from traditional gene-by-gene analysis to large-scale data-driven approaches (Carlson and Choudhuri 2008). The applications of genomics are diverse and impactful, particularly in the field of agriculture. Genetically modified crops with enhanced traits and biofortification are direct outcomes of genomic research (Kumar 2015). Genomics has revolutionized crop improvement by providing tools to understand and manipulate plant genetics at a molecular level. Genomics allows for the isolation and manipulation of genes responsible for important plant traits, enhancing our understanding of plant development and quality traits (Murphy 1998). Next-generation sequencing (NGS) technologies enable whole-genome and transcriptome sequencing, making it feasible to develop genomic resources for various crops, including those with large and complex genomes (Bartoszewski and Malepszy 2012; Edwards 2016). Genomics-Assisted Breeding (GAB) utilizes genomic information for plant breeding, speeding up the process of developing improved crop varieties. Techniques include marker-assisted selection (MAS), genomic selection, and genome-wide association studies (GWAS) (Aglawe et al. 2021; Varshney et al. 2021). Further different approaches like genome editing, phenomics integration, and pan-genomics also open new avenues in the field of agricultural research. Technologies like CRISPR/Cas9, TALENs, and ZFNs allow precise modifications in plant genomes, enabling targeted gene knockouts, single base modifications, and gene stacking (Khan et al. 2024; Jung and Seo 2017; Debbarma et al. 2024; Somashekar 2024). Pan-genome analysis evaluates the genetic diversity within a species, providing insights into structural variants and their impact on agronomic traits, which can be leveraged for crop improvement (Tao et al. 2019). The bottleneck in crop improvement is shifting from genotyping to phenotyping. Advanced phenomics platforms are essential for collecting high-quality phenotypic data, which is crucial for linking genotype to phenotype (Mir et al. 2019; Jin et al. 2021).

Transcriptomics is the study of the complete set of RNA transcripts produced by the genome under specific circumstances or in a specific cell. This field has been revolutionized by next-generation sequencing (NGS) technologies, which allow for high-throughput, single-base resolution analysis of RNA (Ali and Lee 2022; Chen and Tan 2014; Marguerat et al. 2008). RNA sequencing (RNA-seq) captures all

RNA sequences in a sample, providing a comprehensive view of gene expression and revealing novel transcripts and regulatory mechanisms (Ingvarsson et al. 2016; Lowe et al. 2017), while microarrays are used to quantify predetermined RNA sequences, although they are increasingly being superseded by RNA-seq due to the latter's higher resolution and broader scope (Marguerat et al. 2008; Lowe et al. 2017). Transcriptomics has several key applications in crop improvement, particularly in understanding plant stress responses. Transcriptomics helps identify genes involved in plant responses to abiotic (e.g., drought, heat) and biotic (e.g., pathogens) stresses. This knowledge is essential for developing stress-tolerant crop varieties (Yadav et al. 2024; Gupta et al. 2023; Romero et al. 2017; Singh and Pradhan 2020; Zenda et al. 2021; Kido et al. 2022). It provides a comprehensive understanding of the transcriptome changes during different metabolic processes, such as fruit development, ripening, and responses to biotic and abiotic stresses in fruit crops (Singh and Pradhan 2020). By analyzing gene expression profiles, researchers can pinpoint genes that enhance yield and nutritional content. This is particularly important for meeting the food demands of a growing population (Yadav et al. 2024; Zenda et al. 2021; Tahir ul Qamar et al. 2020). Transcriptomics also intersects with epigenetics, where gene expression is influenced by DNA methylation and histone modifications. This adds another layer of control that can be harnessed for crop improvement (Saraswat et al. 2017; Gupta and Salgotra 2022). Transcriptomics has been used to improve crop varieties with high yield, enhanced nutritional value, and resistance to stresses, such as drought, salinity, and insect attacks (Tahir ul Qamar et al. 2020; Feng et al. 2024). Transcriptomics studies are extensively used to understand how crops respond to biotic (e.g., pests, diseases) and abiotic (e.g., drought, salinity) stresses. This assists in identifying stress-responsive genes that can be targeted for developing stress-tolerant crop varieties (Zenda et al. 2021; Vishwanath et al. 2024). By analyzing the transcriptome, researchers can identify differentially expressed genes associated with desirable traits such as higher yield, better quality, and stress tolerance. This facilitates targeted breeding and genetic engineering (Vishwanath et al. 2024; Singh and Pradhan 2020; Naidu et al. 2023). In case of fruit crops, transcriptomics is used to study gene expression during fruit development and ripening, aiding in the improvement of fruit quality and shelf life (Singh and Pradhan 2020). Further, transcriptomic studies in vegetables have revealed gene expression patterns that can be manipulated to enhance nutritional quality, making crops more beneficial for human health (Vishwanath et al. 2024). The data from transcriptomics can be used to construct gene co-expression networks, which help in understanding complex biological processes and identifying key regulatory genes for crop improvement (Zainal-Abidin et al. 2022). Combining transcriptomics with other omics technologies (e.g., genomics, proteomics, metabolomics) provides a comprehensive understanding of plant biology, leading to the development of climate-resilient and nutritionally rich crops (Zenda et al. 2021; Borém and Fritsche-Neto 2014).

Proteomics is a field that studies the entire set of proteins in a biological system, providing insights into protein identification, quantities, interactions, localization, and posttranslational modifications (Vrazas and Katsani 2024; Ravi 2014). Key techniques used in proteomics research include high-throughput and high-resolution analytical techniques, with mass spectrometry playing a leading role (Vrazas and Katsani 2024; Husi and Albalat 2014). Proteomics also provides quantitative, large-scale analysis of proteins present in organisms under specific conditions, offering insights into protein expression profiles, posttranslational modifications, and protein interaction networks (Gupta et al. 2015). Proteomics is increasingly being integrated into crop improvement strategies due to its ability to provide detailed insights into plant biology at the molecular level. Proteomics helps identify proteins involved in plant responses to abiotic (e.g., drought, salinity) and biotic (e.g., pathogens) stresses, which are critical for developing stress-tolerant crops (Das et al. 2015; Mahajan and Zargar 2020; Chauhan et al. 2023; Raja et al. 2017; Mustafa and Komatsu 2021). This includes studying posttranslational modifications that affect protein function under stress conditions (Das et al. 2015; Mustafa and Komatsu 2021). Techniques such as mass spectrometry allows for the quantification of protein expression levels, helping to identify differentially expressed proteins under various conditions. This is crucial for understanding metabolic and defense pathways in plants (Mahajan and Zargar 2020; Tan et al. 2017; Vanderschuren et al. 2013). Proteomics aids in the identification and validation of gene products associated with desirable traits, supporting marker-assisted selection in breeding programs. This can lead to the development of crops with improved yield, quality, and stress resistance (Pedreschi et al. 2013; Kage et al. 2016; Eldakak et al. 2013). Proteomics is also being used to reduce post-harvest losses by identifying markers for quality traits and early detection of physiological or pathogen-related issues (Pedreschi et al. 2013).

Metabolomics is the comprehensive study of metabolites, the small molecules involved in metabolism, within biological systems such as cells, tissues, and biofluids (Dryagina et al. 2020; Wishart et al. 2020; Pushparaj 2019). This field aims to analyze all known and unknown metabolites to understand the metabolic responses to various stimuli, including environmental changes, diseases, and drug treatments. It relies on approaches like mass spectrometry (MS), nuclear magnetic resonance (NMR) spectroscopy and chromatography techniques, etc. Mass spectrometry (MS) is utilized for its ability to identify thousands of compounds in a single sample based on their unique mass-charge ratios (Dryagina et al. 2020; Wishart et al. 2020; Pushparaj 2019). Nuclear magnetic resonance (NMR) spectroscopy allows for the determination of metabolites by analyzing their spectral arrays (Dryagina et al. 2020; Wishart et al. 2020; Pushparaj 2019). Chromatography techniques like liquid chromatography (LC) and gas chromatography (GC), often coupled with MS, are used for enhanced detection capabilities (Wishart et al. 2020 Pushparaj 2019). Metabolomics plays a crucial role in crop improvement by providing insights into plant phenotypes and their underlying biochemical pathways. Metabolomics allows for the detailed phenotyping of crops, aiding in the identification of metabolic markers linked to desirable traits such as stress resistance, nutritional value, and yield (Sakurai 2022; Herrmann and Schauer 2013; Fernie and Schauer 2009; Razzaq et al. 2022). By integrating metabolomics with genomics, researchers can develop functional molecular markers (FMMs) that enhance the precision of MAS, facilitating the selection of traits governed by complex genetic interactions (Herrmann and Schauer 2013; Kage et al. 2016). Metabolomics helps in understanding plant responses to abiotic and biotic stresses by mapping specific metabolites to their genetic pathways, thus enabling the breeding of stress-tolerant crops (Razzaq et al. 2022; Alawiye and Babalola 2021; Singh and Roychoudhury 2023). Metabolomics aids in improving the nutritional content and quality of crops by identifying and manipulating the metabolic pathways responsible for the accumulation of valuable compounds (Pott et al. 2021; Hamany Djande et al. 2020).

The integration of multiple omics approaches represents a powerful strategy for understanding complex biological systems in crops. By combining genomics (studying DNA sequences and genetic variations), transcriptomics (analyzing RNA expression patterns), proteomics (examining protein abundance and modifications), and metabolomics (profiling metabolite dynamics), researchers can construct comprehensive networks of gene expression and regulation that govern important plant traits. This multi-omics integration provides deeper insights into plant development, stress responses, and adaptation mechanisms than single omics approaches alone. Such holistic understanding enables more precise identification of key genetic elements and molecular pathways, facilitating targeted crop improvement strategies for enhanced yield, stress tolerance, and nutritional quality. The synergistic application of these technologies has revolutionized our ability to decipher complex trait architectures and accelerate crop breeding programs through improved selection strategies (Pandita 2023, Yadav et al. 2024).

2 Bibliometric Analysis

Bibliometric analysis of omics research in crop improvement reveals significant trends and patterns in scientific contributions across the field. This quantitative assessment examines publication patterns, citation impacts, research hotspots, collaborative networks, and geographical distributions of research outputs. Analysis indicates a substantial increase in multi-omics studies over the past decade, with key research clusters emerging around stress tolerance, nutrient use efficiency, and crop quality enhancement. Publication trends show a shift from single-omics to integrated approaches, with prominent research institutions and countries leading these advances. Citation patterns highlight the growing impact of meta-omics studies, particularly in areas of practical agricultural applications. Collaboration networks reveal strong international partnerships, though with notable disparities between developed and developing nations. The analysis also identifies emerging research frontiers and potential gaps in current knowledge, providing valuable insights for directing future research efforts and resource allocation in agricultural omics research.

Examining the landscape of omics research in the area of crop improvement through bibliometric analysis unveils different patterns in scientific contributions. This analysis identifies key players from leading nations and research centers to influential scientists and their groundbreaking publications. The data points to three major research focuses, i.e., genomics, gene regulation, and metabolomics which have captured worldwide scientific interest. The field's collaborative spirit is particularly noteworthy, with experts from diverse fields working collaboratively. This cross-pollination of ideas has created a rich ecosystem for innovation. By systematically tracking publication patterns and citation networks, we can identify emerging research frontiers and gauge the momentum behind different research directions. Such insights prove invaluable for both researchers charting their investigative paths and funding bodies seeking to maximize their research investment impact.

The power of bibliometric analysis extends beyond mere performance metrics. While it certainly helps evaluate the scholarly impact of various stakeholders—from individual researchers to entire nations—its true value lies in illuminating the road ahead. By revealing unexplored territories within omics research in field of crop improvement and agriculture, this analytical approach guides the global research community toward promising new avenues of investigation. Additionally, it provides policymakers with data-driven insights for strategic resource allocation and research prioritization. Perhaps most significantly, these analyses catalyze new collaborative ventures. By mapping the interconnections between different scientific disciplines, bibliometric studies facilitate the formation of research partnerships that transcend traditional boundaries. This framework for interdisciplinary collaboration accelerates scientific progress and enables more comprehensive approaches to complex challenges in agriculture research.

This bibliometric study provides a comprehensive overview of the global scientific output and impact of different -omics fields, i.e., genomics, transcriptomics, proteomics, and metabolomics, in the area of crop improvement and agriculture over the past two and a half decades (2000–2025), serving as a valuable resource for researchers, policymakers, and stakeholders in the field of plant sciences and agriculture.

3 Landscape of Scientific Literature Databases

The landscape of scientific literature databases is predominantly characterized by two major platforms: Web of Science (WoS), maintained by Clarivate Analytics, and Scopus, operated by Elsevier. These platforms serve as comprehensive repositories and citation indices, including diverse scholarly outputs comprising peer-reviewed articles, conference proceedings, and academic publications. Comparative analyses have consistently demonstrated Scopus's equivalence in reliability metrics while offering expanded coverage parameters (Harzing and Alakangas 2016; Pranckutė 2021).

For the current study, SCOPUS database was used for the bibliometric analysis. In the process of data collection, the search field utility was accessed in which the field was restricted to "Title" (TITLE), "Abstract" (ABS), and "Keywords" (KEY). The queries were generated using the words "genomics," "transcriptomics," "proteomics," "metabolomics," and "crop" and "agriculture." The possible combinations were queried using "AND" and "OR" logical operators in this process. The initial exclusion criteria on language, year of publication, and document type were applied to filter the dataset. All publications of duration 2000–2025 were considered (to cover early access publications also, if any), Document Types were limited to Article, Review, Book chapter, Book, Editorial, Note, Letter, and Short survey, while Source type were restricted to Journal, Book, and Book series while publications in English language were only considered. The search query was given as TITLE-ABS-KEY ((genomics OR transcriptomics OR proteomics OR metabolomics) AND (crop OR agriculture)) AND PUBYEAR >1999 AND PUBYEAR <2026 AND (LIMIT-TO (LANGUAGE, "English")) AND (LIMIT-TO (SRCTYPE, "b") OR LIMIT-TO (SRCTYPE, "j") OR LIMIT-TO (SRCTYPE, "k")) AND (LIMIT-TO (DOCTYPE, "ar") OR LIMIT-TO (DOCTYPE, "re") OR LIMIT-TO (DOCTYPE, "ch") OR LIMIT-TO (DOCTYPE, "bk") OR LIMIT-TO (DOCTYPE, "ed") OR LIMIT-TO (DOCTYPE, "sh") OR LIMIT-TO (DOCTYPE, "no") OR LIMIT-TO (DOCTYPE, "le")) (Accessed on 31-12-2024). This initial search query resulted in a total of 16,773 publications. Metadata of these publications was downloaded for bibliometric analysis.

4 Research Impact Through Data Analysis

The quantitative assessment of research impact was conducted through four fundamental bibliometric indicators:

1. Publication Metrics.
   - Total Publications (TP): Quantifies research output volume across entities (authors, countries, publishers, journals).
   - Total Citations (TC): Measures cumulative citation frequency of research artifacts.
   - h-index: A composite metric where h publications have received $\geq$h citations,
   - Average Citation Per Publication (ACPP): Calculated as TC/TP, representing mean citation density per publication.

Additionally, by using the metadata from publications, network visualizations utilizing various data attributes were carried out to gain deeper insights. The study employed VOSviewer software (van Eck and Waltman 2010) for bibliometric network visualization following methodology of Biscaro and Giupponi (2014). Two primary network analyses were conducted:

A. Keyword co-occurrence networks were constructed to map scientific networks, where each node represents a keyword, and connecting links indicate the co-occurrence of keywords. Node size correlates with keyword frequency. Thickness of link indicates relationship strength between keyword pairs. Consequently, it offers enhanced clarity regarding the structure of knowledge, the evolution of trending topics, and the trajectories of the scientific domain (Radhakrishnan et al. 2017; Zhuang et al. 2016).

B. Co-authorship networks were generated to examine research collaboration patterns, where, nodes represent individual researchers and node size correlates with number of publications, link represent co-authorship relationships with link weights indicating the number of co-authored articles. Earlier studies have assessed both co-authorship and keyword co-occurrence analyses as network-based approaches (Chen 2004, 2006; Zhang et al. 2012). The integration of these complementary analytical approaches provides a comprehensive framework for understanding both the semantic and social dimensions of scientific research evolution within the field.

5 Publication Outputs

A total of 16,773 documents related to research in field of genomics, transcriptomics, proteomics, and metabolomics in agriculture sector were collected (accessed date: Dec 31, 2024). These documents were classified as article (10,891), review (3084), Book Chapter (1970), Editorial (249), Books (329), and Note (99) (Fig. 1). The yearly publications, citations, and cumulative publications from 2000 to 2023 are presented in chronological order in Fig. 1. A cumulative citation of 622,615 was reported for these publications with h-index 285 and average citation per year 24904.60 and average citation per publication as 37.12.

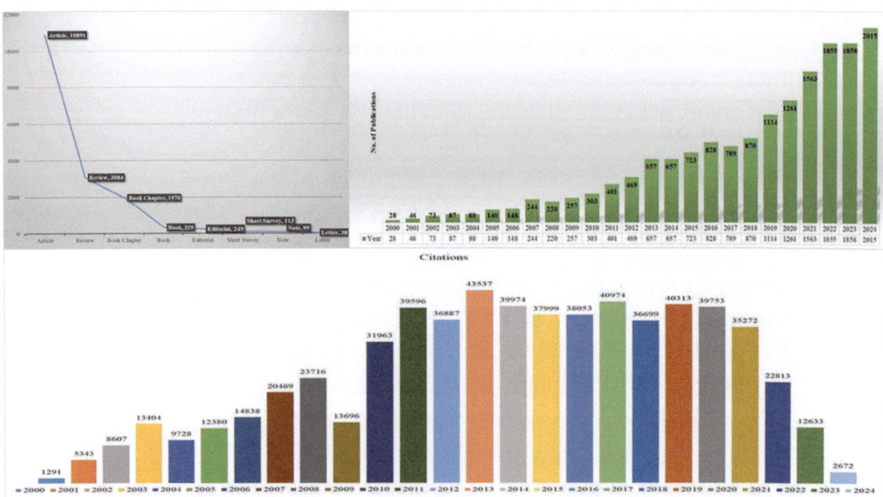

Fig. 1 Publication and citation trends in agricultural meta-omics research (2000–2024). Top panel: Year-wise citation growth showing exponential increase from 2000 to 2024. Middle panel: Annual citation metrics demonstrating consistent growth in research impact. Bottom panel: Publication output by year showing research productivity trends, with notable increases post-2010. The data indicates significant growth in both research output and impact, particularly in the last decade, reflecting the expanding influence of meta-omics approaches in agricultural research. Data sourced from scientific databases and analyzed using bibliometric tools

6 Impact Analysis

Citation metrics serve as quantitative indicators of scholarly impact and knowledge dissemination within scientific disciplines (Lipsman et al. 2014; Zyoud et al. 2015). Analysis of citation data from the SCOPUS database (2000–2025) reveals the most influential publications in agricultural -omics research, with citation frequencies ranging from 3486 to 1651 (Table 1). The highest impact was achieved by research focusing on paleopolyploid soybean genome sequencing (Schmutz et al. 2010). Notably, 40% of the most cited publications concentrate on crop genome sequencing, encompassing soybean (Schmutz et al. 2010), rice (Goff et al. 2002), tomato (Tomato Genome Consortium 2012), and potato (Potato Genome Sequencing Consortium et al. 2011), underscoring genomics' pivotal role in agricultural advancement (Table 1). The remaining high-impact publications address diverse agricultural challenges, including drought stress responses (Chaves et al. 2003), phosphorus acquisition mechanisms (Vance et al. 2003), rhizosphere microbiome interactions (Mendes et al. 2011), antimicrobial resistance (D'Costa et al. 2011), genetic marker applications (Davey et al. 2011), and comparative analysis of ERF gene families in model organisms (Nakano et al. 2006). These influential works predominantly appear in high-impact factor journals, particularly Nature and Science, supporting the correlation between journal prestige and citation frequency (Calcagno et al. 2012; Van Noorden 2017). However, it is important to note that citation metrics may vary across bibliometric databases, with platforms like Google Scholar incorporating nontraditional citation sources, potentially affecting impact rankings (Yang and Meho 2006).

7 Author Statistics and Co-authorship Analysis

The bibliometric analysis of agricultural omics research from 2000 to 2025 revealed a corpus of 16,777 publications authored by 46,013 authors, yielding a mean authorship density of 5 researchers per manuscript. Examination of individual contributions identified Varshney RK as the most prolific author with 164 publications, followed by Fernie AR and Komatsu S contributing 88 and 83 publications, respectively (Table 2). Bibliometric indices demonstrate heterogeneous impact patterns among the top contributors: while Varshney RK is having the highest h-index, Paterson AH exhibited superior citation metrics with 185.80 citations per publication despite a moderate publication count of 50 (Table 2). Further, Kole C reflects least h-index among the top 10 contributors (Table 2). This citation-to-publication ratio indicates that raw publication counts may not directly correlate with research impact in the agricultural omics domain.

Network analysis of co-authorship patterns was conducted using VOSviewer to elucidate collaborative relationships in the field (Fig. 2). The network topology manifests as a heterogeneous structure characterized by densely connected collaborative communities interspersed with peripheral nodes, reflecting varying degrees of research collaboration intensity. This network architecture enables the

Table 1 Highly cited publications in field of -omics areas in crop and agriculture sector (2000–2025)

S. no.	Publication	YoP: year of publication	TC: total citation	ACPY: average citation per year	FWCI[a]
1.	Schmutz J, Cannon SB, Schlueter J, et al. Genome sequence of the palaeopolyploid soybean. Nature. 2010;463(7278):178–83. https://doi.org/10.1038/nature08670. Erratum in: Nature. 2010;465(7294):120	2010	3486	232.40	35.83
2.	Chaves MM, Maroco JP, Pereira JS. Understanding plant responses to drought – from genes to the whole plant. Funct Plant Biol. 2003;30(3):239–264.	2003	2946	133.91	17.52
3.	Goff SA, Ricke D, Lan TH, et al. A draft sequence of the rice genome (*Oryza sativa* L. ssp. *japonica*). Science. 2002;296(5565):92–100	2002	2809	122.13	29.29
4.	Tomato Genome Consortium. The tomato genome sequence provides insights into fleshy fruit evolution. Nature. 2012;485(7400):635–41	2012	2516	193.54	41.76
5.	Vance CP, Uhde-Stone C, Allan DL. Phosphorus acquisition and use: critical adaptations by plants for securing a nonrenewable resource. New Phytol. 2003;157(3):423–447.	2003	2281	103.68	5.42
6.	Mendes R, Kruijt M, de Bruijn I, et al. Deciphering the rhizosphere microbiome for disease-suppressive bacteria. Science. 2011;332(6033):1097–100	2011	1944	138.86	10.68
7.	D'Costa VM, King CE, Kalan L, et al. Antibiotic resistance is ancient. Nature. 2011;477(7365):457–61	2011	1931	137.93	5.65
8.	Davey JW, Hohenlohe PA, Etter PD, Boone JQ, Catchen JM, Blaxter ML. Genome-wide genetic marker discovery and genotyping using next-generation sequencing. Nat Rev. Genet. 2011; 12(7):499–510	2011	1887	134.79	14.63
9.	Nakano T, Suzuki K, Fujimura T, Shinshi H. Genome-wide analysis of the ERF gene family in Arabidopsis and rice. Plant Physiol. 2006;140(2):411–32.	2006	1821	95.84	14.24
10.	Potato Genome Sequencing Consortium; Xu X, Pan S, et al. Genome sequence and analysis of the tuber crop potato. Nature. 2011;475(7355):189–95	2011	1651	117.93	21.60

[a]*FWCI*: field-weighted citation impact shows how well-cited this document is when compared to similar documents. A value greater than 1.00 means the document is more cited than expected according to the average. It takes into account:

The year of publication

Document type

Disciplines associated with its source

The FWCI is the ratio of the document's citations to the average number of citations received by all similar documents over a 3-year window. Each discipline makes an equal contribution to the metric, which eliminates differences in researcher citation behavior

Table 2 Top contributing authors worked in –omics assisted research in crop and agriculture sector

No.	Author	TP	TC	h-index	ACPP
1.	Varshney, R.K.	164	10,724	59	65.39
2.	Fernie, A.R.	88	5983	35	67.99
3.	Komatsu, S.	83	3655	34	44.04
4.	Edwards, D.	67	5400	32	80.60
5.	Paterson, A.H.	50	9290	32	185.80
6.	Batley, J.	48	4343	26	90.48
7.	Pandey, M.K.	47	2611	22	55.55
8.	Siddique, K.H.M.	47	2145	22	45.64
9.	Kole, C.	46	565	11	12.28
10.	Lucini, L.	46	1639	21	35.63

TP total publication, *TC* total citation, *ACPP* average citations per publication, *APP* authors' popular publication, TC of APP

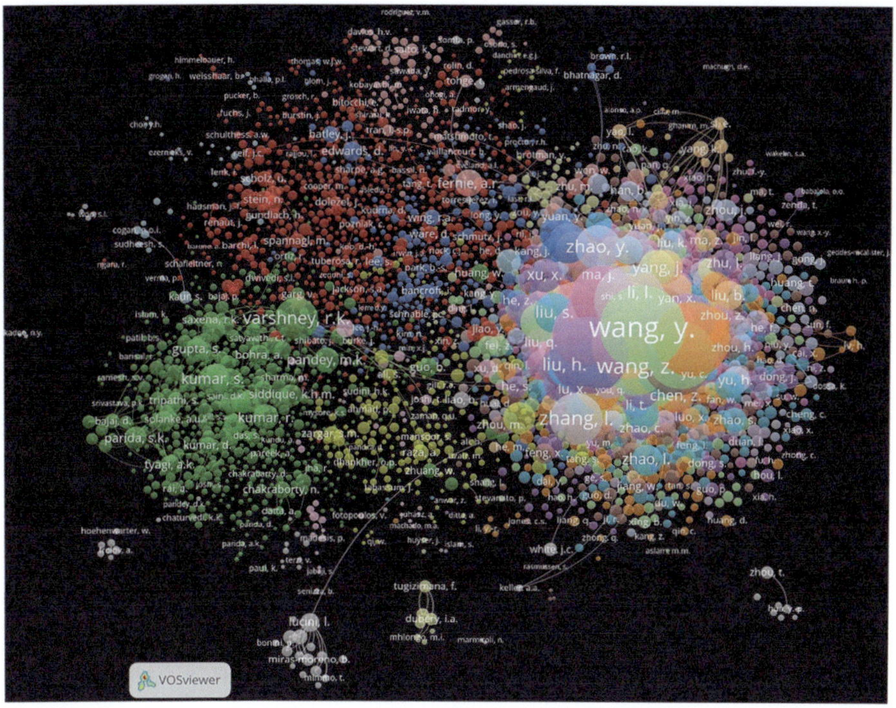

Fig. 2 Co-authorship network visualization of researchers in agricultural meta-omics research (2010–2024). Node sizes represent publication output volume, different colors indicate distinct research clusters, and connecting lines show collaboration relationships between authors. Major research clusters are centered around prominent authors (e.g., Wang, Y. and collaborators), demonstrating the collaborative nature and intellectual structure of the field. Visualization generated using VOSviewer bibliometric analysis software

identification of influential researchers, established research consortia, and potential emerging research trajectories through the analysis of node centrality metrics and cluster formation patterns. The initial dataset comprising 46,013 authors was refined to 3238 authors using a threshold criterion of minimum 5 publications per author under full counting methodology. The largest connected component encompassed 3191 authors, organized into 26 distinct clusters with 79,113 links and a total link strength of 121,804 (Fig. 2). Cluster distribution analysis revealed three predominant groupings: the primary cluster containing 661 authors, followed by cluster 2 and cluster 3 with 600 and 229 authors, respectively (Fig. 2). Topological examination of the network demonstrated that Wang Y emerged as the most prolific node, exhibiting 915 connections with a total link strength of 2841 and contributing 406 documents (Fig. 2). The resultant visualization provides quantitative insights into the collaborative dynamics and intellectual structure of the research domain.

8 Research Hotspots and Keyword Co-occurrence Analysis

Keyword analysis serves as a critical methodological tool for elucidating thematic evolution and research trajectories within scientific domains. In this study we employed VOSviewer software utilizing the full counting algorithm to conduct a comprehensive keyword co-occurrence analysis (Fig. 3). The initial dataset comprised 56,890 keywords which was subsequently refined to 8520 keywords by implementing a minimum occurrence threshold of 5. Network visualization revealed distinct hierarchical clustering patterns and interconnected nodes demonstrating the semantic relationships within the field. Primary nodes were dominated by fundamental "-omics" terminologies specifically "genetics," "genomics," "metabolomics," "metabolism," and regulatory processes such as "gene expression profiling" and "gene expression regulation" (Fig. 3). The network topology exhibited three major clusters:

1. Molecular Biology Cluster: Characterized by genomic and genetic descriptors including "chromosome mapping," "quantitative trait locus," "synteny," and "molecular evolution." Species-specific terms such as "rice," "glycine max," "grain yield," and "peas" were prominently represented, indicating the significance of crop-specific molecular investigations.
2. Metabolic Systems Cluster: Encompassed terms related to physiological processes, including "gene expression profiling and regulation," "flavonoid," "leaf," "plant cell," "chlorophyll," and "dry weight." This cluster demonstrates the integration of metabolic pathway analysis with agricultural phenotypes.
3. Agricultural Systems Cluster: Incorporated broader agricultural parameters including "bacterium," "animals," "soil," "herbicide," "microbial community," and "soil pollutants," reflecting the ecosystem-level applications of omics technologies.

The topological distribution and edge weights within this network visualization provide quantitative insights into the predominant research themes and technological

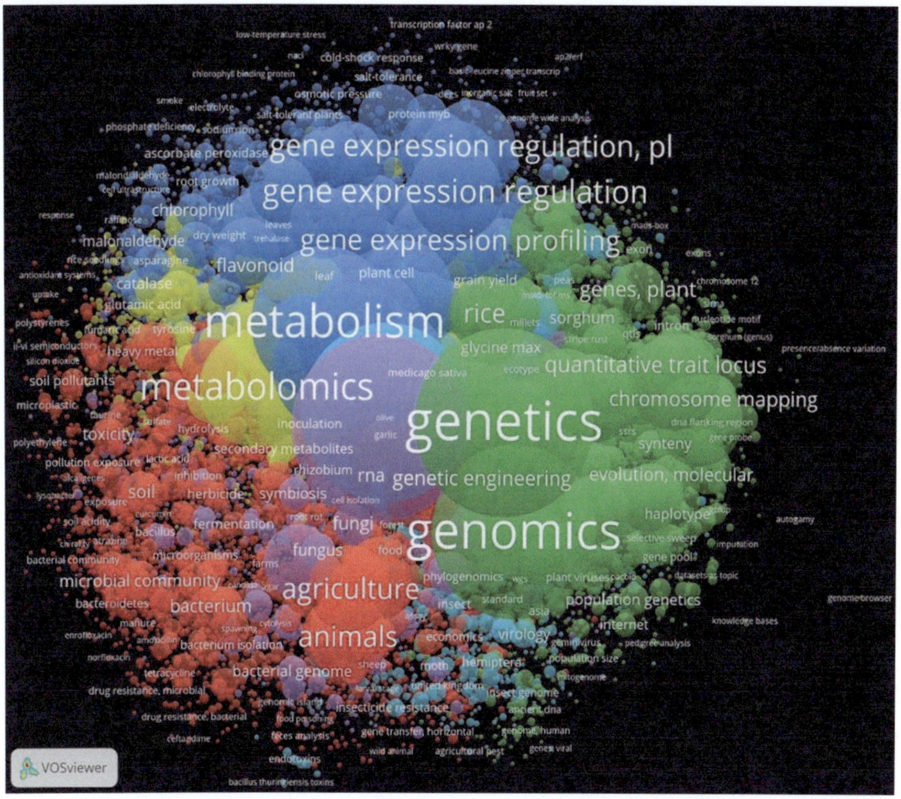

Fig. 3 Keyword co-occurrence network visualization of agricultural omics research (2000–2024). The network map illustrates major research themes and their interconnections, with node size indicating keyword frequency and colors representing different research clusters. Dominant themes include "genomics," "metabolism," "gene expression regulation," and "metabolomics," reflecting the core focus areas. The visualization demonstrates the integration of various omics approaches (blue cluster), their applications in crop genetics and breeding (green cluster), and relationships with soil/microbial studies (red cluster). Generated using VOSviewer, this analysis reveals the conceptual structure and thematic evolution of agricultural omics research over the past two decades

applications within agricultural sciences, while simultaneously demonstrating the interconnected nature of various subdisciplines through bibliometric analysis.

9 Subject Categories' Co-occurrence Analysis

The objective of this study was to elucidate the predominant research domains within agricultural sciences utilizing various -omics methodologies. Analysis of research categories, filtered for those exceeding a threshold of 100 publications, revealed a distinct hierarchical distribution of disciplines (Fig. 4). "Biochemistry, Genetics and Molecular Biology" emerged as the major field, encompassing 9248

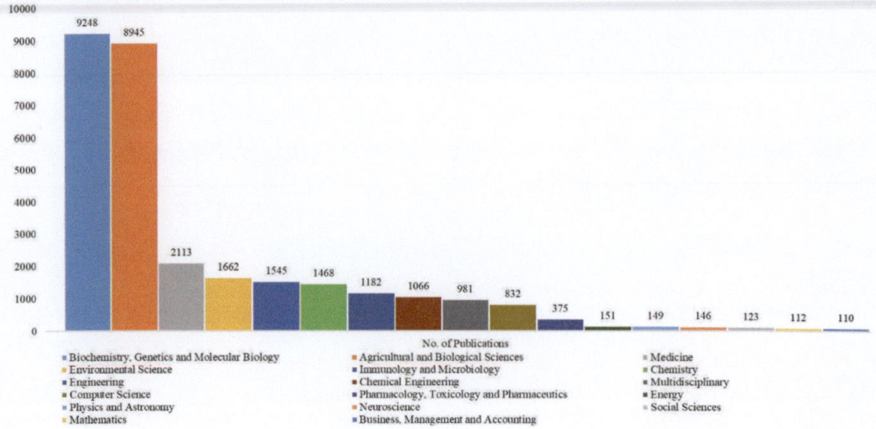

Fig. 4 Distribution of omics-assisted research publications across leading research categories in crop and agriculture sector (2000–2025). The graph shows the dominance of "Biochemistry, Genetics and Molecular Biology" (9238 publications) and "Agricultural and Biological Sciences" (8915 publications) as primary research domains, followed by significant contributions from "Environmental Science" and "Engineering." Other categories including "Immunology and Microbiology," "Chemical Engineering," and interdisciplinary fields demonstrate the broad application of omics approaches in agricultural research. Values represent the number of publications in each category, illustrating the multidisciplinary nature of agricultural omics research

publications, with Agricultural and Biological Sciences following closely at 8945 publications. Other fields included Medicine (2113 publications), Environmental Science (1662 publications), Immunology and Microbiology (1545 publications), and Chemistry (1468 publications), among others (Fig. 4). The substantial representation of both Biochemistry, Genetics and Molecular Biology, and Agricultural and Biological Sciences as core disciplines underscores the interdisciplinary nature of -omics applications in agricultural research, demonstrating the extensive scope and integration of these methodologies across diverse agricultural subdisciplines.

10 Key Journal Ranking and Publisher Analysis

In the current study, it was identified that there were several reputed journals which published research articles in this field. There were 117 journals which published minimum of 25 publications in this area. Figure 5 reflects top journals with minimum 100 publications in this area. Figure 5 presents top ten journals that have published the most research from 2000 to 2025. Frontiers in Plant Science leads with 703 total publications. International Journal of Molecular Sciences follows with 613 publications. BMC Genomics ranks third with 538 publications.

"Frontiers in Plant Science" is most cited plant science journal which advances understanding of plant biology for sustainable food security, functional ecosystems

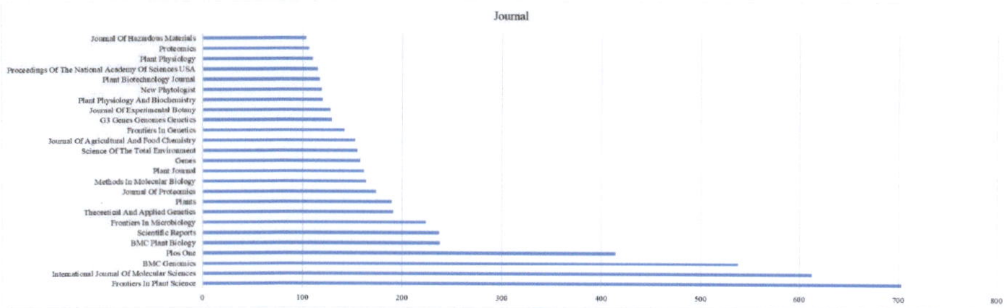

Fig. 5 Leading journals publishing research on omics approaches in agriculture (2000–2025). The horizontal bar graph illustrates publication volumes across major scientific journals, with "Frontiers in Plant Science" and "International Journal of Molecular Sciences" showing the highest publication outputs. The distribution spans specialized journals in plant science, molecular biology, and agricultural research, reflecting the diverse publication venues for agricultural omics studies. Journal rankings are based on the number of publications, demonstrating both the breadth and focus of research dissemination in this field. Data sourced from scientific databases shows publication counts per journal, indicating preferred outlets for agricultural omics research

and human health. Among the top 10 journals, "BMC Genomics" was having highest "ACPP," while "Plants" was having lowest. Further when we look for different "Document Type," then we can observe that they were distributed in three categories, Article, Review, and Editorial. Articles make up the majority of publications across all journals. Reviews vary significantly, while Frontiers in Plant Science has 225 reviews, some journals like Scientific Reports have none. Editorials are less common, with Frontiers in Plant Science having the most (29). Frontiers in Plant Science has the highest total citations (24,177). BMC Genomics follows with (23,683) citations. Plos One has 17,257 citations. In the case of average citations per publication (ACPP), BMC Genomics leads with 44.02 citations per publication followed by Plos One (41.68) and Theoretical and Applied Genetics (41.59 ACPP). The data suggests that while some journals publish more papers (like Frontiers in Plant Science), others might have higher impact per paper (like BMC Genomics). There's also a clear variation in how journals balance different types of content (articles, reviews, and editorials). Interestingly, specialized journals (like BMC Genomics and Theoretical and Applied Genetics) tend to have higher citation rates per paper compared to broader scope journals (like PLoS One), suggesting that focused, field-specific research might generate more citations in this area.

11 Countries Cooperation Analysis

Table 3 illustrates the top 10 countries leading advanced research in the field of omics areas in crop and agriculture sector. These ten countries together account for approximately 67% of the total research publications. China leads with 4634 publications, followed by the United States (4080) and India (3022). These three

Table 3 Top 10 productive countries in omics research in field of crop and agriculture sector (2000–2025)

S. no.	Country/ territory	Total publications	TC: total citation	h-index	ACPP: average citations per publication
1.	China	4634	151,285	159	32.65
2.	United States	4080	245,093	212	60.07
3.	India	3022	77,369	121	25.60
4.	United Kingdom	1154	85,852	140	74.40
5.	Australia	1089	60,428	119	55.49
6.	Germany	1072	73,561	133	68.62
7.	France	824	60,246	115	73.11
8.	Italy	757	40,468	96	53.46
9.	Canada	684	39,668	94	57.99
10.	Spain	591	34,908	94	59.07

countries significantly outpublish the others in the list (Table 3). Based on the total citation and h-index statistics, the United States is producing publications that are highly cited. Despite being second in publications, the United States leads in total citations with 245,093, significantly higher than China's 151,285 (Table 3). In terms of h-index, a measure of both productivity and citation impact, the United States leads with 212. China follows with 159. The United Kingdom ranks third with 140 (Table 3). In the case of average citations per publication (ACPP), the United Kingdom leads with 74.40 citations per publication. France follows closely with 73.11. Germany has ACPP of 68.62. Interestingly, while China leads in total publications, it has a lower ACPP (32.65) compared to most other countries in the list (Table 3). The data reveals different strengths among countries, China and the United States dominate in terms of research output volume. European countries (the United Kingdom, France, Germany) tend to have higher citation rates per publication. Countries like India have high publication numbers but lower citation metrics, suggesting room for increasing research impact (Table 3).

Overall, different -omics approaches, like genomics, transcriptomics, proteomics, and metabolomics has fundamentally transformed our understanding of plant biology and agricultural improvement. The integration of diverse molecular data through high-throughput multi-omics technologies has revolutionized our understanding toward critical plant processes. From developmental stages like growth and aging to agricultural traits such as yield and stress tolerance, these comprehensive analytical approaches have provided valuable insights. Researchers have successfully deployed these powerful tools across major agricultural species, spanning staple grains like wheat, maize, rice, barley, and millet, to important protein sources like soybean and the model legume *Medicago truncatula*, as well as commercial crops such as cotton and tomato (Yang et al. 2021).

12 Conclusion

The comprehensive analysis of omics research in agriculture over the past two and a half decades reveals a transformative impact on crop improvement and agricultural innovation. The field has witnessed remarkable growth, with 16,773 publications garnering 622,615 citations and achieving an h-index of 285, demonstrating both the quantity and quality of research output. The dominance of genomics-related publications, particularly in crop genome sequencing, underscores its fundamental role in advancing agricultural science. The emergence of integrated approaches combining genomics, transcriptomics, proteomics, and metabolomics has created a more holistic understanding of plant systems and their responses to environmental challenges. The dominance of research articles over other publication types indicates a strong focus on original research. Leading countries like China, the United States, and India have emerged as major contributors, though European nations demonstrate higher citation impact per publication, suggesting varying research quality and influence across regions. The analysis of research hotspots through keyword co-occurrence reveals three primary clusters focusing on molecular biology, metabolic systems, and agricultural systems, highlighting the field's breadth and depth. The strong representation of both biochemistry/genetics and agricultural sciences in subject categories demonstrates the successful integration of fundamental and applied research. This convergence has facilitated practical applications in crop improvement, particularly in developing stress-resistant varieties and enhancing nutritional content. Notably, top-cited publications focus heavily on genome sequencing projects and stress response mechanisms, highlighting these areas' fundamental importance. As agriculture faces mounting challenges from climate change and food security demands, the continued evolution of omics technologies, supported by strong international collaboration and interdisciplinary approaches, will be crucial for developing sustainable and resilient agricultural systems.

Acknowledgment DPS is thankful to the Indian Council of Agricultural Research (ICAR) for funding support in the form of Network project on Agricultural Bioinformatics (CABin scheme) at ICAR-IIVR, Varanasi. RP is thankful for the support received under the Institute Project "Soil and Crop health monitoring through sensing, modelling, and big data analytics for resource efficient smart agriculture at field to regional scale (CRSCIARISIL20210034319)."

Conflict of Interest Authors declare no conflict of interest.

References

Aglawe SB, Verma AK, Upadhyay AK (2021) Bioinformatics tools and databases for genomics-assisted breeding and population genetics of plants: a review. Curr Bioinforma 16(6). https://doi.org/10.2174/1574893615999200629123524

Alawiye TT, Babalola OO (2021) Metabolomics: current application and prospects in crop production. Biologia 76:227–239. https://doi.org/10.2478/s11756-020-00574-z

Ali MA, Lee J (2022) Transcriptome profiling: progress and prospects. ISBN:9780323918107
Bartoszewski G, Malepszy S (2012) Plant genomics – its emerging role in crop improvement. Acta Hortic 953:23–29. https://doi.org/10.17660/ActaHortic.2012.953.3
Biscaro C, Giupponi C (2014) Co-authorship and bibliographic coupling network effects on citations. PLoS One 9:e99502. https://doi.org/10.1371/journal.pone.0099502
Borém A, Fritsche-Neto R (2014) Omics in plant breeding. Wiley, Hoboken. Print ISBN:9781118820995
Calcagno V, Demoinet E, Gollner K, Guidi L, Ruths D, de Mazancourt C (2012) Flows of research manuscripts among scientific journals reveal hidden submission patterns. Science 338:1065–1069. https://doi.org/10.1126/science.1227833
Carlson DB, Choudhuri S (2008) Genomics: the state of the science. In: Genomics: fundamentals and applications. CRC Press, Boca Raton
Castillejo MA, Pascual J, Jorrín-Novo JV, Balbuena TS (2023) Proteomics research in forest trees: A 2012–2022 update. Front Plant Sci 14:1130665. https://doi.org/10.3389/fpls.2023.1130665
Chauhan C, Thakur A, Sharma V, Yadav RR, Sivakumar KB, Kumar R, Verma SK, Panwar RK, Tewari SK, Gautam A (2023) Proteomics as a tool for analyzing plant responses to abiotic and biotic stresses. In: Genomics, transcriptomics, proteomics and metabolomics of crop plants. Academic, pp 35–68. https://doi.org/10.1016/B978-0-323-95989-6.00012-7
Chaves MM, Maroco JP, Pereira JS (2003) Understanding plant responses to drought – from genes to the whole plant. Funct Plant Biol 30(3):239–264. https://doi.org/10.1071/FP02076
Chen C (2004) Searching for intellectual turning points: progressive knowledge domain visualization. Proc Natl Acad Sci 101:5303–5310. https://doi.org/10.1073/pnas.0307513100
Chen C (2006) CiteSpace II: detecting and visualizing emerging trends and transient patterns in scientific literature. J Am Soc Inf Sci Technol 57:359–377. https://doi.org/10.1002/asi.20317
Chen H, Tan XF (2014) Excavation of genic resources based on next generation sequencing technologies. Plant Physiol J 50(8):1089–1095
Cortés AJ, Castillejo MÁ, Yockteng R (2023) 'Omics' Approaches for Crop Improvement. Agronomy 13:1401. https://doi.org/10.3390/agronomy13051401
Costa-Neto G, Fritsche-Neto R (2021) Enviromics: bridging different sources of data, building one framework. Crop Breed Appl Biotechnol 21:e393521S12. https://doi.org/10.1590/1984-70332021v21Sa12
D'Costa VM, King CE, Kalan L, Morar M, Sung WW, Schwarz C, Froese D, Zazula G, Calmels F, Debruyne R, Golding GB, Poinar HN, Wright GD (2011) Antibiotic resistance is ancient. Nature 477(7365):457–461. https://doi.org/10.1038/nature10388
Das A, Paudel B, Rohila JS (2015) Potentials of proteomics in crop breeding. In: Al-Khayri J, Jain S, Johnson D (eds) Advances in plant breeding strategies: breeding, biotechnology and molecular tools. Springer, Cham. https://doi.org/10.1007/978-3-319-22521-0_18
Davey JW, Hohenlohe PA, Etter PD, Boone JQ, Catchen JM, Blaxter ML (2011) Genome-wide genetic marker discovery and genotyping using next-generation sequencing. Nat Rev Genet 12(7):499–510. https://doi.org/10.1038/nrg3012
Debbarma J, Bhattacharjee M, Konwar T, Chikkaputtaiah C, Singha DL (2024) Recent tools of genome editing and functional genomics in plants: present and future applications on crop improvement. In: Chen J (ed) CRISPR and plant functional genomics. CRC Press, Boca Raton
Descombes P, Oakeley EJ (2016) Landscape of genomics technology platforms in Switzerland. Chimia 70(12):883–888. https://doi.org/10.2533/chimia.2016.883
Dryagina NV, Kondratyeva EA, Dubrovskii YA, Kondratyev AN (2020) Metabolome of the Brain. Russ Neurol J 25(1):4–12. https://doi.org/10.30629/2658-7947-2020-25-1-4-12
Edwards D (2016) The impact of genomics technology on adapting plants to climate change. In: Edwards D, Batley J (eds) Plant genomics and climate change. Springer, New York. https://doi.org/10.1007/978-1-4939-3536-9_8
Eldakak M, Milad SIM, Nawar AI, Rohila JS (2013) Proteomics: a biotechnology tool for crop improvement. Front Plant Sci 4:35. https://doi.org/10.3389/fpls.2013.00035

Feng Z, Zheng CY, Bo YK, Li YH, Zhu F (2024) Advances in transcriptomics and metabolomics-based exploration of plant responses to drought and herbivorous insects. Chin J Eco-Agric 32(3):369–379. https://doi.org/10.12357/cjea.20230297

Fernie AR, Schauer N (2009) Metabolomics-assisted breeding: a viable option for crop improvement? Trends Genet 25(1):39–48. https://doi.org/10.1016/j.tig.2008.10.010

Goff SA, Ricke D, Lan TH, Presting G, Wang R, Dunn M, Glazebrook J, Sessions A, Oeller P, Varma H et al (2002) A draft sequence of the rice genome (Oryza sativa L. ssp. japonica). Science 296(5565):92–100. https://doi.org/10.1126/science.1068275

Gupta C, Salgotra RK (2022) Epigenetics and its role in effecting agronomical traits. Front Plant Sci 13:925688. https://doi.org/10.3389/fpls.2022.925688

Gupta DB, Shekhar S, Agrawal L (2015) Plant proteomics: technologies and applications. In: Barh D, Khan M, Davies E (eds) PlantOmics: the omics of plant science. Springer, New Delhi. https://doi.org/10.1007/978-81-322-2172-2_8

Gupta S, Verma R, Ravi RK (2023) Multiomics approach for crop improvement under climate change. In: Prakash CS, Fiaz S, Nadeem MA, Baloch FS, Qayyum A (eds) Sustainable agriculture in the era of the OMICs revolution. Springer, Cham. https://doi.org/10.1007/978-3-031-15568-0_2

Hamany Djande CY, Pretorius C, Tugizimana F, Piater LA, Dubery IA (2020) Metabolomics: a tool for cultivar phenotyping and investigation of grain crops. Agronomy 10(6):831. https://doi.org/10.3390/agronomy10060831

Harzing AW, Alakangas S (2016) Google Scholar, scopus and the web of science: a longitudinal and cross-disciplinary comparison. Scientometrics 106:787–804. https://doi.org/10.1007/s11192-015-1798-9

Herrmann A, Schauer N (2013) Metabolomics-assisted plant breeding. In: Weckwerth W, Kahl G (eds) The handbook of plant metabolomics. Wiley-Blackwell. https://doi.org/10.1002/9783527669882.ch13

Houle D, Govindaraju DR, Omholt S (2010) Phenomics: the next challenge. Nat Rev Genet 11:855–866. https://doi.org/10.1038/nrg2897

Husi H, Albalat A (2014) Proteomics. In: Padmanabhan S (ed) Handbook of pharmacogenomics and stratified medicine. Academic, pp 147–179

Ingvarsson PK, Hvidsten TR, Street NR (2016) Towards integration of population and comparative genomics in forest trees. New Phytol 212(2):338–344. https://doi.org/10.1111/nph.14153

Jin X, Zarco-Tejada PJ, Schmidhalter U, Reynolds MP, Hawkesford MJ, Varshney RK (2021) High-Throughput Estimation of Crop Traits: A Review of Ground and Aerial Phenotyping Platforms. IEEE Geosci Remote Sens Mag 9(1):200–231. https://doi.org/10.1109/MGRS.2020.2998816

Jung JH, Seo YW (2017) Challenges in wide implementation of genome editing for crop improvement. J Crop Sci Biotechnol 20:129–135. https://doi.org/10.1007/s12892-017-0019-0

Kage U, Kumar A, Dhokane D, Karre S, Kushalappa AC (2016) Functional molecular markers for crop improvement. Crit Rev Biotechnol 36(5):917–930. https://doi.org/10.3109/07388551.2015.1062743

Khan AA, Iqbal B, Jalal A, Nadeem MA, Baloch FS (2024) Advanced molecular approaches for improving crop yield and quality: a review. J Plant Growth Regul 43:2091–2103. https://doi.org/10.1007/s00344-024-11253-7

Kido A, Ferreira-Neto JRC, Binneck E, da Silva M, da Silva Júnior W, Benko-Iseppon AM (2022) Explore the RNA-sequencing and the next-generation sequencing in crops responding to abiotic stress. In: Sharma P, Yadav D, Gaur RK (eds) Bioinformatics in agriculture: next generation sequencing era. Academic, pp. 161–175

Kumar D (2015) Socioeconomic outcomes of genomics in the developing world. In: Kumar D, Chadwick R (eds) Genomics and society: ethical, legal, cultural and socioeconomic implications. Academic, pp 239–258

Lipsman N, Woodside DB, Lozano AM (2014) Trends in anorexia nervosa research: an analysis of the top 100 most cited works. Eur Eat Disord Rev 22:9–14. https://doi.org/10.1002/erv.2265

Lowe R, Shirley N, Bleackley M, Dolan S, Shafee T (2017) Transcriptomics technologies. PLoS Comput Biol 13(5):e1005457. https://doi.org/10.1371/journal.pcbi.1005457

Mahajan R, Zargar SM (2020) Proteomics as a tool for characterizing the alteration in pathways associated with defense and metabolite synthesis. In: Wani SH, Kumar V (eds) Heat stress tolerance in plants: physiological, molecular and genetic perspectives. Wiley. https://doi.org/10.1002/9781119432401.ch7

Marguerat S, Wilhelm BT, Bähler J (2008) Next-generation sequencing: applications beyond genomes. Biochem Soc Trans 36(5):1091–1096. https://doi.org/10.1042/BST0361091

Mendes R, Kruijt M, de Bruijn I, Dekkers E, van der Voort M, Schneider JH, Piceno YM, DeSantis TZ, Andersen GL, Bakker PA, Raaijmakers JM (2011) Deciphering the rhizosphere microbiome for disease-suppressive bacteria. Science 332(6033):1097–1100. https://doi.org/10.1126/science.1203980

Mir RR, Reynolds M, Pinto F, Khan MA, Bhat MA (2019) High-throughput phenotyping for crop improvement in the genomics era. Plant Sci 282:60–72. https://doi.org/10.1016/j.plantsci.2019.01.007

Murphy DJ (1998) Impact of genomics on improving the quality of agricultural products. In: Copping LG, Dixon GK, Livingstone DJ (eds) Genomics: commercial opportunities from a scientific revolution. CRC Press. https://doi.org/10.1201/9781003423782

Mustafa G, Komatsu S (2021) Plant proteomic research for improvement of food crops under stresses: a review. Mol Omics 17:860–880. https://doi.org/10.1039/D1MO00027F

Muthamilarasan M, Singh NK, Prasad M (2019) Multi-omics approaches for strategic improvement of stress tolerance in underutilized crop species: a climate change perspective. Adv Genet 103:1–38. https://doi.org/10.1016/bs.adgen.2019.01.001

Naidu S, Suman S, Roy A et al (2023) Impact of genomics tools and techniques on crop improvement. In: Mani A, Kushwaha S (eds) Genomics of plant-pathogen interaction and the stress response. CRC Press, Boca Raton

Nakano T, Suzuki K, Fujimura T, Shinshi H (2006) Genome-wide analysis of the ERF gene family in Arabidopsis and rice. Plant Physiol 140(2):411–432. https://doi.org/10.1104/pp.105.073783

Pandita D (2023) Orchestration of omics technologies for crop improvement. In: Hasanuzzaman M (ed) Climate-resilient agriculture, vol 2. Springer, Cham. https://doi.org/10.1007/978-3-031-37428-9_34

Pedreschi R, Lurie S, Hertog M, Nicolaï B, Mes J, Woltering E (2013) Post-harvest proteomics and food security. Proteomics 13(12–13):1772–1783. https://doi.org/10.1002/pmic.201200387

Potato Genome Sequencing Consortium, Xu X, Pan S et al (2011) Genome sequence and analysis of the tuber crop potato. Nature 475(7355):189–195. https://doi.org/10.1038/nature10158

Pott DM, Durán-Soria S, Osorio S et al (2021) Combining metabolomic and transcriptomic approaches to assess and improve crop quality traits. CABI Agric Biosci 2:1. https://doi.org/10.1186/s43170-020-00021-8

Pranckutė R (2021) Web of Science (WoS) and scopus: the Titans of bibliographic information in today's academic world. Publica 9(1):12. https://doi.org/10.3390/publications9010012

Pushparaj PN (2019) Metabolomics. In: Shaik N, Hakeem K, Banaganapalli B, Elango R (eds) Essentials of bioinformatics, volume I. Springer, Cham. https://doi.org/10.1007/978-3-030-02634-9_13

Radhakrishnan S, Erbis S, Isaacs JA, Kamarthi S (2017) Novel keyword co-occurrence network-based methods to foster systematic reviews of scientific literature. PLoS One 12:e0172778. https://doi.org/10.1371/journal.pone.0172778

Raja V, Wani MA, Wani UM, Jan N, John R (2017) Understanding abiotic stress tolerance in plants by proteomics approach. In: Zargar SM, Rai V (eds) Plant OMICS and crop breeding. CRC Press. https://doi.org/10.1201/9781315365930

Ravi I (2014) Proteomics. In: Ravi I, Baunthiyal M, Saxena J (eds) Advances in biotechnology. Springer, New Delhi. https://doi.org/10.1007/978-81-322-1554-7_8

Razzaq A, Wishart DS, Wani SH, Hameed MK, Mubin M, Saleem F (2022) Advances in metabolomics-driven diagnostic breeding and crop improvement. Meta 12(6):511. https://doi.org/10.3390/metabo12060511

Resende RT, Piepho HP, Rosa GJM, Silva-Junior OB, e Silva FF, de Resende MDV, Grattapaglia D (2021) Enviromics in breeding: Applications and perspectives on envirotypic-assisted selection. Theor Appl Genet 134:95–112. https://doi.org/10.1007/s00122-020-03684-z

Romero FM, Marina M, Pieckenstain FL et al (2017) Gaining insight into plant responses to beneficial and pathogenic microorganisms using metabolomic and transcriptomic approaches. In: Kalia V, Saini A (eds) Metabolic engineering for bioactive compounds. Springer, Singapore. https://doi.org/10.1007/978-981-10-5511-9_6

Sakurai N (2022) Recent applications of metabolomics in plant breeding. Breed Sci 72(1):56–65. https://doi.org/10.1270/jsbbs.21065

Salt DE, Baxter I, Lahner B (2008) Ionomics and the study of the plant ionome. Annu Rev Plant Biol 59:709–733. https://doi.org/10.1146/annurev.arplant.59.032607.092942

Saraswat S, Yadav AK, Sirohi P et al (2017) Role of epigenetics in crop improvement: Water and heat stress. J Plant Biol 60:231–240. https://doi.org/10.1007/s12374-017-0053-8

Schmutz J, Cannon SB, Schlueter J et al (2010) Genome sequence of the palaeopolyploid soybean. Nature 463(7278):178–183. https://doi.org/10.1038/nature08670

Singh SK, Pradhan S (2020) Transcriptomics in fruit crops: Present status and future prospects. Indian J Horticult 77(4):563–581

Singh A, Roychoudhury A (2023) Advances in metabolomics research in environmental stress response in plants. In: Biology and biotechnology of environmental stress tolerance in plants: sustainable approaches for enhancing environmental stress tolerance. CRC Press, Boca Raton

Somashekar H (2024) Harnessing genome editing approaches in crop improvement to reshape modern agriculture. In: Chandra K, Chand S, Saini RP, Sharma R (eds) Smart breeding: molecular interventions and advancements for crop improvement. CRC Press, Boca Raton

Tahir ul Qamar M, Faryad A, Bari A, Zahid B, Zhu X, Chen LL (2020) Effectiveness of conventional crop improvement strategies vs. omics. In: Fahad S et al (eds) Environment, climate, plant and vegetation growth. Springer, Cham. https://doi.org/10.1007/978-3-030-49732-3_11

Talukdar D, Sinjushin A (2015) Cytogenomics and mutagenomics in plant functional biology and breeding. In: Barh D, Khan M, Davies E (eds) PlantOmics: the omics of plant science. Springer, New Delhi, pp 113–156. https://doi.org/10.1007/978-81-322-2172-2_5

Tan BC, Lim YS, Lau SE (2017) Proteomics in commercial crops: an overview. J Proteome 169:176–188. https://doi.org/10.1016/j.jprot.2017.05.018

Tao Y, Zhao X, Mace E, Henry R, Jordan D (2019) Exploring and exploiting pan-genomics for crop improvement. Mol Plant 12(2):156–169. https://doi.org/10.1016/j.molp.2018.12.016

Tomato Genome Consortium (2012) The tomato genome sequence provides insights into fleshy fruit evolution. Nature 485(7400):635–641. https://doi.org/10.1038/nature11119

van Eck NJ, Waltman L (2010) Software survey: VOSviewer, a computer program for bibliometric mapping. Scientometrics 84:523–538. https://doi.org/10.1007/s11192-009-0146-3

Van Noorden R (2017) The science that's never been cited. Nature 552:162–165. https://doi.org/10.1038/d41586-017-08404-0

Vance CP, Uhde-Stone C, Allan DL (2003) Phosphorus acquisition and use: critical adaptations by plants for securing a nonrenewable resource. New Phytol 157(3):423–447. https://doi.org/10.1046/j.1469-8137.2003.00695.x

Vanderschuren H, Lentz E, Zainuddin I, Gruissem W (2013) Proteomics of model and crop plant species: status, current limitations and strategic advances for crop improvement. J Proteome 93:5–19. https://doi.org/10.1016/j.jprot.2013.05.036

Varshney RK, Bohra A, Yu J, Graner A, Zhang Q, Sorrells ME (2021) Designing future crops: genomics-assisted breeding comes of age. Trends Plant Sci 26(6):631–649. https://doi.org/10.1016/j.tplants.2021.03.010

Vishwanath PP, Bidaramali V, Lata S et al (2024) Transcriptomics: illuminating the molecular landscape of vegetable crops: a review. J Plant Biochem Biotechnol. https://doi.org/10.1007/s13562-023-00871-z

Vrazas V, Katsani KR (2024) Proteomics. In: Liu D (ed) Handbook of molecular biotechnology. CRC Press, Boca Raton

Wishart DS, Godzien J, Gil-De-La-Fuente A et al (2020) Metabolomics. In: New developments in mass spectrometry. Royal Society of Chemistry, London

Wu S, Ning F, Zhang Q, Wu X, Wang W (2017) Enhancing omics research of crop responses to drought under field conditions. Front Plant Sci 8:174. https://doi.org/10.3389/fpls.2017.00174

Yadav AK, Shree B, Lakhwani D, Singh AK (2024) Omics technologies for crop improvement. In: Gupta MK, Mondal S, Katara P, Singh RL (eds) Integrative omics: concept, methodology, and application. Academic, pp 293–313

Yang K, Meho LI (2006) Citation analysis: a comparison of Google Scholar, Scopus, and Web of Science. Proc Am Soc Inf Sci Technol 43:1–15. https://doi.org/10.1002/meet.14504301185

Yang Y, Saand MA, Huang L, Abdelaal WB, Zhang J, Wu Y, Li J, Sirohi MH, Wang F (2021) Applications of multi-omics technologies for crop improvement. Front Plant Sci 12:563953. https://doi.org/10.3389/fpls.2021.563953

Zainal-Abidin RA, Harun S, Vengatharajuloo V, Tamizi AA, Samsulrizal NH (2022) Gene co-expression network tools and databases for crop improvement. Plants 11(13):1625. https://doi.org/10.3390/plants11131625

Zenda T, Liu S, Dong A, Li J, Wang Y, Liu X, Wang N, Duan H (2021) Omics-facilitated crop improvement for climate resilience and superior nutritive value. Front Plant Sci 12:774994. https://doi.org/10.3389/fpls.2021.774994

Zhang J, Xie J, Hou W, Tu X, Xu J, Song F, Wang Z, Lu Z (2012) Mapping the knowledge structure of research on patient adherence: knowledge domain visualization based co-word analysis and social network analysis. PLoS One 7:e34497. https://doi.org/10.1371/journal.pone.0034497

Zhang L, Chen F, Zeng Z, Xu M, Sun F, Yang L, Bi X, Lin Y, Gao Y, Hao H et al (2021) Advances in metagenomics and its application in environmental microorganisms. Front Microbiol 12:3847. https://doi.org/10.3389/fmicb.2021.766136

Zhuang Y, Zhang L, Du Y, Chen G (2016) Current patterns and future perspectives of best management practices research: a bibliometric analysis. J Soil Water Conserv 71:98A–104A. https://doi.org/10.2489/jswc.71.3.98A

Zyoud SH, Al-Jabi SW, Sweileh WM, Al-Khalil S, Alqub M, Awang R (2015) Global methaemoglobinaemia research output (1940–2013): a bibliometric analysis. Springerplus 4:1–7. https://doi.org/10.1186/s40064-015-1431-7

MIX
Papier aus verantwortungsvollen Quellen
Paper from responsible sources
FSC® C105338

If you have any concerns about our products,
you can contact us on
ProductSafety@springernature.com

In case Publisher is established outside the EU,
the EU authorized representative is:
Springer Nature Customer Service Center GmbH
Europaplatz 3, 69115 Heidelberg, Germany

Printed by Libri Plureos GmbH
in Hamburg, Germany